Paul Baginski, Benjamin Fine, Anja Moldenhauer, Gerhard Rosenberger, and Vladimir Shpilrain (Eds.)

Elementary Theory of Groups and Group Rings, and Related Topics

De Gruyter Proceedings in Mathematics

Elementary Theory of Groups and Group Rings, and Related Topics

Proceedings of the Conference held at Fairfield University and at the Graduate Center, CUNY, November 1-2, 2018

Edited by
Paul Baginski, Benjamin Fine, Anja Moldenhauer, Gerhard Rosenberger, and Vladimir Shpilrain

DE GRUYTER

Editors
Prof. Dr. Paul Baginski
Department of Mathematics
Fairfield University
1073 North Benson Road
Fairfield
CT 06430
USA
pbaginski@fairfield.edu

Prof. Dr. Benjamin Fine
Department of Mathematics
Fairfield University
1073 North Benson Road
Fairfield
CT 06430
USA
fine@fairfield.edu

Dr. Anja Moldenhauer
Heidberg 22
22301 Hamburg
Germany
a.moldenhauer@gmx.net

Prof. Dr. Gerhard Rosenberger
Fachbereich Mathematik
Bereich AZ
University of Hamburg
Bundesstrasse 55
20146 Hamburg
Germany
gerhard.rosenberger@math.uni-hamburg.de

Prof. Dr. Vladimir Shpilrain
Department of Mathematics
The City College of New York
NAC 8/133
Convent Ave at 138th Street
New York
NY 10031
USA
shpil@groups.sci.ccny.cuny.edu

ISBN 978-3-11-063673-4
e-ISBN (PDF) 978-3-11-063838-7
e-ISBN (EPUB) 978-3-11-063709-0

Library of Congress Control Number: 2019953906

Bibliographic information published by the Deutsche Nationalbibliothek
The Deutsche Nationalbibliothek lists this publication in the Deutsche Nationalbibliografie; detailed bibliographic data are available on the Internet at http://dnb.dnb.de.

Typesetting: VTeX UAB, Lithuania
Printing and binding: CPI books GmbH, Leck

www.degruyter.com

Introduction

This volume contains the proceedings of the conference on the elementary theory of groups and group rings and related topics held at Fairfield University and City University of New York on November 1–2, 2018. The conference was held in cooperation with the New York Group Theory Seminar. The conference was held primarily to highlight the extensive research activity in the area since the proofs of the Tarski conjectures by Kharlampovich and Myasnikov and independently by Sela. The conference was also to honor the memory of Gilbert Baumslag who for many years was the guiding inspiration for the New York Group Theory Seminar. This seminar was initiated by Wilhelm Magnus and has run continuously for over 50 years. The conference also honored Ben Fine and Anthony Gaglione on their 70th birthdays. Both have been long time participants in the seminar and have done a great deal of research work in the area of the conference. In addition to the submitted papers, we have included a remembering Gilbert section. Gilbert Baumslag was one of the leading infinite group theorists for much of the second half of the twentieth century. He was also a mentor and guide to both his own students and to many others in the Group Theory community. We have asked several people who worked closely with Gilbert to write both working with him and the impact of his work. Many of these remembrances are related to the talks given at the memorial service for Gilbert held at City University. These talks can be viewed at https://www.youtube.com/watch?v=kIPw79wqdHo&list=PLeYOJGSDgPSr9Elws8z2mj25DvVwQt9DK&index=5.

The Editors

Proceedings of the elementary theory of groups and group rings

Paul Baginski
Ben Fine
Anja Moldenhauer
Gerhard Rosenberger
Vladimir Shpilrain

https://doi.org/10.1515/9783110638387-201

Conference booklet and talks

A conference celebrating the enduring success of the New York Group Theory Seminar

In memory of the work of long time seminar leader Gilbert Baumslag

And to celebrate the 70th birthday of long time participants, Benjmain Fine and Tony Gaglione

Fairfield University – November 1, 2018
Graduate Center CUNY – November 2, 2018

The New York Group Theory seminar which now runs Friday afternoons at the City University Graduate Center is the world's longest running Group Theory seminar. For over 60 years, the Group Theory Seminar has been crucial to the research efforts and accomplishments of the New York group Theory community. This will be a 2-day conference to honor the achievements of the seminar and in memory of the long time seminar leader, Gilbert Baumslag, who died 2 years ago. Through his students and his own work, Gilbert inspired a great deal of the research done by the seminar's participants, many of whom have been attending the seminar throughout their professional lives. The conference will also celebrate the 70th birthday of two of the long time participants, Ben Fine and Tony Gaglione, both of whom have worked closely with Gilbert and with many other people associated with the seminar.

Over the past 20 years, there has been tremendous progress in both infinite group theory and the interactions of group theory with logic and cryptography. Much of this material came together in the monumental work that resulted in a proof, by Kharlampovich and Myasnikov and, independently, by Sela, of the Tarski conjectures on free groups.

To show an appreciation of these achievements that arose from the Group Theory seminar and to honor on their 70th birthdays two people, Benjamin Fine and Anthony Gaglione, who have made significant contributions to the area, there will be a 2-day conference on November 1–2, 2018. Thursday, November 1, it will be at Fairfield University in Fairfield, Connecticut, while November 2, 2018, it will be at the Graduate Center of CUNY at 34th Street and 5th Avenue in Manhattan. On Thursday evening there will be a dinner and birthday party, free of charge to conference participants, hosted by Fairfield University. The speakers are predominantly people associated with the Group Theory Seminar.

Conference dinner

There will be a conference dinner and celebration on Thursday night at 6 PM at the Kelley Center Auditorium hosted by Fairfield University. This is free to conference participants.

https://doi.org/10.1515/9783110638387-202

Festschrift volume
We are planning a Festschrift volume and invite papers.

Transportation
Fairfield, Connecticut and New York City are connected by the Metro North rail line (New Haven Branch). In New York City, the local station is Grand Central Station at 42nd Street. It is a 15–20 minute walk from Grand Central to the Graduate Center. In Fairfield, the local station is Fairfield Station (not Fairfield Metro!). Fairfield University is a 15–20 minute walk from the station. Tickets may be bought in the station from kiosks or, for a hefty surcharge, on the train. Fares differ whether the trip is peak or nonpeak.

Organizing committee
Paul Baginski (Fairfield)
Ben Fine (Fairfield)
Gerhard Rosenberger (University of Hamburg)
Vladimir Shpilrain (CUNY Graduate Center)

Elementary Theory of Group Rings and Related Topics. A Conference in Honor of Gilbert Baumslag and the 70th Birthdays of Ben Fine and Tony Gaglione
November 1, 2018 – Fairfield University, November 2, 2018 – CUNY Graduate Center

Tentative schedule
Thursday, November 1 – Alumni House Fairfield University
8:00–8:30 – Coffee and Breakfast
8:30–8:45 – Opening Remarks
8:50–9:20 – Martin Kreuzer – “Gröbner basis methods in group theory”
9:30–10:10 – Walter Neumann – “Applications of Leighton’s theorem”
10:10–10:40 – Coffee
10:40–11:10 – Alexander Ushakov – “Quadratic equations in free metabelian groups”
11:20–11:50 – Dennis Spellman – “Some model theory of the Heisenberg group”
12:00–2:00 – Lunch
2:00–2:30 – Alexei Myasnikov – “On elementary theories of one-relator groups”
2:40–3:10 – Neha Hooda – “Rational subgroups and invariants of F_4 and G_2”
3:20–3:50 – Jon Gryak – “Solving the conjugacy decision problem via machine learning”
3:50–4:15 – Coffee
4:15–4:45 – Volker Diekert – “The elementary theory of free inverse monoids: about decidable and undecidable fragments”
4:45–5:15 – Anthony Clement – “Some properties of the Baumslag groups G(m, n)”

5:25–5:55 – Barry Mittag – “On nested infinite radicals”
6:00–Onward – Drinks, dinner and party – Alumni House Fairfield University

Friday – November 2 – CUNY Graduate Center – Science Center – Fourth Floor
10:00–10:15 – Opening remarks
10:20–10:50 – Al Thaler – “On Gilbert Baumslag and the Group Theory Seminar”
11:00–11:30 – Olga Karlampovich – “First-order properties of group rings and limit groups”
11:40–12:10 – Ilya Kapovich – “Attracting trees of ‘random’ automorphisms of free groups”
12:10–2:00 – Lunch
2:00–2:30 – Mahmood Sohrabi – “Bi-interpretability of the ring of integers Z with the pure groups SL(n, Z), GL(n, Z), and T(n, Z), n > 2”
2:40–3:10 – Lev Shneerson – “On conditions for polynomial growth in inverse semigroups”
3:10–3:40 – Coffee
3:40–4:10 – Gerhard Rosenberger – “On vector-valued Hecke forms”
4:20–4:50 – Mariana Bonanome – “A sampling of remarkable groups: Thompson’s, self-similar, lamplighter and Baumslag-Solitar”
4:50–5:20 – Zoran Sunic – “Rewriting in Thompson’s Group F”
5:30–6:00 – Ben Steinberg – TBA
6:30–Onward – Dinner at the Persian Grill.

Remembrances of Gilbert

Gilbert Baumslag was one of the leading infinite group theorists of the second half of the twentieth century. Through his own work, his students, and his mentoring, he had a profound effect on both the direction and interests in group theory. There is an ongoing seminar on his work given at Cornell University and various talks from this seminar can be found on the internet. The year he organized on combinatorial group theory at MSRI in 1989 ushered in much of the interest on automatic groups and hyperbolic groups. We note the paper by Alonso, Gersten, Shapiro, and Short explaining Gromov's ideas on hyperbolic groups and which came out of that conference as one of foundational works in geometric group theory. Gilbert's work and influence has been extremely wide, from classical combinatorial group theory, involving embedding theorems, and theory of nilpotent and one-relator groups to automatic and hyperbolic groups; the development of algebraic geometry over group with Myasnikov and Remeslennikov, which was instrumental in the proof of the Tarski theorems; the use of the variety of group representations with Peter Shalen; the development of group-based cryptography. In addition to all this work in theoretical group theory, there was also his work on the Magnus project and all that it entailed in terms of computer implementations of group theoretic procedures. Since group theoretic procedures do not necessarily terminate, the Magnus project work as well as the work on group-based cryptography, veered greatly into computer science applications.

Ben Fine and Gerhard Rosenberger

Ben Fine began working with Gilbert through the Group Theory seminar as soon as I completed my own thesis work under Wilhelm Magnus. Gilbert was extremely helpful and always willing to listen to work and to give ideas. After the Friday seminar talk and either before or during dinner, a great many ideas for theorems and proofs flowed freely – most coming from Gilbert. He was fiercely protective of his own work as any great mathematician should be, but he was an extremely generous colleague who was willing to help anyone who needed it. Up until his death, he had an active group working with him consisting of both his own students and others. Gerhard Rosenberger and Ben Fine began working with Gilbert in the mid 1980s when some his work on the variety of group representations with Peter Shalen and our work with Jim Howie on essential representations overlapped. While Ben worked in the CAISS lab, lunchtime conversations with Gilbert led to a great deal more work particularly in group-based cryptography which got to Gerhard via email. We especially like material we did on provable password security. He became our co-author, along with Martin Kreuzer, on our book "A Course in Mathematical Cryptography," published by DeGruyter in 2015.

In 2000, I (Ben Fine) mentioned to Gilbert that the Magnus framework would be ideal with statistical evaluations, where the problems of whether the algorithms ter-

https://doi.org/10.1515/9783110638387-203

minated or not did not appear. In Gilbert's inimitable style, he said "great idea. You go ahead and do it." That began a 15-year project at CAISS developing CAISS-Stat that involved the programming work of Xiaowei Xu, Yeagor Brjukhov, P. C. Wong, and others as well as the work of Gilbert and myself. We developed what I felt was an excellent package which if further developed properly would have rivaled SAS and SPSS. What we developed up to a point was excellent, but we learned just how difficult it is to remove bugs from a multiuser software package. We also learned how difficult it is to actually market a new idea – no matter how good it is. I also learned how enthusiastic Gilbert could become. We ran several very nice and instructive seminars on CAISS-Stat. Through all of this work, it was more that just mathematics. We had conversations about all conceivable topics (we agreed on all things political). I considered Gilbert my good friend and mentor and I miss him.

Chuck Miller

Looking us up individually on MathSciNet, one finds a graphic showing that we were each others most frequent collaborator. We each had a number of other collaborators, and Gilbert was a lot more productive than I was. But we had a productive and enjoyable collaboration over many years. Our collaboration began with a mistaken example. In about 1974, I was a junior faculty member at Princeton, and Gilbert had returned to CUNY the year before. We saw each other fairly often in the NY Group Theory Seminar. One day he phoned me to tell me about a group that he and Frank Cannonito had constructed because he knew I would be interested.

Their group was (supposedly) locally free but could not be imbedded in a finitely presented group. He described the construction, but I could not immediately see how it worked. After a day or so thinking about this, I came up with a proof that every locally free group could be imbedded in a finitely presented group. So their purported example could not exist. It was also clear the method applied in more general circumstances where all the isomorphisms between finitely generated subgroups can be recursively enumerated. One simply realizes all of these isomorphisms in a large HNN extension which is then embeddable in a finitely presented group. After a few phone calls and mail messages, Gilbert was convinced. At his suggestion, we joined forces and eventually published four papers on this and related topics. Years earlier, in 1963–1964, I was a masters degree student in a subject on finitely presented groups that Gilbert taught at NYU.

There were several appealing aspects of Gilbert's subject. Some tools used were free groups, amalgamated free products, and HNN extensions. Basic questions were: when are groups finitely presented or finitely generated and the same for subgroups? One learned these constructions can be used to make new groups with interesting properties; for instance, the non-hopfian Baumslag-Solitar group BS(2, 3) they had recently concocted. Also one studied decision problems and showed the Adian–Rabin theorem that most properties are not recognizable. I found the connections with logic and the ability to construct things of particular interest. After that year, I moved to Illinois which had strengths in algebra and logic, and eventually Bill Boone was my PhD supervisor.

So Gilbert and that one subject had a strong influence on my mathematical interests and career. In 1976, I moved to Melbourne, but managed to visit the US fairly regularly. In the fall of 1979, I was on sabbatical resident at IAS when we began another collaborative effort on the homology of finitely presented groups with Eldon Dyer. We showed that any suitably given sequence of abelian groups could be realized as the homology sequence of a finitely presented group. Although this is a remarkable result, it is frustratingly short of a characterization. Like all the collaborations I have been involved in, this had elements of social enjoyment as well as mathematical pleasure.

Over the next 10 years, I was heavily involved in administration in Melbourne and visits to New York were less frequent. When I did came to New York, Gilbert often had a project in mind that he hoped we might do something toward. Resulting from one such project was our short paper "Some odd finitely presented groups" which we later jokingly called our most underrated paper. In it, we constructed a nontrivial finitely presented group which maps onto its direct square. So it is non-hopfian. A related group has a finitely generated derived group which is isomorphic to its direct square.

Starting in 1991, Gilbert became quite involved in the project to develop the Magnus computational group theory package. During much of that year, I was in New York and was also involved. Among other things, I wrote a collection of programs for manipulating words and subgroups in free groups. I also began collaborations with Gilbert and Hamish Short and eventually Martin Bridson.

After 2000, my obligations in Melbourne lessened and I visited New York more regularly. Gilbert was still concerned with the Magnus project and funding, but he managed time for lots of small research projects. He had a talent for suggesting projects one could make progress on in finite time. The results were generally interesting if not all that deep. He was also very generous with help for students and colleagues. For one such project, we were trying to better understand a group constructed by B. H. Neumann in the 1930s. But we only talked about this group while riding on the crosstown bus. We could just pick up where we had left off on a previous trip. This was good sport. With all the mathematical jargon, we were sure other passengers thought we were very strange.

Gilbert was a good friend and a very talented mathematician. We enjoyed each others company and doing mathematics together. He was remarkably generous with ideas and particularly helpful to students.

There is another aspect of his personality that I admired. When dealing with someone at say a counter in an office or a teller or an official, he would chat with them in an understanding way that indicated appreciation of their circumstance. It often put them at ease and made their lives seem a bit brighter, as though he was spreading a little joy.

Martin Bridson

Gilbert Baumslag was my friend. We shared passions – for cricket, for arguing about how the world should be, for a measure of carousing, and above all for mathematics, especially the mathematics of group theory, about which he taught me a very great deal. For many years, starting in 1989, it was through conversation that he nurtured my appreciation of the elegance and achievements of pre-geometric group theory. But his great influence on our subject is in large part a function of his ability to express his many insights with such clarity in his writing. He was a master of explaining his solution to hard problems by exhibiting concise examples, and he had a well-honed skill for illuminating key ideas without stinting on rigor.

It is remarkable to reflect on how often Gilbert's papers opened doors to new worlds that held many wonders. One such paper that has been of particular significance in my life is Gilbert's article with Jim Roseblade on the subgroups of a direct product of two non-abelian free groups. Gilbert and Jim showed that, on the one hand, the array of finitely generated subgroups is strikingly diverse, while on the other hand, the only finitely presented subgroups are the obvious ones. This insight, and the homological techniques used to establish it, were the first key steps in what became a remarkably rich theory of subdirect products of hyperbolic and related groups, a subject that was the scene of many adventures for Gilbert, myself, Jim Howie, Chuck Miller, and Hamish Short (in various combinations) over 15 years.

Gilbert and I built many groups together but only published four papers. This is an inadequate indication of how much his mathematics influenced mine. A more accurate picture emerges if one scans the list of references in my other papers: he is there in many of them, for a variety of reasons – sometimes for his work on isoperimetric inequalities, subdirect products or equations over groups, but also for his seminal work on metabelian and nilpotent groups, profinite completions, and non-hopfian groups. His insights have populated my universe with great characters that I would not otherwise have known. My son, James, is a gifted cricketer in whom Gilbert took a particular delight. After Gilbert died, I pinned an email from Gilbert next to a picture on my office wall showing James in the act of bowling. I had shared that picture with Gilbert and his typically brief reply was one of the last things he wrote to me: "I envy you James. Love to you both, Gilbert."

> "We are all completely mad," he would often say. But oh what a life-affirming madness it was. I miss him greatly.

Martin Bridson, Oxford, January 2019

Anthony Gaglione and Dennis Splellman

Our Reflections on Gilbert Baumslag:

Dennis recalls attending a talk of Paul Schupp in which he was discussing groups with a certain property – call it P. (Dennis does not remember what P was!) Schupp's theorem was that finitely presented (or maybe finitely generated?) groups are generically – P. He justified the importance of his theorem with the assertion that unless you are Gilbert Baumslag, concrete examples of such groups are notoriously difficult to find. Such was the respect for Gilbert's power among his contemporaries. Not only does Gilbert's prolific outpouring of significant research merit recognition but also his encouragement of others.

Dennis also recalls the late Seymour Lipschutz relating to him Seymour's discussing a result of his with Gilbert and Gilbert replying that the result implies a solution to the conjugacy problem for cyclically-pinched one relator groups and that Seymour must publish. Thus, also in his encouragement of others, Gilbert helped advance the frontiers of our science.

In the New York Group Theory Seminar, Magnus had posed the question of whether or not the surface groups are residually finite. In a paper published in 1962, Gilbert established the stronger result that the surface groups are residually free. Here, he introduced the technique of big powers. For this, we are forever in his debt as we use that very technique over and over again in our own work. Implicit in Gilbert's 1962 paper is the fact that the surface groups are not only residually free but even freely discriminated in the sense that one can distinguish finitely many distinct elements by mapping into a free group. This paper along with a paper published in 1967 by his brother Benjamin was the basis for our result (independently proven by Remeslennikov) that, among finitely presented non-abelian groups, the freely discriminated ones are precisely the universally free examples. Incidentally, having mentioned the New York Group Theory Seminar, we would be remiss in not pointing out that that seminar's storied longevity is in no small measure due to Gilbert's stewardship for many years. Consulting Hanna Neumann's classic monograph on "Varieties of Groups," we see that Gilbert along with Bernhard, Hanna and Peter M. Neumann introduced (what we shall now call) varietally discriminating groups. Given a nontrivial variety of groups V, a group G in V is said to discriminate V provided given finitely many words $w(x_1, \ldots, x_n)$ such that none of the equations $w(x_1, \ldots, x_n) = 1$ is a law in V, there are elements $g_1, \ldots, g_n$ in G such that simultaneously $w(g_1, \ldots g_n) = 1$ for all the W. A group G is then varietally discriminating provided it discriminates the variety it generates. Suppose c and r are integers with $c > 2$ and let N be free of rank r in the variety N_c of groups nilpotent of class at most c. Then, since N is universally equivalent to the free group of countably infinite rank in N_C, N discriminates N_c. From that, it follows that N is varietally discriminating. Below are listed three papers in which we had collaborated with Gilbert (and for which we are indebted for his clear prose.) We want here to discuss one of these only; namely, the one on the nondiscrimination of nilpotent groups (see [3]). That would appear to contradict our discussion of free nilpotent groups above but here Gilbert was instrumental in promoting a different concept of discriminating group.

(It is somewhat amusing that Gilbert confessed to having forgotten his decades earlier definition joint with the Neumanns!) The newer definition of discriminating group is equivalent to the following: A group is discriminating provided it discriminates its direct square. From that, one immediately deduces that a discriminating group is universally equivalent to its direct square. With Gilbert's guidance, using the technique of Malcev completions (at which Gilbert was a master), we proved that no nilpotent group is discriminating unless it is abelian. (Torsion-free abelian groups are easily proven to be discriminating.) An ad hoc argument, without using the above theorem, that the groups N alluded to earlier cannot be discriminating in the newer sense follows. The fact that the centralizer of any element outside the $(c+1)$-st term of the upper central series is abelian is captured by a universal sentence which is true in N but false in $N \times N$. The theorem on nondiscrimination of nilpotent groups was proven independently using centralizer dimensions by Myasnikov et al. Although we have for decades studied the intersection of group theory and model theory, we both started our careers in the commutator calculus. Tony recalls that on a visit to the U.S. Naval Academy, Gilbert suggested to us that we study that ranks of the lower central quotients of free polynilpotent groups. This study ultimately led to the publication of a paper in which we gave recurrence relations to effectively compute the ranks of the lower central quotients of any free abelian by nilpotent group.

References

[1] G. Baumslag, O. Bogopolski, B. Fine, A. M. Gaglione, G. Rosenberger, and D. Spellman, On some finiteness properties infinite groups, Algebra Colloquium, Vol. 15 (2008), 1–22.

[2] G. Baumslag, B. Fine, A. M. Gaglione, and D. Spellman, Reflections on discriminating groups, Journal of Group Theory, Vol. 10, (2007), 87–99.

[3] G. Baumslag, B. Fine, A. M. Gaglione, and D. Spellman, A note on the nondiscrimination of nilpotent groups and Malcev completions, Combinatorial Group Theory, Discrete Groups, and Number theory, AMS Contemporary Mat. Series Vol. 421, (2006), 29–34.

Doug Troeger

CAISS was founded to bring together the Science and Engineering communities at City College for the sharing of gains in mathematics, computer science, and related disciplines, and exploring the practical extensions of these gains. Gilbert Baumslag was the founding director of CAISS.

Building on the software package MAGNUS developed under Baumslag's leadership in the 1990s, the primary goals were to apply algebra and group theory to a wide array of applications, and to extend and expand the approach to computer algebra embodied in MAGNUS.

CAISS' ties to computer science were formalized with Baumslag's transfer to the Department of Computer Science in 2007. He was the key to recruiting N. Fazio, R. Gennaro, L. Gurvits, and W.Skeith to join the computer science faculty, with simultaneous appointment to CAISS. CAISS remains a vital center at City College; R. Gennaro currently serves as director. Projects illustrating Baumslag's role in bridging computer science and mathematics initiated by CAISSin the years 2000–2011 included efforts to:

1. Combine Axiom, MetaPrl, and MAGNUS to create software joining theorem proving and symbolic computation (A Next-Generation Computational Science Platform, proposal to NSF ITR, with S. Artemov, G. Baumslag, S. Cleary, T. Daly, J. Hickey, A. Nogin, W. Sit, and D. Troeger [2003]);
2. Port MAGNUS to the Sharp SL5600, a Linux-based PDA (G. Baumslag, T. Daly, J. Niu);
3. Use finitely presented groups as a replacement for social security numbers, for access control in data bases and electronic communications, leveraging computations carried out on hand-held devices running MAGNUS (Social Security Numbers, Electronic Communication and Finitely Presented Groups, proposal to NSF ITR, with G. Baumslag, S. Cleary, T. Daly, A. Kawaguchi, C. Miller [2003]; Universal Passwords and Cryptography Based on Group Theory, proposal to NSF ITR, with G. Baumslag, S. Cleary, T. Daly, A. Kawaguchi, and M. Anshel [2004]);
4. Extending AXIOM to allow indefinite computing in arbitrary mathematical domains (Parametric Computations in Axiom: Towards Indefinite Symbolic Computing, proposal to NSF CCF, with G. Baumslag, S. Cleary, T. Daly, W. Sit, and D. Troeger [funded 2006–2008]);
5. Enhance MAGNUS to (a) take full advantage of parallel computation on multiple CPUs – specifically on CAISS' 132-node Beowulf cluster, (b) integrate with the general purpose computer algebra system SAGE, (c) exploit functional programming to provide computational tools for working with infinitely presented groups, and (d) support a new interface for visual programming (Interactive Computation with Finitely Presented Groups, proposal to NSF, with G. Baumslag, P. Brinkman, Y. Bryukhov, C. Miller, and D. Troeger [2008]);
6. Use the MAGNUS front-end as a zero learning curve interface to a variety of backends, such as GAP, a linear algebra package, and a statistics package, resulting in CAISS-Stat (G. Baumslag, Y. Bryukhov, T. Daly, M. Dean, B. Fine, X. Xu, and M. Zyman);
7. Use finitely presented groups as a source of computer games, leading to Expacon (G. Baumslag, Y. Bruykhov);
8. Use the group randomizer system from MAGNUS for challenge-response security (Challenge Response Password Security Using Combinatorial Group Theory, G. Baumslag, B. Fine, Y. Bryukhov, and D. Troeger, Groups – Complexity – Cryptology, Volume 2, Issue 1, pp. 67–81 [2010]);
9. Use noncommutative groups for homomorphic encryption (Provable Security from Group Theory and Applications, proposal to NSF TC, N. Fazio, W. Skeith, V. Shpilrain, and G. Baumslag [funded 2011–2016]).

CAISS contributed to the bridging of the mathematics and computer science departments as well by organizing a number of conferences and workshops of interest to both and held on the City College campus. These included Software for the Working Mathematician [2003], Axiom Conference [2005], Computation and Complexity [2006], Visualization Day [2008], Security and Privacy Day [2011], and Faces of Modern Cryptography [2011]. In addition, CAISS maintained a small Linux laboratory for the use of undergraduates, facilitating their participation in many of the Center's projects. T. Daly and R. Bruykhov offered courses on open-source programming using this lab, which was also used to host summer research programs for talented high school students.

The Gilbert Group

Definition 0.1. *The Baumslag Group*: advisees and protegees of Gilbert Baumslag during his last few decades; Kati Bencsath, Mariana Bonanome, Anthony Clement, Peggy Dean, Sal Liriano, Steve Majewicz, Gretchen Ostheimer, Marcos Zyman

1. Beginnings

Peggy: I first met Gilbert when I was an undergraduate and he was my Abstract Algebra professor in 1975. What joy that class was! Gilbert led us into wonderland and let us explore. I also took non-Euclidean geometry with Gilbert, and again he guided us through strange and marvelous places. Gilbert became my unofficial advisor at that time, and it was he who explained that I must go to graduate school, and helped me make it happen.

Gretchen: Gilbert was a mentor to me. My advisor took me to meet him in 1993 to get feedback about a problem we had identified as a possible thesis. Gilbert said it was a hard problem, maybe too hard. And then, when I solved it and started writing it up, I discovered that Gilbert himself had solved it many years prior! We joked that that's one liability of having such a huge body of work it was impossible for him to remember even his own results! It all worked out well: I was able to generalize the result and to begin my career as an independent researcher.

Steve: Every Friday, the Graduate Center offers various seminars. One of them was the Group Theory Seminar. I think it was in 1998 that I attended my first seminar. I remember being one of several students who would join the professors for coffee and cookies before the seminar (you can guess who some of the other students were – the *Baumslag Group*). I was not very sociable at the time, so I didn't interact with anyone. One person in the crowd with an accent that I was unfamiliar with attracted a lot of attention, including my own Gilbert. He was very comical and intelligent, and seemed to be the kind of person that I would like to learn from. I did not know if he taught at the GC, nor did I know his teaching style. Luckily for me, he was the speaker at the seminar. He gave a fascinating and intriguing talk in the area of combinatorial group theory. I immediately became interested in learning more about the subject, and I wanted Gilbert to teach me.

Anthony: I met Gilbert Baumslag between 2000 and 2001, when I took a group theory course he was teaching. I was working toward becoming an algebraic topologist. After my first class with Gilbert, I immediately changed my mind. Gilbert's teaching was smooth and captivating. He made group theory not only accessible, but even somewhat easy for me to follow. His style of joking throughout his lectures succeeded in making his audience feel completely relaxed and engaged as we learned.

Marianna: In 2001, I took a combinatorial group theory course at the CUNY Graduate Center and was fortunate to have Gilbert as my professor. What a magical time! I was finding my way as a graduate student and walking into Gilbert's class was like coming home.

Marcos: I am forever honored and humbled to have crossed paths with Gilbert Baumslag. In 2001, I audited a course he gave on combinatorial group theory at the CUNY Graduate Center. I had never seen such remarkable teaching in my life. It was because of him that I became a PhD student. Ever since that initial encounter, Gilbert's teaching and advice marked me in a profound way, both mathematically and personally. I never stopped attending his lectures from then on.

2. Teaching

Marianna: Gilbert was dynamic, supportive and full of humor in the classroom. He wrangled his students into group theoretic explorations requiring patience, skill, and creativity. He made every topic new at each revisiting and we his students happily revisited with him. He made me understand that it is okay to pull apart an idea, a theorem, a proof, a construction, again and again for weeks and years at a time. For years, I continued to take every class I could with Gilbert, even after graduating, sometimes taking the same class several times. I still return to my notes from these classes. They are prized possessions. Gilbert's spirit lives on in all of our teaching styles. The Baumslag Group, strives to inject joy, awe, and appreciation of mathematics in our classrooms; it is our way to honor our dear friend.

Peggy: Twenty years later, having left graduate school for a job, the time was right to go back and finish my degree. Gilbert was now on the CUNY GC faculty. I started taking his topics in group theory course, every semester for about 4 semesters. About midway through each lecture, Gilbert would ask us his signature question: Are you feeling weak, or are you feeling strong?" After my course work was complete, I kept sitting in on his courses along with other members of the Baumslag Group. There were often more unregistered students than registered!

Marcos: Gilbert was the embodiment of generosity. He loved discussing and sharing ideas with everyone. In the spring of 2007, Gilbert taught a course in finitely generated solvable groups at the CUNY Graduate Center. He gave his notes to some members of the Baumslag Group to organize them, fill in details, produce examples, expand them, and add our own thoughts. This is how I came to learn about semidirect products, wreath products, HNN-extensions, and, of course, finitely generated solvable groups. The notes eventually became a book. This is an example of who Gilbert was. He was all about offering projects, stimulating ideas, and teaching wonderful mathematics. He did all this because he enjoyed it, with boundless energy, and a remarkable sense of humor.

Anthony: Following that first class with Gilbert, I started reading some of his papers, even though I could not fully understand most of them. I met with Gilbert several times afterwards, and he directed me to read a particular paper, "On generalised free products." This was the beginning of Gilbert becoming my unofficial advisor. A year or two later, I was still unsure whether he was my official advisor. I did not wish to disrupt the momentum we had established; we had been men at work. One afternoon when Gilbert was speaking with one of his colleagues I overheard him saying, "my student Anthony"; then I knew I was officially Gilbert's student.

Steve: Gilbert had a way of explaining difficult subject matter that was understandable for all of the students. He made the course enjoyable by throwing a few jokes around here and there. And he would always ask the audience if they had questions. One day, I met with Gilbert after class to ask for guidance. At that time, I was interested in nilpotent groups and exponential A-groups. Little did I know that Gilbert wrote a book on nilpotent groups. This was certainly meant to be! Gilbert took me under his wing and gave me a thesis project.

Doing Mathematics

Anthony: Gilbert Baumslag was a very kind, generous, and loving teacher and advisor. I met with Gilbert, religiously, every week for about 5 years. During my last year of meeting with Gilbert, I was having some difficulty completing my PhD thesis problem. He consoled and encouraged me by saying "There are lots of people out there with PhDs in Mathematics. I know that you know much more than many of them." That statement, coming from Gilbert, bolstered my self-confidence. Gilbert then left for Europe for 3 months. During his absence, I worked with renewed vigor and inspiration. Just before Gilbert returned, I solved and completed the problem. Gilbert will always have a special place in my heart. I am extremely grateful to have spent some time on this earth with him.

Marcos: Interacting with Gilbert was electric. His sheer passion for mathematics and for teaching sustained me throughout graduate school and sustains me still. I often remember those Friday evenings, on our way to dinner after the Group Theory seminar, walking around Manhattan with Gilbert and other seminar goers. Gilbert would discuss math in the street, over dinner, everywhere; always motivated by deep intellectual curiosity. His presence is still felt.

Peggy: Gilbert became my advisor for my PhD thesis. It took me forever. Gilbert continues to be my inspiration for mathematics to this very day. Because of Gilbert, Marianna and I joined forces to write a book on "A Sampling of Remarkable Groups." When I lay awake at night struggling with some problem, I always thought, "What would Gilbert do?"

Marianna: I had the privilege of working as a research assistant for Gilbert at his lab at CCNY for several years and he was also my thesis advisor for a year. He had a knack for knowing exactly what challenge to pose to me. I recall him writing problems on the board in his office (playing with elements from free products with amalgamation) for me to solve. He would leave me to struggle, chat with someone else, then come back to check on me, always with his white ceramic mug in hand (never a disposable cup in sight!) Gilbert generously doled out ideas, matching them to interest and ability. When I told him I had an idea to use methods from quantum computation to solve decision problems in combinatorial group theory, he encouraged me to find a physicist to advise and guide me, which I did. Gilbert continued to guide me mathematically and pushed me to innovate. Peggy and I still argue: "What would Gilbert do?" What we don't argue about anymore is "He must have made a mistake here!"

Steve: Friday was the best day of the week, the day that I would go to the GC for a class taught by Gilbert, go to a seminar and sit with Gilbert, and/or discuss group theory with Gilbert. He was a person that you just wanted to hang out with. I can still hear him say "Every finitely generated torsion-free nilpotent group has a poly-infinite cyclic and central series. Make sure that you can prove this in your sleep." After completing my PhD, I had the idea of writing a book on nilpotent groups, written for both students and professors in the field. Gilbert gave me some ideas. He emphasized that it would take a lot of work and a long time to complete. He advised me that no matter how good the book is, there will always be critics. I pursued the project with Anthony and Marcos. During the years that we worked on it, Gilbert helped us fill in some of the missing pieces from the literature. His support, encouragement, and guidance was truly heartening, especially at times when the work became tiresome and endless. The book, entitled 'The Theory of Nilpotent Groups,' was finally published in 2017. Unfortunately, Gilbert passed on before seeing it. It is dedicated to him. I often think about Gilbert and how he has touched the lives of so many people. God bless him.

Gretchen: In 2005, I started working with Gilbert and Chuck Miller on a question about metabelian groups, and then we just kept on working. Eventually, we published a paper about intersections of subgroups and then we moved our attention to decompositions of nilpotent groups. We were trying to figure out under what conditions it could be decided whether a given nilpotent group was decomposable. My last meeting with Gilbert and Chuck together was in August 2014. We were able to finally see our way to an algorithm for deciding decomposability in the case of finitely generated torsion-free nilpotent groups. Gilbert got his diagnosis that month, and was gone within 2 months. It turned out there were quite a few details yet to be worked out, and the work was finally published in 2016. Gilbert was a friend. We didn't talk about personal stuff that much. But Gilbert knew when I was going through difficult personal stuff. He got it, he just got it that life is hard sometimes and that's the way it is, and knowing that Gilbert got it was always a great solace to me. I really don't know how I got so lucky to have been able to work with Gilbert for almost 20 years. It is certainly one of the things in this life for which I feel most grateful. Maybe it wasn't luck, but rather Gilbert's incredible generosity. He gave to so many of us, regardless of our academic stature, that which was most precious: his treasure trove of questions, his time, and his serious attention.

Addendum

Peggy: My *current husband*, David, attended my defense. He knew nothing of mathematics, and at one point Gilbert looked over at him and asked "David, are you understanding any of this?" Gretchen, who was sitting next to David, said "He's been taking notes this whole time!" Here are the results of those notes (to be sung to the tune of "Old McDonald"):

The Group Theory Group Anthem

Gilbert Baumslag had a group non-Abelian.
And this group was torsion-free, torsion, torsion-free.
Reprise:
Torsion here, and free free there,
A torsion portion, and it's free,
We are on a modding spree.
Gilbert Baumslag had a group non-Abelian.
And this group was nilpotent, nil, nil pot-pot-tent.
With a nil-nil here and potent there,
Nil-a-nil and pot-pot-tent,
An isomorph? Ah there it went.
Reprise
And this group was para-free, para-para free,
With para here and free-free there,
Para free isn't really free,
It's a fact, o can't you see?
With a nil-nil here and potent there,
Reprise
And this group has an Ore Domain, Ore Ore Domain.
With an Ore here and Domain there,
Ore Ore, Do-Do main,
This Ore Domain is one big pain.
With para here and free-free there,
A + B equals sub-prime P

With a nil-nil here and potent there,
Reprise
And this group has a skew fields, skew-skew, skew-skew field.
With a skew-skew here and a field-field there.
Here a skew, there a field,
This skew fields is no big deal.
With an Ore here and Domain there,
With para here and free-free there, B + B equals sub-prime P
With a nil-nil here and potent there,
Reprise
And then there is a wreath product. Wreath-wreath, prod-prod-duct.
With wreath-wreath here and a product there,
Here a wreath, there a duct,
Which makes us ask,
So why a duct?
With a skew-skew here and a field-field there.
With an Ore here and Domain there,
With para here and free-free there,
P to the second power equals B.
With a nil-nil here and potent there,
Reprise
And a Marcos here and a Gretchen there,
And a Kati here and a Peggy there,
And Delaram and Anthony,
Marianna, Shu, and Niu,
I like this group at any price,
For this group is very nice [don't be fooled that's a technical term!]
Here a wreath, there a duct,
With a skew-skew here and a field-field there.
With an Ore here and Domain there,
With para here and free-free there,
B + A = P + [B B B]
With a nil-nil here and potent there,
Reprise

Ilya Kapovich

The 4 years (1992–1996) that I was a student of Gilbert Baumslag at the CUNY Graduate Center, formed and defined me as a mathematician. Gilbert played a crucial role in this process. I learned many invaluable lessons from Gilbert during that time: How to think about mathematics, how to write mathematics, how to present and talk about mathematics, how to teach mathematics, and ultimately, how to be a mathematician.

Gilbert gave me the time and the freedom necessary to discover what kind of mathematics most agreed with my talents and inclinations, and helped me chose a project with the right mix of geometry, algebra and combinatorics. I also learned from Gilbert that every graduate student is different, and there is no single formula or standard for advising them. My most intense mathematical interactions with Gilbert occurred when I was working on my first several papers. The process of writing the first of these papers, on small cancellation theory, was particularly agonizing (as I thought at the time). I must have gone through a hundred drafts of just the Introduction. After giving each draft to Gilbert, I would get it back densely covered in red ink comments. Every time I thought to myself: Is this pain ever going to end? Gilbert would always smile and say: "Ilya, don't worry, we will get there, it is getting better!" Looking at that paper now, I think that Gilbert should have made me through a hundred iterations more! He worked hard to break some of my bad writing habits and to replace them by good ones. He taught me the importance of mathematical discipline and clarity, of avoiding complicated notation, overly long (as he said, Tolstovian) sentences and paragraphs, of keeping the main arguments short, clear, and well articulated, of breaking long proofs into shorter lemmas and propositions, and so on. These lessons define my mathematical writing style to this day, and I have tried to impart them on my own students. Gilbert taught me to direct my writing toward graduate students and mathematicians just entering the field, and not worry too much about trying to impress the experts and the snobs. A crucial conversation, also from the time of writing that small cancellation theory paper, concerned how to decide which mathematics is important. I told Gilbert that I was concerned that in my 20-plus page paper the crucial argument boiled down to a single page with a couple of pictures. Gilbert said: "Ilya, don't worry!" He said that he would tell me a big mathematical secret, which was that in any paper, no matter how long, there is usually a single short place where a crucial argument occurs. If a paper has two such places, then the paper is uncommonly good, and if it has three then, as he put it, "the paper is Inventiones level work." I learned from Gilbert that at the end you have to trust your own intuition about what kind of mathematics is interesting and important, and not be afraid to pose your own questions and problems. While I was struggling with the Introduction for one of my papers, Gilbert showed me the paper that he was working on the time, called "Musings on Magnus." When I read his Introduction, I was blown away: Not only did he use the pronoun "I" twice in the opening sentence, the first word of that sentence was "I." At the time, I could not imagine doing something so bold and brave in a mathematical paper. I learned from Gilbert the importance of storytelling in writing and presenting mathematics, and in my own papers I have been trying to tell stories ever since then.

So far, I had given Gilbert seven "mathematical PhD grandchildren," with two more on the way. In part through them, Gilbert's mathematical legacy and his genius continue to live on. I am proud and honored to have been Gilbert Baumslag's student.

Ilya Kapovich, Professor, Hunter College of CUNY.

Delarm Kahrobaei

I graduated in 2004 under direction of Distinguished Professor Baumslag, writing a thesis related to residual solvability. Afterwards, I went to the University of St. Andrews (UK) as an assistant professor. Then for 12 years I was at CUNY, where I got my Full Professorship in 2015, and became a faculty member at the PhD Program in Computer Science. I am currently the chair of cyber security at the University of York (UK) in the Department of Computer Science. I also hold an adjunct professorship position at NYU, Department of Computer Science.

Gilbert Baumslag really helped me to pursue my passion to be a mathematician and computer scientist. He gave me encouragement and support. He also helped me to become a better person.

He encouraged me to dream big, and some of my major accomplishments in part are due to him. He told me examples of successful women mathematicians such as Hanna Neumann, who proved great theorems while she was raising children.

He was great at fostering an environment that created opportunities for other people. For example, he organized many conferences and seminars at City College and the Graduate Center focused on many aspects of group theory and some in cryptography. At these conferences/seminars, I met my key collaborators such as Vladimir Shpilrain.

Following the model put forward by Gilbert, I have since organized numerous conferences including many which that were designed to encourage women mathematicians and computer scientists.

He also encouraged me to take on leadership roles. For instance, he inspired me to create and be a director of York Interdisciplinary Centre for Cyber Security (University of York, UK, www.cs.york.ac.uk/security).

He was a superb lecturer and mentor. The first conference talk I ever gave was at the Albany Group Theory Conference. He coached me for over a month. He said I am a mathematical descendent of Philip Hall and I must give beautiful lectures.

Scientifically. I learned a lot from his mentorship to pay attention to details and also his interests in the interdisciplinary fields between combinatorial algebra and cryptography. This made me define a new career path to be a better scholar.

He had many students, and learning from his mentorship skills, I have had 8 PhD students (8 academic grandchildren for him), whom all graduated with degrees in mathematics and computer science and have gotten prestigious positions both in academia and industry.

I will miss him very much.

Professor Delaram Kahrobaei, Chair of Cyber Security Department of Computer Science (CSE 039), University of York, U.K. Professor of Computer Science (Adjunct), New York University, U.S.A. Doctoral Faculty in the PhD Program in Computer Science (Adjunct) Graduate Center, The City University of New York, U.S.A. http://www-users.cs.york.ac.uk/~delaram/.

Taken on January 23, 2004, Dinner with Gilbert Baumslag after the NY Group Theory Seminar
Left to right: Bob Gilman, Gilbert Baumslag, Michael Thompson's girlfriend, Michael Thompson, Anthony E. Clement, Indira Chatterji and Murray Elder

Delaram Kahrobaei and Gilbert Baumslag

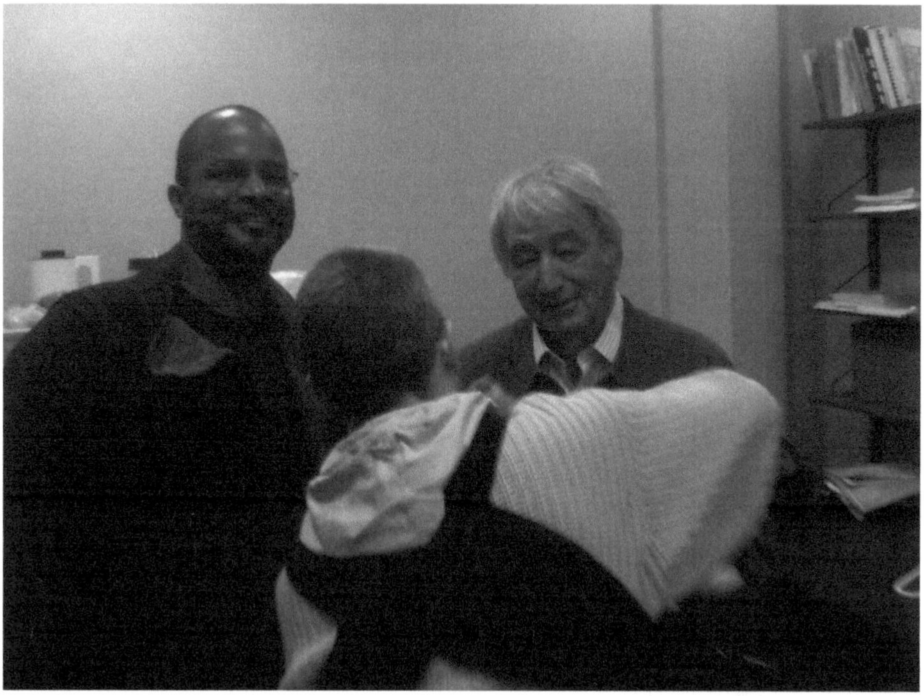

Taken on February 4th, 2011 at Gilbert Baumslag's office at the Graduate Center, NY
Left to right: Anthony E. Clement (Gilbert's Student), Gretchen Ostheimer (collaborator) and Gilbert Baumslag

Kati Bencsath and Gilbert Baumslag

Young Gilbert Baumslag

Left to right: Anthony Gaglione, Alexei Miasnikov, Ben Fine, Gerhard Rosenberger, guest speaker, Dennis Spellman, Gilbert Baumslag, and Delaram Kahrobaei

Contents

Sergey Bakulin and Alexei Miasnikov

Explicit universal axioms for Kaplansky groups

Abstract: In this paper, we give an explicit set of universal axioms in the pure group theory language for the class of groups whose group algebras over an arbitrary field have no zero divisors. This answer a question posed by Fine, Gaglione, Rosenberger, and Spellman.

Keywords: Kaplansky conjecture, Kaplansky groups, universal axioms

MSC 2010: 20B07

1 Introduction

In [1], Fine, Gaglione, Rosenberger, and Spellman introduced a class $\mathcal{K}$ of groups whose group algebras over fields do not have zero divisors (they called them Kaplansky groups) and proved that the class $\mathcal{K}$ is universally axiomtazible in group theory language. Their argument does not provide any explicit set of universal axioms, so they posed an open problem to find such a set. In this paper, we provide an explicit and natural set of universal axioms for $\mathcal{K}$.

There is a well-known *Kaplansky's zero divisors conjecture* in group theory, that states that for any torsion free group G its group algebra over a field does not have zero divisors [2]. The conjecture was confirmed for some particular classes of groups, but is still open in general. Of course, if the conjecture is true then $\mathcal{K}$ is axiomatized by the following set of universal axioms:

$$\{\forall x(x^n = 1 \rightarrow x = 1) \mid n = 1, 2, 3, \dots\}.$$

It is interesting to come up with an explicit set of universal axioms for $\mathcal{K}$ independently whether the conjecture is true or not. Furthermore, it may happen (if the conjecture fails) that there are groups G such that the group algebras FG are domains (have no zero divisors) for some fields F and not so for others. To this end, for a field F by $\mathcal{K}_F$ we denote the class of groups G whose group algebras FG are domains. In a similar fashion, if p is a prime number or zero by $\mathcal{K}_p$ we denote the class of groups G such that FG is a domain for any field of characteristic p.

To get an universal axiomatization of $\mathcal{K}$, we first describe an explicit set of universal axioms $\mathcal{A}_F$ for the class $\mathcal{K}_F$ for every field F. We show that if the Diophantine

Acknowledgement: The results of this paper were obtained with support of Russian Science Foundation grant (project no. 19-11-00209).

Sergey Bakulin, Alexei Miasnikov, Department of Mathematics, Stevens Institute of Technology, K, 1 Castle Point Terrace #217, Hoboken, NJ 07030, USA, e-mail: amiasnikov@gmail.com

https://doi.org/10.1515/9783110638387-001

problem is decidable in F, then the set $\mathcal{A}_F$ is recursive. Furthermore, we show that if F_p is a prime field of characteristic p (so $F_0 = \mathbb{Q}$) and $\bar{F}_p$ is its algebriac closure then the set of axioms $\mathcal{A}_{\bar{F}_p}$ also gives a recursive set of universal axioms for the class $\mathcal{K}_p$. Now the union $\mathcal{A} = \cup_p \mathcal{A}_{\bar{F}_p}$ is an explicit set of universal axioms for the class $\mathcal{K}$. Note that in the notation above

$$\mathcal{K} = \bigcap\{\mathcal{K}_F \mid F \text{ is a finite field}\}$$

(see, e. g., [3] for the proof) and

$$\mathcal{K}_0 = \mathcal{K}_{\mathbb{C}} = \mathcal{K}_{\bar{\mathbb{Q}}} = \bigcap\{\mathcal{K}_F \mid [F : \mathbb{Q}] < \infty\}$$

(see below).

2 Universal axioms for $\mathcal{K}_F$

In this section, we describe a system of universal axioms for the class $\mathcal{K}_F$ for an arbitrary field F.

Fix a field F, a group G, and numbers $n, m \in \mathbb{N}$. Denote $I_n = \{1, \dots, n\}$. Let $g = (g_1, \dots, g_n) \in G^n$ and $h = (h_1, \dots, h_m) \in G^m$ be two tuples of elements of G such that $g_i \neq g_j$ for $i \neq j \in I_n$, and $h_p \neq h_q$ for $p \neq q \in I_m$. Consider two elements (where $\alpha_i, \beta_j \in F$)

$$u = \Sigma_{i=1}^n \alpha_i g_i, \quad v = \Sigma_{j=1}^m \beta_j h_j$$

from the group algebra FG. The support of u is the set $\{g_1, \dots, g_n\}$. Now, the condition $uv = 0$, $u \neq 0$, $v \neq 0$ is equivalent to the following:

(1) There is a partition

$$I_n \times I_m = \bigcup_{k=1}^{s} E_k$$

of the set $I_n \times I_m$ such that:

(2) For any $k = 1, \dots, s$, for any $(i, j), (p, q) \in E_k$,

$$g_i h_j = g_p h_q.$$

(3) For any $k = 1, \dots, s$,

$$\Sigma_{(i,j)\in E_k} \alpha_i \beta_j = 0.$$

(4) $\prod_{i=1}^n \alpha_i \prod_{j=1}^m \beta_j \neq 0$.

Given a partition, say P, as in (1), the condition (2) is about elements in G, and the conditions (3) and (4) about elements in F. If one denotes by $Q_{P,k}$, the polynomial

$$Q_{P,k}(x_1,\ldots,x_n,y_1,\ldots,y_m) = \Sigma_{(i,j)\in E_k} x_i y_j$$

and by $Q_{n,m}$ the polynomial

$$Q_{n,m}(x_1,\ldots,x_n,y_1,\ldots,y_m) = \prod_{i=1}^{n} x_i \prod_{j=1}^{m} y_j$$

then the condition (3) states that $\bar{\alpha} = (\alpha_1,\ldots,\alpha_n)$ and $\bar{\beta} = (\beta_1,\ldots,\beta_m)$ are solutions of the polynomial equation $Q_{P,k}(\bar{x},\bar{y}) = 0$ in F for $k = 1,\ldots,s$ (here, $\bar{x} = (x_1,\ldots,x_n)$ and $\bar{y} = (y_1,\ldots,y_m)$), while the condition (4) states that $Q_{n,m}(\bar{\alpha},\bar{\beta}) \neq 0$. Since F is a field the latter inequality can be written in terms of equations introducing a new variable, say z. Namely, consider an integral polynomial

$$Q_{n,m}(\bar{x},\bar{y},z)^* = Q_{n,m}(\bar{x},\bar{y})z - 1.$$

It is clear that $Q_{n,m}(\bar{\alpha},\bar{\beta}) \neq 0$ in F if and only if $Q_{n,m}(\bar{\alpha},\bar{\beta},\gamma)^* = 0$ in F for some $\gamma \in F$. Denote the following system of integral polynomial equations by $S_P(\bar{x},\bar{y},z)$:

$$(\bigwedge_{k=1}^{s} Q_{P,k}(\bar{x},\bar{y}) = 0) \wedge (Q_{n,m}(\bar{x},\bar{y},z)^* = 0)$$

In fact, the coefficients in the system $S_P(\bar{x},\bar{y},z) =$ are all equal to 1, so one may consider it as a polynomial system over any unitary commutative ring, in particular, a field F. It follows that $\bar{\alpha},\bar{\beta}$ satisfy (3) and (4) if and only if $S_P(\bar{\alpha},\bar{\beta},\gamma) = 0$ for some $\gamma \in F$.

Let $\mathcal{P}_{n,m,F}$ be the set of all partitions of $I_n \times I_m$ for which the system $S_P = 0$ has a solution in F.

Now we deal with the condition (2). Let $P = (E_1,\ldots,E_s)$ be a partition of $I_n \times I_m$. Put

$$\Phi_{P,n,m}(\bar{x},\bar{y}) = \bigwedge_{k=1}^{s} \bigwedge_{(i,j),(p,k)\in E_k} (x_i y_j = x_p y_q).$$

Clearly, for any $\bar{g} \in G^n$, $\bar{h} \in G^m$ the following holds:

$$G \models \Phi_{P,n,m}(\bar{g},\bar{h}) \iff \bar{g} \text{ and } \bar{h} \text{ satisfy (2)}.$$

We need also the following formula that says that all components in $\bar{g}$ and $\bar{h}$ are pairwise distinct:

$$\Delta_{m,n}(\bar{x},\bar{y}) = \bigwedge_{1\le i<j\le n} (x_i \neq x_j) \bigwedge_{1\le i<j\le m} (y_i \neq y_j)$$

Put

$$\Psi_{P,n,m}(\bar{x},\bar{y}) = \Delta_{m,n}(\bar{x},\bar{y}) \wedge \Phi_{P,n,m}(\bar{x},\bar{y}).$$

The discussion above is summarized in the following lemma.

Lemma 1. *Let $P \in \mathcal{P}_{n,m,F}$ and $\bar{g} \in G^n$, $\bar{h} \in G^m$ such that*

$$G \models \Psi_{P,n,m}(\bar{g}, \bar{h}).$$

Then for any solution $(\bar{\alpha}, \bar{\beta}, \gamma)$ of the system $S_P(\bar{x}, \bar{y}, z) = 0$ in F (and such a solution does exist) the elements $u = \Sigma_{i=1}^{n} \alpha_i g_i$ and $v = \Sigma_{j=1}^{m} \beta_j h_j$ are nonzero and $uv = 0$ in FG.

Notice that the set $\mathcal{P}_{n,m,F}$ is finite so one can consider the formula

$$\Psi_{n,m,F} = \exists \bar{x} \exists \bar{y} \Big(\bigvee_{P \in \mathcal{P}_{n,m,F}} \Psi_{P,n,m}(\bar{x}, \bar{y}) \Big).$$

Lemma 2. *Let G be a group. Then $\Psi_{n,m,F}$ holds in G if and only if there are two nonzero elements $u, v \in FG$ such that $|\mathrm{supp}(u)| = n$, $|\mathrm{supp}(v)| = m$ and $uv = 0$.*

Proof. Follows from Lemma 1. □

Theorem 1. *The set*

$$\mathcal{A}_F = \{ \neg \Psi_{n,m,F} \mid n, m \in \mathbb{N}, n, m \geq 2 \}$$

is a set of universal axioms of the class $\mathcal{K}_F$. Furthermore,

(1) *if F is countable with decidable equality problem, then the set $\mathcal{A}_F$ is computably enumerable;*
(2) *if F is countable with decidable Diophantine problem, then the set $\mathcal{A}_F$ is computable.*

Proof. The first statement of the theorem follows from Lemma 2.

If the field F is countable and the equality problem in F is decidable, one can computably enumerate all systems $S_P(\bar{x}, \bar{y}, z) = 0$ which have a solution in F, by trying all possible $\alpha \in F^n$, $\beta \in F^m$, $\gamma \in F$ and checking whether or not the element $S_P(\bar{\alpha}, \bar{\beta}, \gamma)$ is equal to 0 in F. This proves the statement (1).

If the Diophantine problem in F is decidable, then the set of those systems $S_P(\bar{x}, \bar{y}, z) = 0$ which have solutions in F is computable. This proves (2). □

Corollary 1. *Let p be a prime number or zero, F_p the prime field of characteristic p and $\bar{F}_p$ its algebraic closure. Then the set $\mathcal{A}_{\bar{F}_p}$ is a computable set of universal axioms of the class $\mathcal{K}_p$.*

Proof. Let F be an arbitrary field of characteristic p. Then F is contained in its algebraic closure $\bar{F}$, hence for any partition P of the set $I_n \times I_m$ as above, if the system $S_P(\bar{x}, \bar{y}, z) = 0$ has a solution in F then it has a solution in $\bar{F}$. This implies that $P_{n,m,F} \subseteq S_{n,m,\bar{F}}$, so $\mathcal{A}_F \subseteq \mathcal{A}_{\bar{F}}$. Note that the system $S_P(\bar{x}, \bar{y}, z) = 0$ has a solution in $\bar{F}$ if and only if it has a solution in $\bar{F}_p$ (or any other algebraically closed field of characteristic p). Hence $\mathcal{A}_{\bar{F}} = \mathcal{A}_{\bar{F}_p}$. It follows that $\mathcal{A}_{\bar{F}_p}$ is a set of universal axioms of the class $\mathcal{K}_p$. The Diophantine problem in the field $\bar{F}_p$ is decidable, that is, there is an algorithm that for any P as above decides whether or not the system $S_P(\bar{x}, \bar{y}, z) = 0$ has a solution in $\bar{F}_p$. Hence, the set $\mathcal{A}_{\bar{F}_p}$ is computable, as claimed. □

The following result answers the question posed in [1].

Theorem 2. *The following are sets of universal axioms of $\mathcal{K}$:*

$$\mathcal{A}_{\text{fin}} = \bigcup\{\mathcal{A}_F \mid F \text{ is a finite field}\},$$
$$\mathcal{A}_{\text{acl}} = \bigcup\{\mathcal{A}_{\bar{F}_p} \mid p \text{ is a prime or zero}\}.$$

Proof. As we mentioned above, $\mathcal{K} = \mathcal{K}_{\text{fin}}$. Therefore, by Theorem 1 the set $\mathcal{A}_{\text{fin}}$ is, indeed, a set of universal axioms of $\mathcal{K}$. Similarly, $\mathcal{K} = \cap_p \mathcal{K}_p$ so by Corollary 1 A_{acl} is also a set of universal axioms of $\mathcal{K}$. □

3 Axioms for $\mathcal{K}_{\mathbb{Z}}$

In this section, we provide a different set of universal axioms for the class $\mathcal{K}_{\mathbb{Z}}$.

Let G be a group. Every element $u \in \mathbb{Z}G$ can be written in the form

$$u = (g_1 + \cdots + g_n) - (h_1 + \cdots + h_m), \tag{1}$$

where $g_i, h_j \in G$, and $g_i \neq h_j$ for any $i \in I_n$, $j \in I_m$. In this case, we say that the tuples $g = (g_1, \ldots, g_n)$ and $h = (h_1, \ldots, h_m)$ represent the element $u \in \mathbb{Z}G$. Note that if $m, n > 0$ then $u \neq 0$.

The following formula

$$\Upsilon_{1,n,m}(x, y) = \bigwedge_{i=1}^{n} \bigwedge_{j=1}^{m} x_i \neq y_j, \tag{2}$$

holds on tuples g, h if and only if these tuples give a representation of the type (1).

It was shown by Passman in [4] (Lemma 13.1.2) that $\mathbb{Z}G$ has zero divisors if and only if there exist a nonzero element $u \in \mathbb{Z}G$ such that $u^2 = 0$. Now for u given as in (1) we express the condition $u^2 = 0$ in terms of the tuples g and h.

For arbitrary tuples $a = (a_1, \ldots, a_n)$ and $b = (b_1, \ldots, b_m)$ denote by $a \cup b$ the concatenation of a and b, and by ab the tuple $(a_1b_1, \ldots, a_nb_m)$, whose components are "formal products" a_ib_j ordered in an arbitrary but fixed order. Let $u = (g_1 + \cdots + g_n) - (h_1 + \cdots + h_m) \in \mathbb{Z}G$ be a nonzero element such that $u^2 = 0$. Then

$$\begin{aligned} u^2 &= ((g_1 + \cdots + g_n) - (h_1 + \cdots + h_m))^2 \\ &= \sum_{i,j=1}^{n} g_ig_j + \sum_{i,j=1}^{n} h_ih_j - \left(\sum_{i,j=1}^{n} g_ih_j + \sum_{i,j=1}^{n} h_ig_j \right) = 0 \end{aligned}$$

So, the element u^2 is represented by a pair of tuples $gg \cup hh$ and $gh \cup hg$. Taking the augmentation homomorphism $\mathbb{Z}G \to \mathbb{Z}$, which sends $G \to 1$, and applying it to u^2, one can conclude that $m = n$. Furthermore, $u^2 = 0$ if and only if some permutation of the tuple $gg \cup hh$ gives the tuple $gh \cup hg$. This is equivalent to the condition that these tuples satisfy in G the following formula:

$$\Upsilon_{2,n}(x,y) = \bigvee_{\varphi \in S_{2n^2}} \bigwedge_{i=1}^{2n^2} x_i = y_{\varphi(i)}, \tag{3}$$

where S_{2n^2} is the symmetric group on $2n^2$ symbols.

For $n \geq 1$, put

$$\Upsilon_n = \exists x \exists y \Upsilon_{1,n,n}(x,y) \bigwedge \Upsilon_{2,n}(xx \cup yy, xy \cup yx). \tag{4}$$

The discussion above gives the following result.

Lemma 3. *The sentence Υ_n holds in a group G if and only if there exists a nonzero element $u \in \mathbb{Z}G$ with representation(1) such that $u^2 = 0$.*

The following result follows from Lemma 3.

Theorem 3. *The class $\mathcal{K}_{\mathbb{Z}}$ is axiomatizable by the set of universal axioms*

$$\{\neg \Upsilon_n \mid n = 1, 2, 3, \ldots\}.$$

Bibliography

[1] Fine B, Gaglione A, Rosenberger G, Spellman D. Orderable groups, elementary theory, and the Kaplansky conjecture. Groups Complex Cryptol. 2018;10(1):43–52.
[2] Kaplansky I. Problems in the Theory of Rings. Am Math Mon. 1970;77:445–54. (Revisited).
[3] Malman B. Zero-divisors and Idempotents in Group Rings, Master's theses in mathematical sciences, ISSN 1404-6342, Lund University, 2014.
[4] Passman DS. The Algebraic Structure of Group Rings. Pure and Applied Mathematics. New York: Wiley-Interscience [John Wiley & Sons]; 1977.

Gilbert Baumslag, Martin R. Bridson, and Charles F. Miller III

Isoperimetric and isodiametric functions of group extensions

Abstract: We establish a general upper bound for the isoperimetric and isodiametric functions of an extension of one finitely presented group by another.

Keywords: Isoperimetric and isodiametric functions, group extensions

MSC 2010: 20F10

This note[1] was written to record a conversation between the three authors that took place in Bryant Park, New York, in 1992. That conversation was about the geometry of the word problem in finitely presented groups, more specifically isoperimetric and isodiametric inequalities.

If $\langle A \mid R \rangle$ is a finite presentation for a group G, then a word w in the free group $F = F(A)$ will equal the identity in G if and only if it can be expressed in F as a product of conjugates of the defining relations and their inverses,

$$w =_F \prod_{i=1}^{M(w)} \theta_i r_i \theta_i^{-1}$$

with $r_i \in R^{\pm 1}$. A function $f : \mathbb{N} \to \mathbb{N}$ is said to be an *isoperimetric function* for G if for all $n \in \mathbb{N}$ and all words of length $|w| \leq n$ that represent the identity in G, one can find such an expression with $M(w) \leq f(n)$. And $g : \mathbb{N} \to \mathbb{N}$ is said to be an *isodiametric function* for G if for all such words there is an equality in which the lengths of the conjugating elements θ_i are no greater than $g(n)$. The dependence of these definitions on the chosen presentation of G can be removed by working modulo the equivalence relation $f_1 \simeq f_2$ that equates functions if $f_1 \preceq f_2$ and $f_2 \preceq f_1$, where $f_1 \preceq f_2$ means that there is a constant $C > 0$ such that $f_1(n) \leq Cn + f_2(Cn + C)$ for all positive integers n.

When we say that (f, g) is a pair of isoperimetric–isodiametric functions for G, we assume that the above bounds on $M(w)$ and $|\theta_i|$ hold simultaneously. Isoperimetric

1 Edited slightly in July 2019 by the second and third authors to be included in a memorial volume for the first author, who died in October 2014.

Gilbert Baumslag, Department of Mathematics, City College of New York, Convent Ave, and 138th St., New York, NY 10031, USA, e-mail: gilbert@groups.sci.ccny.cuny.edu
Martin R. Bridson, Mathematical Institute, University of Oxford, ROQ, Woodstock Road, Oxford OX2 6GG, United Kingdom, e-mail: bridson@maths.ox.ac.uk
Charles F. Miller III, Department of Mathematics and Statistics, University of Melbourne, Melbourne 3010, Australia, e-mail: c.miller@ms.unimelb.edu.au

https://doi.org/10.1515/9783110638387-002

and isodiametric functions are often studied using geometric methods but the arguments in this note are purely algebraic.

Theorem (Bryant Park lemma). *Suppose that*

$$1 \to N \to G \to Q \to 1$$

is an exact sequence of finitely presented groups. Let f be an isoperimetric function for N, and let (g,h) be a pair of isoperimetric–isodiametric functions for Q. Then there is a positive integer constant c and an isoperimetric function $\delta_G(n)$ for G such that

$$\delta_G(n) \le c^n + g(n) + g(n)c^{h(n)+1} + f(c^n + g(n)c^{h(n)+1}).$$

If the exact sequence splits, so that G is a semidirect product $N \rtimes Q$, then $\delta_G(n) \le c^n + g(n) + f(c^n)$.

If the action of Q on N is trivial, then $\delta_G(n) \le n^2 + g(n) + cg(n)h(n) + f(cg(n))$.

Proof. We fix finite presentations

$$N = \langle a_1, \dots, a_\alpha \mid r_1 = 1, \dots, r_\rho = 1\rangle$$

and

$$Q = \langle b_1, \dots, b_\beta \mid s_1 = 1, \dots, s_\sigma = 1\rangle.$$

From these, one can derive a finite presentation for G (see, for instance, [4]). Indeed, for suitable words u_i $(i = 1, \dots, \sigma)$ and x_{ij}, y_{ij} $(i = 1, \dots, \alpha, j = 1, \dots, \beta)$ on the a_i, the group G can be presented as

$$\begin{aligned}&\langle a_1, \dots,\ a_\alpha, b_1, \dots, b_\beta \mid r_1 = 1, \dots, r_\rho = 1,\ s_1 = u_1, \dots, s_\sigma = u_\sigma,\\ &b_i^{-1} a_j b_i = x_{ij}\ (i = 1, \dots, \alpha, j = 1, \dots, \beta),\\ &b_i a_j b_i^{-1} = y_{ij}\ (i = 1, \dots, \alpha, j = 1, \dots, \beta)\rangle.\end{aligned}$$

In the special case of a split extension, all of the words u_i can be chosen to be 1 and so the corresponding relations are $s_i = 1$.

Consider an arbitrary word w on the generators of G, of length $|w| = n$, such that $w =_G 1$. We want to find an upper bound for the number of uses of the defining relations needed to prove $w =_G 1$.

Let k denote the maximum of the lengths of all the words u_i, x_{ij}, and y_{ij}. We will deal only with the case $k \ge 2$. In case $k = 1$, somewhat better bounds can be derived.

Using the relations $b_i^{-1} a_j b_i = x_{ij}$ and $b_i a_j b_i^{-1} = y_{ij}$, we can successively move all of the b_i symbols to the right over all the a_j until we obtain a word of the form $w_a w_b =_G w$ where w_a is a word on the a_j and w_b is a word on the b_i. To move a b_i from left to right over m of the a_j requires m uses of the defining relations and replaces these a_j by at most mk symbols. In w, there are at most n occurrences of the letters a_j and at most n of the b_i. Considering the worst case in which all the b_i are at the left to start, one calculates that $|w_a| \leq nk^n$ and that at most

$$n(1 + k + \cdots + k^{n-1}) = n\frac{k^n - 1}{k - 1} \leq nk^n$$

uses of the defining relations are required. Observe that the number of symbols in w_b is at most the number of b_i in w and so $|w_b| \leq n$.

Now Q is naturally the quotient of G and so $w_b =_Q 1$. Thus in the free group F_b on the b_i we have an equation of the form

$$w_b =_{F_b} z_1 s_{j_1}^{\epsilon_1} z_1^{-1} \cdots z_q s_{j_q}^{\epsilon_q} z_q^{-1}$$

where $\epsilon = \pm 1$ and where we may assume $q \leq g(|w_b|) \leq g(n)$ and $|z_i| \leq h(n)$ for $i = 1, \ldots q$. (In the special case of a split extension, we must have $w_b =_G 1$, and this can be proved with $q \leq g(n)$ uses of the defining relations.)

In G, we have defining relations $s_j = u_j$ where u_j is a word on the a_i. Thus

$$z_i s_{j_i}^{\epsilon_i} z_i^{-1} =_G z_i u_{j_i}^{\epsilon_i} z_i^{-1}.$$

Now, using the relations $b_i^{-1} a_j b_i = x_{ij}$ and $b_i a_j b_i^{-1} = y_{ij}$, we can successively pass the symbols of z_i from left to right over a_j symbols in $u_{j_i}^{\epsilon_i}$ and freely reduce to obtain $z_i u_{j_i}^{\epsilon_i} z_i^{-1} =_G t_i$ where t_i is a word on the a_j. Calculating as before, we find $|t_i| \leq |u_{j_i}| k^{|z_i|}$ and that the number of relations needed is at most

$$|u_{j_i}|(1 + k + \cdots + k^{|z_i|-1}) = |u_{j_i}| \frac{k^{|z_i|} - 1}{k - 1} \leq |u_{j_i}| k^{|z_i|}.$$

Let $\overline{w_b} = t_1 \cdots t_q$. Using the estimates $|u_{j_i}| \leq k$, $|z_i| \leq h(n)$ and $q \leq g(n)$, it follows that $|\overline{w_b}| \leq g(n)k^{h(n)+1}$. Moreover, the proof of $w_b =_G \overline{w_b}$ requires at most $g(n) + g(n)k^{h(n)+1}$ uses of the defining relations. (Here, the initial $g(n)$ relations are required to replace the s_{j_i} by the corresponding u_{j_i}.)

Combining this with earlier estimates, it follows that

$$|w_a \overline{w_b}| \leq nk^n + g(n)k^{h(n)+1}$$

and that the proof of $w =_G w_a \overline{w_b}$ requires at most

$$nk^n + g(n) + g(n)k^{h(n)+1}$$

uses of the defining relations. Using the isoperimetric function f for N, it follows that the proof that $w_a \overline{w_b} =_G 1$ requires at most $f(nk^n + g(n)k^{h(n)+1})$ uses of the defining relations. Adding the various contributions, the proof that $w =_G 1$ therefore requires at most

$$nk^n + g(n) + g(n)k^{h(n)+1} + f(nk^n + g(n)k^{h(n)+1})$$

uses of the defining relations.

Observe that $nk^n \le k^{2n} = (k^2)^n$ and $k^{n+1} \le k^{2n}$ provided $n \ge 1$. Thus, setting $c = k^2$, the above isoperimetric estimate can be replaced by

$$c^n + g(n) + g(n)c^{h(n)+1} + f(c^n + g(n)c^{h(n)+1}).$$

(In the case of a split extension, the discussion simplifies to give an isoperimetric upper bound of $nk^n + g(n) + f(nk^n)$, and hence of $c^n + g(n) + f(c^n)$.)

This completes the proof of the theorem. □

The hypotheses of the following corollary hold, for instance, when N is a finitely generated nilpotent group [3] and Q is an asynchronously combable group [2], [1].

Corollary. *Suppose that*

$$1 \to N \to G \to Q \to 1$$

is an exact sequence of finitely presented groups. Assume:
(1) *that an isoperimetric function f for N is bounded above by some polynomial n^p;*
(2) *that an isoperimetric function g for Q is bounded above by a simple exponential c_0^n, where $c_0 \ge 1$ is a constant; and*
(3) *that, simultaneously, the isodiametric function h for Q is bounded by $c_1 n$, where $c_1 \ge 1$ is a constant.*

Then there is a positive integer constant C such that C^n is an isoperimetric function for G.

Proof. According to the theorem, G has an isoperimetric function

$$\delta(n) \le c^n + c_0^n + c_0^n c^{c_1 n+1} + (c^n + c_0^n c^{c_1 n+1})^p.$$

If we let c_3 be the maximum of c and c_0, we have

$$\begin{aligned}\delta(n) &\le c_3^{c_1 n+4} + (c_3^{c_1 n+3})^p \\ &\le c_3^{(p+1)c_1 n+3p+4} \le c_3^{(4p+5)c_1 n}.\end{aligned}$$

Set $C = c_3^{(4p+5)c_1}$. □

Bibliography

[1] Bridson MR. Combings of semidirect products and 3-manifold groups. Geom Funct Anal. 1993;3(3):263–78.

[2] Epstein DBA, Cannon wJW, Holt DF, Levy SVF, Paterson MS, Thurston WP. Word processing in Groups. Boston–London: Jones and Barlett; 1992.

[3] Gersten SM. Isoperimetric and Isodiametric Functions of Finite Presentations. In: Niblo G, Roller M, editors. Geometric Group Theory. London Math. Soc. Lecture Notes. vol. 181. Cambridge: Cambridge Univ. Press; 1993. p. 79–96.

[4] Hall P. Finiteness conditions for soluble groups. Proc Lond Math Soc (3). 1954;4:419–36.

Martin R. Bridson

The homology of groups, profinite completions, and echoes of Gilbert Baumslag

Abstract: We present novel constructions concerning the homology of finitely generated groups. Each construction draws on ideas of Gilbert Baumslag. There is a finitely presented acyclic group U such that U has no proper subgroups of finite index and every finitely presented group can be embedded in U. There is no algorithm that can determine whether or not a finitely presentable subgroup of a residually finite, biautomatic group is perfect. For every recursively presented abelian group A, there exists a pair of groups $i : P_A \hookrightarrow G_A$ such that i induces an isomorphism of profinite completions, where G_A is a torsion-free biautomatic group that is residually finite and superperfect, while P_A is a finitely generated group with $H_2(P_A, \mathbb{Z}) \cong A$.

Keywords: Homology of groups, undecidability, Grothendieck pairs

MSC 2010: 20J05, 20F10, 20E18, 20F65

1 Introduction

Gilbert Baumslag took a great interest in the homology of groups. Famously, with Eldon Dyer and Chuck Miller [10], he proved that an arbitrary sequence of countable abelian groups (A_n), with A_1 and A_2 finitely generated, will arise as the homology sequence $H_n(G, \mathbb{Z})$ of some finitely presented group G, provided that the A_n can be described in an untangled recursive manner. This striking result built on Gilbert's earlier work with Dyer and Alex Heller [9]. A variation on arguments from [10] and [9] yields the following result, which will be useful in our study of profinite completions of discrete groups. Recall that a group G is termed *acyclic* if $H_n(G, \mathbb{Z}) = 0$ for all $n \geq 1$.

Theorem A. *There is a finitely presented acyclic group U such that:*
(1) *U has no proper subgroups of finite index;*
(2) *every finitely presented group can be embedded in U.*

A recursive presentation $(X \mid R)_{\mathrm{Ab}}$ of an abelian group is said to be *untangled* if the set R is a basis for the subgroup $\langle R \rangle$ of the free abelian group generated by X. The following corollary can be deduced from Theorem A using the Baumslag–Dyer–Miller construction; see Section 3.

Note: For Gilbert Baumslag, in memoriam.

Martin R. Bridson, Mathematical Institute, Andrew Wiles Building, ROQ, Woodstock Road, Oxford, United Kingdom, e-mail: bridson@maths.ox.ac.uk

https://doi.org/10.1515/9783110638387-003

Corollary B. *Let $\mathcal{A} = (A_n)_n$ be a sequence of abelian groups, the first of which is finitely generated. If the A_n are given by a recursive sequence of recursive presentations, each of which is untangled, then there is a finitely presented group $Q_{\mathcal{A}}$ with no proper subgroups of finite index and $H_n(Q_{\mathcal{A}}, \mathbb{Z}) \cong A_{n-1}$ for all $n \geq 2$.*

In [14], Gilbert and Jim Roseblade used homological arguments to prove that every finitely presented subdirect product of two finitely generated free groups is either free or of finite index. This insight was the germ for a large and immensely rich body of work concerning residually free groups and subdirect products of hyperbolic groups, with the homology of groups playing a central role. The pursuit of these ideas has occupied a substantial part of my professional life [21, 24, 25, 26] and also commanded much of Gilbert's attention in the latter part of his career [7, 8, 13, 6]. A cornerstone of this program is the 1-2-3 theorem, which Gilbert and I proved in our second paper with Chuck Miller and Hamish Short [8].

The proof of the following theorem provides a typical example of the utility of the 1-2-3 theorem. It extends the theme of [7, 8], which demonstrated the wildness that is to be found among the finitely presented subgroups of automatic groups. It also reinforces the point made in [23] about the necessity of including the full input data in the effective version of the 1-2-3 theorem [26]. The proof that the ambient biautomatic group is residually finite relies on deep work of Wise [48, 49] and Agol [1] as well as Serre's insights into the connection between residual finiteness and cohomology with finite coefficient modules [46, Section I.2.6].

Theorem C. *There is no algorithm that can determine whether or not a finitely presentable subgroup of a residually finite, biautomatic group is perfect.*

To prove this theorem, we construct a recursive sequence (G_n, H_n) where G_n is a biautomatic group given by a finite presentation $\langle X \mid R_n \rangle$ and $H_n < G_n$ is the subgroup generated by a finite set S_n of words in the generators X; the cardinality of S_n and R_n does not vary with n. The construction ensures that H_n is finitely presentable, but a consequence of the theorem is that there is no algorithm that can use this knowledge to construct an explicit presentation of H_n. An artefact of the construction is that each G_n has a finite classifying space $K(G_n, 1)$.

Besides picking up on the themes of Gilbert mentioned above, Theorem C also resonates with a longstanding theme in his work, often pursued in partnership with Chuck Miller, whereby one transmits undecidability phenomena from one context to another in group theory by building groups that encode the appropriate phenomenon by means of graphs of groups, wreath products, directly constructed presentations, or whatever else one can dream up. This is already evident in his early papers, particularly [5].

I have discussed three themes from Gilbert Baumslag's oeuvre: **(i)** decision problems and their transmission through explicit constructions; **(ii)** homology of groups; and **(iii)** subdirect products of free and related groups. To these I add two more (neglecting others): **(iv)** a skill for constructing explicit groups that illuminate important phenomena, inspired in large part by his formative interactions with Graham Higman, Bernhard Neumann, and Wilhelm Magnus, and **(v)** an enduring interest in residual finiteness and nilpotence, with an associated interest in profinite and pronilpotent completions of groups.

In the 1970s, Gilbert and his students, particularly Fred Pickel [42, 33], explored the extent to which finitely generated, residually finite groups are determined by their finite images (equivalently, their profinite completions; see Section 2.5). He maintained a particular focus on residually nilpotent groups, motivated in particular by a desire to find the right context in which to understand parafree groups. In his survey [4], he writes: "More than 35 years ago, Hanna Neumann asked whether free groups can be characterized in terms of their lower central series. Parafree groups grew out of an attempt to answer her question." It is a theme that he returned to often; see [4]. I was drawn to the study of profinite completions later, by Fritz Grunewald [20]. As I have become increasingly absorbed by it, Gilbert's illuminating examples and provocative questions have been invigorating.

I shall present one result concerning profinite completions here and further results in the sequel to this paper [19].

The proof of the following result combines a refinement of Corollary B and parts of the proof of Theorem C with a somewhat involved spectral sequence argument. It extends arguments from Section 6 of [22] that were developed to answer questions posed by Gilbert in [4].

Recall that the profinite completion $\widehat{G}$ of a group G is the inverse limit of the directed system of finite quotients of G. A *Grothendieck pair* [32] is a pair of residually finite groups $\iota : A \hookrightarrow B$ such that the induced map of profinite completions $\hat{\iota} : \widehat{A} \to \widehat{B}$ is an isomorphism. Recall also that a group G is termed *superperfect* if $H_1(G, \mathbb{Z}) = H_2(G, \mathbb{Z}) = 0$.

Theorem D. *For every recursively presented abelian group A there exists a Grothendieck pair $P_A \hookrightarrow G_A$ where G_A is a torsion-free biautomatic group that is residually-finite, superperfect and has a finite classifying space, while P_A is finitely generated with $H_2(P_A, \mathbb{Z}) \cong A$.*

Note that A need not be finitely generated here; for example, A might be the group of additive rationals $\mathbb{Q}$, or the direct sum of the cyclic groups $\mathbb{Z}/p\mathbb{Z}$ of all prime orders.

The diverse background material that we require for the main results is gathered in Section 2.

2 Preliminaries

I shall assume that the reader is familiar with the basic theory of homology of groups ([15] and [27] are excellent references) and the definitions of small cancellation theory [38] and hyperbolic groups [31]. Recall that a classifying space $K(G, 1)$ for a discrete group G is a CW-complex with fundamental group G and contractible universal cover. $H_n(G,\mathbb{Z}) = H_n(K(G,1),\mathbb{Z})$. One says that G is of *type* F_n if there is a classifying space $K(G,1)$ with finite n-skeleton. Finite generation is equivalent to type F_1 and finite presentability is equivalent to type F_2.

2.1 Fiber products

Associated to a short exact sequence of groups $1 \to N \to G \xrightarrow{\eta} Q \to 1$ one has the *fiber product*

$$P = \{(g,h) \mid \eta(g) = \eta(h)\} < G \times G.$$

The restriction to P of the projection $G \times G \to 1 \times G$ has kernel $N \times 1$ and can be split by sending $(1,g)$ to (g,g). Thus $P \cong N \rtimes G$ where the action is by conjugation in G.

1-2-3 Theorem ([8]). *Let $1 \to N \to G \xrightarrow{\eta} Q \to 1$ be a short exact sequence of groups. If N is finitely generated, G is finitely presented, and Q is of type F_3, then the associated fiber product $P < G \times G$ is finitely presented.*

The *effective 1-2-3 theorem*, proved in [25], provides an algorithm that, given the following data, will construct a finite presentation for P: a finite presentation $G = \langle A \mid S\rangle$ is given, with a finite generating set for N (as words in the generators A), a finite presentation $\mathcal{P}$ for Q, a word defining $\eta(a)$ for each $a \in A$, and a set of generators for $\pi_2\mathcal{P}$ as a $\mathbb{Z}Q$-module.

The proof of Theorem C shows that one cannot dispense with this last piece of data, while Theorem D shows that the 1-2-3 theorem would fail if one assumed only that Q was finitely presented.

By definition, a generating set A for G defines an epimorphism $\mu : F \to G$, where F is the free group on A. We can choose a different presentation $Q = \langle A \mid R\rangle$ such that the identity map on A defines the composition $\eta \circ \mu : F \to Q$. The following lemma is easily checked.

Lemma 2.1. *With the above notation, the fiber product $P < G \times G$ is generated by the image of $\{(a,a),(r,1) \mid a \in A, r \in R\} \subset F \times F$.*

We shall also need an observation that is useful when computing with the LHS spectral sequences associated to $1 \to N \to G \to Q \to 1$ and to $1 \to N \times 1 \to P \to 1 \times G \to 1$. In the first case, the term $H_0(Q,H_1N)$ arises, which by definition is the group of coinvariants for the action of G on N by conjugation, i. e. $N/[N,G]$. The second spectral sequence contains the term $H_0(G,H_1N)$; here the action of $g \in G$ is induced by conjugation of (g,g) on $N \times 1 < G \times G$, so $H_0(G,H_1N)$ is again $N/[N,G]$. More generally, because the action of G on N is the same in both cases we have the following.

Lemma 2.2. *In the context described above, $H_0(Q,H_kN) \cong H_0(G,H_kN)$ for all $k \geq 0$.*

2.2 Universal central extensions

A *central extension* of a group Q is a group $\widetilde{Q}$ equipped with a homomorphism $\pi : \widetilde{Q} \to Q$ whose kernel is central in $\widetilde{Q}$. Such an extension is *universal* if given any other central extension $\pi' : E \to Q$ of Q, there is a unique homomorphism $f : \widetilde{Q} \to E$ such that $\pi' \circ f = \pi$.

The standard reference for universal central extensions is [40] pages 43–47. The properties that we need here are the following, which all follow easily from standard facts (see [17] for details and references).

Proposition 2.3.

(1) *Q has a universal central extension $\widetilde{Q} \to Q$ if and only if Q is perfect. (If it exists, $\widetilde{Q} \to Q$ is unique up to isomorphism over Q.)*

(2) *There is a short exact sequence*

$$1 \to H_2(Q, \mathbb{Z}) \to \widetilde{Q} \to Q \to 1.$$

(3) $H_1(\widetilde{Q}, \mathbb{Z}) = H_2(\widetilde{Q}, \mathbb{Z}) = 0.$

(4) *If Q has no nontrivial finite quotients, then neither does $\widetilde{Q}$.*

(5) *For $k \geq 2$, if Q is of type F_k, then so is $\widetilde{Q}$.*

(6) *If Q has a compact 2-dimensional classifying space $K(Q, 1)$, then $\widetilde{Q}$ is torsion-free and has a compact classifying space.*

The following result is Corollary 3.6 of [17]; the proof relies on an argument due to Chuck Miller.

Proposition 2.4. *There is an algorithm that, given a finite presentation $\langle A \mid R\rangle$ of a perfect group G, will output a finite presentation $\langle A \mid \overline{R}\rangle$ defining a group $\widetilde{G}$ such that the identity map on the set A induces the universal central extension $\widetilde{G} \to G$. Furthermore, $|\overline{R}| = |A|(1 + |R|)$.*

2.3 Applications of the Lyndon–Hochshild–Serre spectral sequence

Besides the Mayer–Vietoris sequence, the main tool that we draw on in our calculations of homology groups is the Lyndon–Hochshild–Serre spectral sequence associated to a short exact sequence of groups $1 \to N \to G \to Q \to 1$. The E^2 page of this spectral sequence is $E^2_{pq} = H_p(Q, H_q(N, \mathbb{Z}))$, and the sequence converges to $H_n(G, \mathbb{Z})$; see [27], page 171. A particularly useful region of the spectral sequence is the corner of the first quadrant, from which one can isolate the 5-term exact sequence

$$H_2(G, \mathbb{Z}) \to H_2(Q, \mathbb{Z}) \to H_0(Q, H_1(N, \mathbb{Z})) \to H_1(G, \mathbb{Z}) \to H_1(Q, \mathbb{Z}) \to 0. \tag{1}$$

From this, we immediately have the following.

Lemma 2.5. *Let $1 \to N \to G \to Q \to 1$ be a short exact sequence of groups. If $H_1(G,\mathbb{Z}) = H_2(G,\mathbb{Z}) = 0$, then $H_2(Q,\mathbb{Z}) \cong H_0(Q,H_1N)$.*

The following calculations with the LHS spectral sequence will be needed in the proofs of our main results.

Lemma 2.6. *If $1 \to N \to G \to Q \to 1$ is exact and N is acyclic, then $G \to Q$ induces an isomorphism $H_n(G,\mathbb{Z}) \to H_n(Q,\mathbb{Z})$ for every n.*

Proof. In the LHS spectral sequence, the only nonzero entries on the second page are $E^2_{n0} = H_n(Q,\mathbb{Z})$, so $E^2 = E^\infty$ and $H_n(G,\mathbb{Z}) \to E^\infty_{n0} = H_n(Q,\mathbb{Z})$ is an isomorphism. □

In the following lemmas, all homology groups have coefficients in the trivial module $\mathbb{Z}$ unless stated otherwise.

Lemma 2.7. *Let $1 \to N \to B \overset{\eta}{\to} C \to 1$ be a short exact sequence of groups. If $H_1N = H_2B = 0$ and $\eta_* : H_3B \to H_3C$ is the zero map, then $H_0(C,H_2N) \cong H_3C$.*

Proof. The hypothesis $H_1N = 0$ implies that on the E^2-page of the LHS spectral sequence, the terms in the second row E^2_{*1} are all zero. Thus all of the differentials emanating from the bottom two rows of the E^2-page are zero, so $E^3_{p0} = E^2_{p0}$ for all $p \in \mathbb{N}$ and $E^3_{0q} = E^2_{0q}$ for $q \leq 2$. Hence the only nonzero differential emanating from place $(3,0)$ is on the E^3-page, and this is $d_3 : H_3C \to H_0(C,H_2N)$. The kernel of d_3 is E^∞_{30}, the image of $\eta_* : H_3B \to H_3C$, which we have assumed to be zero. And the cokernel of d_3 is E^∞_{02}, which injects into H_2B, which we have also assumed is zero. Thus $d_3 : H_3C \to H_0(C,H_2A)$ is an isomorphism. □

Lemma 2.8. *Let $P = A \rtimes B$. If $H_1A = H_2B = 0$, then $H_2P \cong H_0(B,H_2A)$.*

Proof. By hypothesis, on the E^2-page of the LHS spectral sequence for $1 \to A \to P \to B \to 1$, the only nonzero term E^2_{pq} with $p+q = 2$ is $E^2_{02} = H_0(B,H_2A)$. It follows that $H_2P \cong E^\infty_{02} = E^4_{02}$. And since $E^2_{21} = H_2(B,H_1A) = 0$, we also have $E^3_{02} = E^2_{02} = H_0(B,H_2A)$. As B is a retract of P, for every n the natural map $H_nP \to H_nB$ is surjective, so all differentials emanating from the bottom row of the spectral sequence are zero. In particular, $d_3 : H_3B \to H_0(B,H_2A)$ is the zero map, and hence $E^4_{02} = E^3_{02} = H_0(B,H_2A)$. □

Lemma 2.9. *Let $1 \to N \to B \overset{\eta}{\to} C \to 1$ be a short exact sequence of groups. Suppose that H_1N is finitely generated, $H_1B = H_2C = 0$ and C has no nontrivial finite quotients. Then $H_1N = 0$.*

Proof. As H_1N is finitely generated, its automorphism group is residually finite. Thus, since C has no finite quotients, the action of C on H_1N induced by conjugation in B must be trivial and $H_0(C,H_1N) = H_1N$. From the LHS spectral sequence, we isolate the exact sequence $H_2C \to H_0(C,H_1N) \to H_1B$. The first and last groups are zero by hypothesis, so $H_1N = H_0(C,H_1N) = 0$. □

2.4 An adapted version of the Rips construction

Eliyahu Rips discovered a remarkably elementary construction [45] that has proved to be enormously useful in the exploration of the subgroups of hyperbolic and related groups. There are many refinements of his construction in which extra properties are imposed on the group constructed. The following version is well adapted to our needs.

Proposition 2.10. *There exists an algorithm that, given an integer $m \geq 6$ and a finite presentation $\mathcal{Q} \equiv \langle X \mid R\rangle$ of a group Q, will construct a finite presentation $\mathcal{P} \equiv \langle X \cup \{a_1, a_2\} \mid \tilde{R} \cup V\rangle$ for a group Γ so that:*

(1) *$N := \langle a_1, a_2\rangle$ is normal in Γ;*
(2) *Γ/N is isomorphic to Q;*
(3) *$\mathcal{P}$ satisfies the small cancellation condition $C'(1/m)$, and*
(4) *Γ is perfect if Q is perfect.*

Proof. The original argument of Rips [45] proves all but the last item. In his argument, one chooses a set of reduced words $\{u_r \mid r \in R\} \cup \{v_{x,i,\varepsilon} \mid x \in X^{\varepsilon},\ i = 1,2,\ \varepsilon = \pm 1\}$ in the free group on $\{a_1, a_2\}$, all of length at least $m\max\{|r| : r \in R\}$, that satisfies $C'(1/m)$. Then $\tilde{R} = \{ru_r \mid r \in R\}$ and V consists of the relations $xa_ix^{-1}v_{x,i,\varepsilon}$ with $x \in X$, $i = 1,2$, and $\varepsilon = \pm 1$. Such a choice can be made algorithmic (in many different ways).

To ensure that (4) holds, one chooses the words $v_{x,i,\varepsilon}$ to have exponent sum 0 in a_1 and a_2. Such a choice ensures that the image of N in $H_1\Gamma$ is trivial, so if $\Gamma/N \cong Q$ is perfect then so is Γ. One way to arrange that the exponent sums are zero is by a simple substitution: choose $\tilde{R} \cup V$ as above and then replace each occurrence of a_1 by $a_1a_2a_1^{-2}a_2^{-1}a_1$ and each occurrence of a_2 by $a_2a_1a_2^{-2}a_1^{-1}a_2$. If the original construction is made so that the presentation is $C'(1/5m)$, then this modified presentation will be $C'(1/m)$. □

Remark 2.11. Dani Wise [48] proved that metric small cancellation groups can be cubulated and, building on work of Wise [49], Agol [1] proved that cubulated hyperbolic groups are virtually compact special in the sense of Haglund and Wise [35]. In particular, the group Γ constructed in Proposition 2.10 is residually finite (cf. [47] and [35]). It also follows from Agol's theorem, via Proposition 3.6 of [34], that virtually compact special hyperbolic groups are good in the sense of Serre [46], meaning that for every finite $\mathbb{Z}G$-module M and $p \geq 0$, the map $H^p(\widehat{G}, M) \to H^p(G, M)$ induced by the inclusion of G into its profinite completion $G \hookrightarrow \widehat{G}$, is an isomorphism. We shall need this remark in our proof of Theorem D.

2.5 Profinite completions and Grothendieck pairs

Throughout, $\widehat{G}$ denotes the profinite completion of a group G. By definition, $\widehat{G}$ is the inverse limit of the directed system of finite quotients of G. The natural map $G \to \hat{G}$ is injective if and only if G is residually finite. A Grothendieck pair is a monomorphism

$u : P \hookrightarrow G$ of residually finite groups such that $\hat{u} : \hat{P} \to \hat{G}$ is an isomorphism but P is not isomorphic to G. The existence of nontrivial Grothendieck pairs of finitely presented groups was established by Bridson and Grunewald in [20] following an earlier breakthrough by Platonov and Tavgen in the finitely generated case [43].

The following criterion plays a central role in [43], [2], and [20].

Proposition 2.12. *Let $1 \to N \to H \to Q \to 1$ be an exact sequence of groups with fiber product P. Suppose H is finitely generated, Q is finitely presented, and $H_2(Q, \mathbb{Z}) = 0$. If Q has no proper subgroups of finite index, then the inclusion $P \hookrightarrow H \times H$ induces an isomorphism of profinite completions.*

It follows easily from the universal property of profinite completions that if $\widehat{G} \cong \widehat{H}$ then G and H have the same finite images. For finitely generated groups, the converse is true [44, pp. 88–89]. Asking for $P \hookrightarrow G$ to be a Grothendieck pair is more demanding than asking simply that there should be an abstract isomorphism $\widehat{P} \cong \widehat{G}$. To see this, we consider a pair of groups constructed by Gilbert Baumslag [3].

Proposition 2.13. *Let $G_1 = (\mathbb{Z}/25) \rtimes_{\alpha} \mathbb{Z}$ and let $G_2 = (\mathbb{Z}/25) \rtimes_{\alpha^2} \mathbb{Z}$, where $\alpha \in \mathrm{Aut}(\mathbb{Z}/25)$ is multiplication by 6.*

(1) $G_1 \not\cong G_2$.

(2) $\widehat{G_2} \cong \widehat{G_1}$.

(3) *No homomorphism $G_1 \to G_2$ or $G_2 \to G_1$ induces an isomorphism between $\widehat{G_1}$ and $\widehat{G_2}$.*

Proof. For $i = 1, 2$, let A_i be the unique $\mathbb{Z}/25 < G_i$. Each monomorphism $\phi : G_1 \to G_2$ restricts to an isomorphism $\phi : A_1 \to A_2$ and induces a monomorphism $G_1/A_1 \to G_2/A_2$. This last map cannot be an isomorphism: choosing a generator $t \in \mathbb{Z} < G_1$ so that $t^{-1}at = a^6$ for every $a \in A_1$ (writing the group operation in A multiplicatively), we have $\phi(t)^{-1}\alpha\phi(t) = \alpha^6$ for all $\alpha \in A_2$, whereas $\tau^{-1}\alpha\tau = \alpha^{\pm 11}$ for each $\tau \in G_2$ such that τA_2 generates G_2/A_2. This proves (1).

With effort, one can prove that G_1 and G_2 have the same finite quotients by direct argument after noting that any finite quotient $G_i \to Q$ that does not kill A_i must factor through $G_i \to A_i \rtimes (\mathbb{Z}/5k)$ for some k. Baumslag [3] gives a more elegant and instructive proof of (2).

As G_1 and G_2 are residually finite, any map $\phi : G_1 \to G_2$ that induces an isomorphism $\widehat{\phi} : \widehat{G}_1 \to \widehat{G}_2$ must be a monomorphism. The argument in the first paragraph shows in this case the image of ϕ will be a proper subgroup of finite index in G_2. If the index is $d > 1$, then the image of $\widehat{\phi}$ will have index d in $\widehat{G}_2$. The same argument is valid with the roles of G_1 and G_2 reversed, so (3) is proved. ☐

2.6 Biautomatic groups

The theory of automatic groups grew out of investigations into the algorithmic structure of Kleinian groups by Cannon and Thurston, and it was developed thoroughly in the book by Epstein et al. [29]; see also [11]. Let G be a group with finite generating set A and let A^* be the set of all finite words in the alphabet $A^{\pm 1}$. An *automatic structure* for G is determined by a normal form $\mathcal{A}_G = \{\sigma_g \mid g \in G\} \subseteq A^*$ such that $\sigma_g = g$ in G. This normal form is required to satisfy two conditions: first, $\mathcal{A}_G \subset A^*$ must be a *regular language*, that is, the accepted language of a finite state automaton; and second, the edge-paths in the Cayley graph $\mathcal{C}(G, A)$ that begin at $1 \in G$ and are labeled by the words σ_g must satisfy the following *fellow-traveler condition*: there is a constant $K \geq 0$ such that for all $g, h \in G$ and all integers $t \leq \max\{|\sigma_g|, |\sigma_h|\}$,

$$d_A\big(\sigma_g(t), \sigma_h(t)\big) \leq K d_A(g, h),$$

where d_A is the path metric on $\mathcal{C}(G, A)$ in which each edge has length 1, and $\sigma_g(t)$ is the image in G of the initial subword of length t in σ_g.

A group is said to be *automatic* if it admits an automatic structure. If G admits an automatic structure with the additional property that for all integers $t \leq \max\{|\sigma_g|, |\sigma_h|\}$,

$$d_A\big(a.\sigma_g(t), \sigma_h(t)\big) \leq K d_A(ag, h),$$

for all $g, h \in G$, and $a \in A$, then G is said to be *biautomatic*. Biautomatic groups were first studied by Gersten and Short [30]. Automatic and biautomatic groups form two of the most important classes studied in connection with notions of nonpositive curvature in group theory; see [18] for a recent survey.

The established subgroup theory of biautomatic groups is considerably richer than that of automatic groups. Biautomatic groups have a solvable conjugacy problem, whereas this is unknown for automatic groups. Groups in both classes enjoy a rapid solution to the word problem, and have classifying spaces with finitely many cells in each dimension. The isomorphism problem is open in both classes. No example has been found to distinguish between the two classes.

2.7 Some groups without finite quotients

Graham Higman [36] gave the first example of a finitely presented group that has no nontrivial finite quotients. Many others have been discovered since, including the group

$$B_p = \langle a, b, \alpha, \beta \mid ba^{-p}b^{-1}a^{p+1},\ \beta\alpha^{-p}\beta^{-1}\alpha^{p+1},\ [bab^{-1}, a]\beta^{-1},\ [\beta\alpha\beta^{-1}, \alpha]b^{-1}\rangle.$$

This presentation is aspherical for $p \geq 2$; see [20]. B_2 is a quotient of the 4-generator finitely presented group H that Baumslag and Miller concocted in [12]. There is a surjection $H \to H \times H$, from which it follows that H (and hence B_2) cannot map onto a nontrivial finite group: for if Q were such a group, then the number of distinct epimorphisms would satisfy $|\mathrm{Epi}(H, Q)| < |\mathrm{Epi}(H \times H, Q)|$, which is nonsense if H maps onto $H \times H$.

3 Proof of Theorem A and Corollary B

The proof of the following lemma is based on similar arguments in [9] and [10].

Lemma 3.1. *Let Π be a property of groups that is inherited by direct limits and suppose that every finitely presented group G can be embedded in a finitely presented group G_Π that has property Π. Let Π' be a second such property. Then there exists a group $U^\dagger = K \rtimes \mathbb{Z}$ such that:*
(1) *$U^\dagger$ is finitely presented;*
(2) *$U^\dagger$ contains an isomorphic copy of every finitely presented group;*
(3) *K has property Π and property Π'.*

Proof. Let U_0 be a finitely presented group that contains an isomorphic copy of every finitely presented group. The existence of such groups was established by Higman [37]. By hypothesis, there is a finitely presented group V that contains U_0 and has property Π, and there is a finitely presented group W that contains V and has property Π'. Consider the following chain of embeddings, where the existence of the embedding into $U_1 \cong U_0$ comes from the universal property of U_0,

$$U_0 < V < W < U_1. \tag{2}$$

We fix an isomorphism $\phi : U_1 \to U_0$ and define $U^\dagger$ to be the ascending HNN extension $\langle U_1, t \mid t^{-1}ut = \phi(u)\ \forall u \in U_1\rangle$. Let K be the normal closure of U_1 in $U^\dagger$ and note that this is the kernel of the natural retraction $U^\dagger \to \langle t\rangle$. Note, too, that $t^{-i}U_1t^i < U_0$ for all positive integers i. It follows that for each positive integer d, we can express K as an ascending union

$$K = \bigcup_{i\geq d} t^i U_1 t^{-i} = \bigcup_{i\geq d-1} t^i U_0 t^{-i}.$$

From (2), we deduce that K is the direct limit of each of the ascending unions $\bigcup_i t^i V t^{-i}$ and $\bigcup_i t^i W t^{-i}$. The first union has property Π, while the second has property Π'. □

3.1 Proof of Theorem A

Every finitely presented group can be embedded in a finitely presented group that has no finite quotients; see [16] for explicit constructions. And it is proved in [9] that every finitely presented group can be embedded in a finitely presented acyclic group. It is clear that having no nontrivial finite quotients is preserved under passage to direct limits, and acyclicity is preserved because homology commutes with direct limits. Thus Lemma 3.1 provides us with a finitely presented group $U^\dagger = K \rtimes \mathbb{Z}$ such that K is acyclic and has no nontrivial finite quotients.

Let B be a finitely presented acyclic group that has no nontrivial finite quotients and let $\tau \in B$ be an element of infinite order; we can take B to be B_p from Section 2.7, for example. Let $U = U^\dagger *_C B$ be the amalgamated free product in which $\langle \tau \rangle$ is identified with $C := 1 \times \mathbb{Z} < U^\dagger$.

As K is acyclic, by Lemma 2.6, $U^\dagger \to C$ induces an isomorphism $H_*(U^\dagger, \mathbb{Z}) \cong H_*(C, \mathbb{Z})$. In particular, $H_n(U^\dagger, \mathbb{Z}) = 0$ for $n \geq 2$, and in the Mayer–Vietoris sequence for $U = U^\dagger *_C B$ the only potentially nonzero terms are

$$0 \to H_2(U, \mathbb{Z}) \to H_1(C, \mathbb{Z}) \to H_1(U^\dagger, \mathbb{Z}) \oplus H_1(B, \mathbb{Z}) \to H_1(U, \mathbb{Z}) \to 0.$$

$H_1(B, \mathbb{Z}) = 0$ and $H_1(C, \mathbb{Z}) \to H_1(U^\dagger, \mathbb{Z})$ is an isomorphism, so we deduce that U is acyclic.

Each subgroup of finite index $S < U$ will intersect both $U^\dagger$ and B in a subgroup of finite index. Since neither has any proper subgroups of finite index, S must contain both $U^\dagger$ and H. Hence $S = U$. □

3.2 Proof of Corollary B

Theorem E of [10] (see also [39]) states that if $\mathcal{A} = (A_n)$ is as described in Corollary B then there is a finitely generated, recursively presented group $G_{\mathcal{A}}$ with $H_n(G_{\mathcal{A}}, \mathbb{Z}) \cong A_n$ for all $n \geq 1$.

By the Higman embedding theorem [37], $G_{\mathcal{A}}$ can be embedded in the universal finitely presented group U constructed in the preceding proof. We form the amalgamated free product of two copies of U along $G_{\mathcal{A}}$,

$$Q_{\mathcal{A}} := U *_{G_{\mathcal{A}}} U.$$

Note that because $G_{\mathcal{A}}$ is finitely generated, $Q_{\mathcal{A}}$ is finitely presented. As in the preceding proof, since the factors of the amalgam have no proper subgroups of finite index, neither does $Q_{\mathcal{A}}$.

The Mayer–Vietoris sequence for this amalgam yields, for all $n \geq 2$, an exact sequence (where the $\mathbb{Z}$ coefficients have been suppressed):

$$H_n U \oplus H_n U \to H_n Q_{\mathcal{A}} \to H_{n-1} G_{\mathcal{A}} \to H_{n-1} U \oplus H_{n-1} U.$$

Thus, since U is acyclic, $H_n Q_{\mathcal{A}} \cong H_{n-1} G_{\mathcal{A}} \cong A_{n-1}$ for all $n \geq 2$. □

Theorem E in [10] is complemented by a number of "untangling results" which avoid the untangled condition that appears in that theorem and in our Corollary B. The following is a special case of what is established in the proof of [10, Theorem G].

Proposition 3.2. *For every recursively presented abelian group A, there exists a finitely generated, recursively presented group G such that $H_1(G, \mathbb{Z}) = 0$ and $H_2(G, \mathbb{Z}) \cong A$.*

Exactly as in the proof of Corollary B, we deduce the following.

Corollary 3.3. *For every recursively presented abelian group A, there exists a finitely presented group Q_A with no proper subgroups of finite index such that $H_1(Q_A, \mathbb{Z}) = H_2(Q_A, \mathbb{Z}) = 0$ and $H_3(Q_A, \mathbb{Z}) \cong A$.*

4 Proof of Theorem C

The seed of undecidability that we need in Theorem C comes from the following construction of Collins and Miller [28].

Theorem 4.1. [28] *There is an integer k, a finite set X, and a recursive sequence (R_n) of finite sets of words in the letters $X^{\pm 1}$ such that:*

(1) *$|R_n| = k$ for all n, and $|X| < k$;*
(2) *all of the groups $Q_n \cong \langle X \mid R_n \rangle$ are perfect;*
(3) *there is no algorithm that can determine which of the groups Q_n are trivial;*
(4) *when Q_n is nontrivial, the presentation $\mathcal{Q}_n \equiv \langle X \mid R_n \rangle$ is aspherical.*

We apply our modified version of the Rips algorithm (Proposition 2.10) to the presentations $\mathcal{Q}_n$ from Theorem 4.1 to obtain a recursive sequence of finite presentations $(\mathcal{P}_n)$ of perfect groups (Γ_n). By applying the algorithm from Proposition 2.4 to these presentations, we obtain a recursive sequence of finite presentations $(\widetilde{\mathcal{P}}_n)$ for the universal central extensions $(\widetilde{\Gamma}_n)$. By Proposition 2.3(3), $\widetilde{\Gamma}_n$ is perfect. We define $G_n = \widetilde{\Gamma}_n \times \widetilde{\Gamma}_n$, with the obvious presentation $\mathcal{E}_n$ derived from $\mathcal{P}_n$.

In more detail, with the notation established in Proposition 2.10 and Proposition 2.4, if $\mathcal{Q}_n \equiv \langle X \mid R_n \rangle$ then $\mathcal{P}_n = \langle X, a_1, a_2 \mid R_n \cup V_n \rangle$ and $\widetilde{\mathcal{P}}_n = \langle X, a_1, a_2 \mid \overline{R_n \cup V_n} \rangle$ while

$$\mathcal{E}_n \equiv \langle X_1, X_2, a_{11}, a_{12}, a_{21}, a_{22} \mid C,\ S_{1,n},\ S_{2,n} \rangle,$$

where X_1 and X_2 are two copies of X corresponding to the two factors of $\widetilde{\Gamma}_n \times \widetilde{\Gamma}_n$ and C is a list of commutators forcing each $x_1 \in X_1 \cup \{a_{11}, a_{12}\}$ to commute with each $x_2 \in X_2 \cup \{a_{21}, a_{22}\}$, and $S_{i,n}$ $(i = 1, 2)$ is the set of words obtained from $\overline{R_n \cup V_n}$ by replacing the ordered alphabet (X, a_1, a_2) with (X_i, a_{i1}, a_{i2}). Note that the generating set of $\mathcal{E}_n$ does not vary with n, and nor does the cardinality of the set of relators. The map $X \cup \{a_1, a_2\} \to Q_n$ that kills a_1 and a_2 and is the identity on X extends to give the composition of the universal central extension of Γ_n and the map $\Gamma_n \to Q_n$ in the Rips construction:

$$\widetilde{\Gamma}_n \to \Gamma_n \to Q_n. \tag{3}$$

By construction, the kernel of this map is the preimage $\widetilde{N}_n < \widetilde{\Gamma}_n$ of $N_n = \langle a_1, a_2 \rangle < \Gamma_n$. In particular, since the kernel of $\widetilde{\Gamma}_n \to \Gamma_n$ is finitely generated (isomorphic to

$H_2(\Gamma_n, \mathbb{Z})$), we see that $\widetilde{N}_n$ is finitely generated. Thus for each n we have a short exact sequence

$$1 \to \widetilde{N}_n \to \widetilde{\Gamma}_n \to Q_n \to 1 \tag{4}$$

with $\widetilde{N}_n$ finitely generated, $\widetilde{\Gamma}_n$ finitely presented (indeed it has a finite classifying space), and Q_n as in Theorem 4.1. In particular, since Q_n is of type F_3, the 1-2-3 theorem tells us that the fiber product $P_n < \widetilde{\Gamma}_n \times \widetilde{\Gamma}_n = G_n$ associated to this short exact sequence is finitely presentable. And Lemma 2.1 tells that P_n is generated by

$$\{(x,x), (a_1,1), (a_2,1), (r,1) \mid x \in X, r \in R_n\}.$$

At this stage, we have constructed the desired recursive sequence of pairs of groups $(P_n \hookrightarrow G_n)_n$ with an explicit presentation for the perfect group G_n and an explicit finite generating set for P_n. The inclusion $P_n \hookrightarrow G_n$ is defined by $(x,x) \mapsto x_1x_2$, $(a_i,1) \mapsto a_{1i}$, etc. Our next task is to prove that there is no algorithm that can determine for which n the group P_n is perfect.

Claim: The recursively enumerable set $\{n \mid P_n \text{ is perfect}\} \subset \mathbb{N}$ is not recursive.

The claim will follow if we can argue that P_n is perfect if and only if Q_n is the trivial group. If $Q_n = 1$, then $P_n = G_n$, and we constructed G_n to be perfect. If $Q_n \neq 1$, then by Theorem 4.1(4), the presentation $\mathcal{Q}_n$ is aspherical, that is, the presentation 2-complex K for $\mathcal{Q}_n$ is a classifying space $K(Q_n, 1)$. In this case, $H_2(Q_n, \mathbb{Z}) = H_2(K, \mathbb{Z})$ is free abelian. As $H_1(Q_n, \mathbb{Z}) = H_1(K, \mathbb{Z}) = 0$, the rank of $H_2(Q_n, \mathbb{Z})$ is $\nu_2 - \nu_1$, where ν_2 is the number of generators on $\mathcal{Q}_n$ (1-cells in K) and ν_2 is the number or relators (2-cells). Theorem 4.1(1) tells us that $H_2(Q_n, \mathbb{Z}) \neq 0$, so we will be done if we can prove that $H_1(P_n, \mathbb{Z}) \cong H_2(Q_n, \mathbb{Z})$.

From the 5-term exact sequence for $1 \to \widetilde{N}_n \to \widetilde{\Gamma}_n \to Q_n \to 1$, we have

$$H_2(\widetilde{\Gamma}_n, \mathbb{Z}) \to H_2(Q_n, \mathbb{Z}) \to H_0(Q_n, H_1\widetilde{N}_n) \to H_1(\widetilde{\Gamma}_n, \mathbb{Z}).$$

The first and last terms are zero, by Proposition 2.3(2), so $H_2(Q_n, \mathbb{Z}) \cong H_0(Q_n, \widetilde{N}_n)$. On the other hand, from the 5-term exact sequence for $P_n = \widetilde{N}_n \rtimes \widetilde{\Gamma}_n$ we have $H_0(\widetilde{\Gamma}_n, H_1\widetilde{N}_n) \cong H_1(P_n, \mathbb{Z})$. As in Lemma 2.2, we observe that $H_0(\widetilde{\Gamma}_n, H_1\widetilde{N}_n) = H_0(Q_n, H_1\widetilde{N}_n)$, so $H_1(P_n, \mathbb{Z}) \cong H_2(Q_n, \mathbb{Z})$. This completes the proof of the claim.

In order to complete the proof of Theorem C, we must explain why G_n is biautomatic and residually finite. First, Neumann and Reeves [41] proved that all finitely generated central extensions of hyperbolic groups are biautomatic; Γ_n is hyperbolic and, therefore, $\widetilde{\Gamma}_n$ is biautomatic. And the direct product of two biautomatic groups is biautomatic, so G_n is biautomatic. The residual finiteness of $\widetilde{\Gamma}_n$ (and hence G_n) is a deeper fact, depending on the work of Wise and Agol: we saw in Remark 2.11 that Γ_n is residually finite and good in the sense of Serre; if A is a finitely generated abelian group and G is a finitely generated residually finite group that is good, then for any central extension $1 \to A \to E \to G \to 1$, the group E is residually finite; see [46, Section I.2.6] and [34, Corollary 6.2]; thus Γ_n is residually finite. □

5 Proof of Theorem D

We restate Theorem D, for the convenience of the reader.

Theorem 5.1. *For every recursively presented abelian group A there exists a Grothendieck pair $P_A \hookrightarrow G_A$ where G_A is a torsion-free biautomatic group that is residually finite, has a finite classifying space and is superperfect, while P_A is finitely generated with $H_2(P_A, \mathbb{Z}) \cong A$.*

Proof. Corollary 3.3 provides us with a finitely presented group Q that has no finite quotients, with $H_1(Q_A, \mathbb{Z}) = H_2(Q_A, \mathbb{Z}) = 0$ and $H_3(Q_A, \mathbb{Z}) \cong A$. As in the proof of Theorem C, we apply Proposition 2.10 to obtain a short exact sequence

$$1 \to N \to \Gamma_A \xrightarrow{p} Q_A \to 1$$

where Γ_A is a metric small cancellation group and N is finitely generated. The argument in the final two paragraphs of the proof of Theorem C shows that the universal central extension $\tilde{\Gamma}_A$ is biautomatic (by [41]) and that it is residually finite (by virtue of the connection between specialness and goodness in the sense of Serre). The asphericity of the small cancellation presentation for Γ_A implies, in the light of Proposition 2.3, that $\tilde{\Gamma}_A$ has a finite classifying space $K(\tilde{\Gamma}_A, 1)$.

Let $\eta : \tilde{\Gamma}_A \to Q_A$ be the composition of the central extension $\tilde{\Gamma}_A \to \Gamma_A$ and $p : \Gamma_A \to Q_A$ and let $P_A < G_A := \tilde{\Gamma}_A \times \tilde{\Gamma}_A$ be the fiber product associated to the short exact sequence

$$1 \to \tilde{N} \to \tilde{\Gamma}_A \xrightarrow{\eta} Q_A \to 1. \tag{5}$$

Lemma 2.1 assures us that P_A is finitely generated. Thus we will be done if we can show that $P_A \hookrightarrow G_A$ induces an isomorphism of profinite completions and that $H_2(P_A, \mathbb{Z}) \cong A$. The first of these assertions is a special case of Lemma 2.12, since $\widehat{Q}_A = 1$ and $H_2(Q_A, \mathbb{Z}) = 0$. The second assertion relies on a comparison of the LHS spectral sequences associated to (5) and $1 \to \tilde{N} \to P_A \to \tilde{\Gamma}_A \to 1$. The key points are isolated in the lemmas in Section 2.3. Using these lemmas, we conclude our argument as follows.

From Lemma 2.2, we have

$$H_0(\tilde{\Gamma}_A, H_2\tilde{N}) = H_0(Q_A, H_2\tilde{N}), \tag{6}$$

where the first group of coinvariants is for the action induced by conjugation in P_A and the second is induced by conjugation in $\tilde{\Gamma}_A$.

From Lemma 2.8, we have

$$H_2(P_A, \mathbb{Z}) \cong H_0(\tilde{\Gamma}_A, H_2N). \tag{7}$$

Lemma 2.9 applies to the short exact sequence (5), yielding $H_1\tilde{N} = 0$. And we *claim* that Lemma 2.7 also applies to (5), yielding $H_0(Q_A, H_2\tilde{N}) \cong H_3(Q_A, \mathbb{Z})$. By combining this isomorphism with (6) and (7), we have $H_2(P_A, \mathbb{Z}) \cong H_3(Q_A, \mathbb{Z})$, as desired.

It remains to justify the claim that Lemma 2.7 applies to (5). Specifically, we must argue that $\eta : \widetilde{\Gamma}_A \to Q_A$ induces the zero map on $H_3(-, \mathbb{Z})$. By construction, η factors through $\widetilde{\Gamma}_A \to \Gamma_A$. The homology of Γ_A can be calculated from the standard 2-complex of its aspherical presentation, so $H_k(\Gamma_A, \mathbb{Z}) = 0$ for all $k > 2$, and hence the composition

$$H_3(\widetilde{\Gamma}_A, \mathbb{Z}) \to H_3(\Gamma_A, \mathbb{Z}) \to H_3(Q_A, \mathbb{Z})$$

is the zero map. □

Bibliography

[1] Agol I. The virtual Haken conjecture. Doc Math. 2013;18:1045–87. With Appendix by Agol I, Groves D and Manning J.

[2] Bass H, Lubotzky A. Nonarithmetic superrigid groups: counterexamples to Platonov's conjecture. Ann Math. 2000;151:1151–73.

[3] Baumslag G. Residually finite groups with the same finite images. Compos Math. 1974;29:249–52.

[4] Baumslag G. Parafree groups. In: Infinite groups: geometric, combinatorial and dynamical aspects. Prog Math. vol. 248. Basel: Birkhäuser; 2005. p. 1–14.

[5] Baumslag G, Boone WW, Neumann BH. Some unsolvable problems about elements and subgroups of groups. Math Scand. 1959;7:191–201.

[6] Baumslag G, Bridson MR, Holt DJ, Miller CF III. Finite presentation of fibre products of metabelian groups. J Pure Appl Algebra. 2003;181:15–22.

[7] Baumslag G, Bridson MR, Miller CF III, Short H. Subgroups of automatic groups and their isoperimetric functions. J Lond Math Soc. 1997;56:292–304.

[8] Baumslag G, Bridson MR, Miller CF III, Short H. Fibre products, non-positive curvature, and decision problems. Comment Math Helv. 2000;75:457–77.

[9] Baumslag G, Dyer E, Heller A. The topology of discrete groups. J Pure Appl Algebra. 1980;16:1–47.

[10] Baumslag G, Dyer E, Miller CF III. On the integral homology of finitely presented groups. Topology. 1983;22:27–46.

[11] Baumslag G, Gersten SM, Shapiro M, Short H. Automatic groups and amalgams. J Pure Appl Algebra. 1991;76:229–316.

[12] Baumslag G, Miller CF III. Some odd finitely presented groups. Bull Lond Math Soc. 1988;20:239–44.

[13] Baumslag G, Myasnikov A, Remeslennikov V. Algebraic geometry over groups. I. Algebraic sets and ideal theory. J Algebra. 1999;219:16–79.

[14] Baumslag G, Roseblade J. Subgroups of direct products of free groups. J Lond Math Soc. 1984;30:44–52.

[15] Bieri R. Homological dimension of discrete groups. 2nd ed. Queen Mary College Mathematical Notes. Queen Mary College, Department of Pure Mathematics: London; 1981.

[16] Bridson MR. Controlled embeddings into groups that have no nontrivial finite quotients. Geom Topol Monogr. 1998;1:99–116.

[17] Bridson MR. Decision problems and profinite completions of groups. J Algebra. 2011;326:59–73.

[18] Bridson MR. Semihyperbolicity. In: Hagen M, Webb R, Wilton H, editors. Beyond Hyperbolicity. London Math Soc Lecture Note Series. vol. 454. Cambridge: Camb. Univ. Press; 2019. p. 25–64.

[19] Bridson MR. The homology of discrete groups and centres of profinite completions, in preparation.
[20] Bridson MR, Grunewald FJ. Grothendieck's problems concerning profinite completions and representations of groups. Ann Math (2). 2004;160:359–73.
[21] Bridson MR, Miller CF III. Structure and finiteness properties of subdirect products of groups. Proc Lond Math Soc (3). 2009;98:631–51.
[22] Bridson MR, Reid AW. Nilpotent completions of groups, Grothendieck pairs, and four problems of Baumslag. Int Math Res Not. 2015;8:2111–40.
[23] Bridson MR, Wilton H. On the difficulty of presenting finitely presentable groups. Groups Geom Dyn. 2011;5:301–25.
[24] Bridson MR, Howie J, Miller CF III, Short H. The subgroups of direct products of surface groups. Geom Dedic. 2002;92:95–103.
[25] Bridson MR, Howie J, Miller CF III, Short H. Subgroups of direct products of limit groups. Ann Math. 2009;170:1447–67.
[26] Bridson MR, Howie J, Miller CF III, Short H. On the finite presentation of subdirect products and the nature of residually free groups. Am J Math. 2013;135:891–933.
[27] Brown KS. Cohomology of groups. Graduate Texts in Mathematics. vol. 87. Berlin–Heidelberg–New York: Springer; 1982.
[28] Collins DJ, Miller CF III. The word problem in groups of cohomological dimension 2. In: Campbell CM, Robertson EF, Smith GC, editors. Groups St. Andrews in Bath. Lond Math Soc Lect Note Ser. vol. 270. 1998. p. 211–8.
[29] Epstein DBA, Cannon JW, Holt DF, Levy SVF, Paterson MS, Thurston WP. Word processing in groups. Boston, MA: Jones and Bartlett Publishers; 1992.
[30] Gersten SM, Short HB. Rational subgroups of biautomatic groups. Ann Math. 1991;134:125–58.
[31] Gromov M. Hyperbolic groups. In: Gersten SM, editor. Essays in Group Theory. MSRI Publ. vol. 8. New York: Springer; 1987. p. 75–263.
[32] Grothendieck A. Representations linéaires et compatifications profinie des groupes discrets. Manuscr Math. 1970;2:375–96.
[33] Grunewald FJ, Pickel PF, Segal D. Polycyclic groups with isomorphic finite quotients. Ann Math. 1980;111:155–95.
[34] Grunewald F, Jaikin-Zapirain A, Zalesskii PA. Cohomological goodness and the profinite completion of Bianchi groups. Duke Math J. 2008;144:53–72.
[35] Haglund F, Wise D. Special cube complexes. Geom Funct Anal. 2008;17:1551–620.
[36] Higman G. A finitely generated infinite simple group. J Lond Math Soc. 1951;26:61–4.
[37] Higman G. Subgroups of finitely presented groups. Proc R Soc Lond Ser A. 1961;262:455–75.
[38] Lyndon RC, Schupp PE. Combinatorial Group Theory. Ergebnisse der Mathematik und ihrer Grenzgebiete. vol. 89. Heidelberg: Springer; 1977.
[39] Miller CF III. Decision problems for groups: survey and reflections. In: Algorithms and classification in combinatorial group theory. MSRI Publ. vol. 23. New York: Springer; 1992. p. 1–59.
[40] Milnor J. Introduction to Algebraic K-Theory. Ann Math Stud. vol. 72. Princeton: Princeton University Press; 1971.
[41] Neumann WD, Reeves L. Central extensions of word hyperbolic groups. Ann Math. 1997;145:183–92.
[42] Pickel PF. Finitely generated nilpotent groups with isomorphic quotients. Trans Am Math Soc. 1971;160:327–41.
[43] Platonov VP, Tavgen OI. Grothendieck's problem on profinite completions and representations of groups. K-Theory. 1990;4:89–101.
[44] Ribes L, Zalesskii PA. Profinite Groups. Ergeb Math. vol. 40. Berlin, New York: Springer; 2000.

[45] Rips E. Subgroups of small cancellation groups. Bull Lond Math Soc. 1982;14:45–7.
[46] Serre J-P. Cohomologie galoisienne. Fifth ed. Lecture Notes in Mathematics. vol. 5. Berlin: Springer; 1994.
[47] Wise DT. A residually finite version of Rips's construction. Bull Lond Math Soc. 2003;35:23–9.
[48] Wise DT. Cubulating small cancellation groups. GAFA. 2004;14:150–214.
[49] Wise DT. The structure of groups with a quasiconvex hierarchy. Preprint, McGill 2011. cf. MR2558631.

Anthony E. Clement

Some properties of the Baumslag groups $G(m, n)$

Abstract: In his 1969 paper, "A noncyclic one-relator group all of whose finite quotients are cyclic," G. Baumslag showed that every finite quotient of $G = \langle a, b \mid a = [a, a^b]\rangle$ rewritten as $G(1, 2) = \langle a, b \mid b^{-1}a^{-1}bab^{-1}ab = a^2\rangle$ is cyclic and as a result presented then yet another example of a one-relator group which is not residually finite. This paper describes the structure and present some properties of the Baumslag groups $G(m, n) = \langle a, b \mid b^{-1}a^{-1}ba^m b^{-1}ab = a^n \mid m \neq 0, n \neq 0 \in \mathbb{Z}\rangle$.

Keywords: Baumslag groups, one-relator group, ascending union, Freiheitssatz Magnus breakdown, generalized free products, cyclic amalgam, isomorphic copies, Baumslag–Solitar groups

MSC 2010: 20E06, 20F05

1 Preliminaries

The Baumslag groups have presentation of the form:

$$\begin{aligned} G(m, n) &= \langle a, b \mid [a^m, a^b] = a^{(n-m)} \mid m \neq 0, n \neq 0 \in \mathbb{Z}\rangle \\ &= \langle a, b \mid b^{-1}a^{-1}ba^m b^{-1}ab = a^n \mid m \neq 0, n \neq 0 \in \mathbb{Z}\rangle. \end{aligned}$$

This paper examines the structure of $G(m, n)$ and confirms for each $n = m+1$ that it contains an embedded copy of the Baumslag–Solitar group [1] $B(m, n) = \langle a, b \mid a^{-1}b^m a = b^n \mid m \neq 0, n \neq 0 \in \mathbb{Z}\rangle$ for the same $n = m + 1$, by virtue of possessing a normal subgroup N that is an ascending union of generalized free products of isomorphic copies of $B(m, n)$ with cyclic amalgam at each stage.

The Baumslag group $G(1, 2)$ first appeared in [2] as $G = \langle a, b \mid a = [a, a^b]\rangle$ in which G. Baumslag showed that G fails to be residually finite. In order to address some properties about $G(1, 2)$, we need to state an important theorem that plays a central role in one-relator group theory.

Theorem 1.1 (W. Magnus' Freiheitssatz). *Let G be a group with a single defining relator, that is, $G = \langle x_1, \ldots, x_q; r\rangle$. Suppose that r is cyclically reduced, that is, the first and last letters in r are not inverses of each other. If each of $x_1, \ldots, x_q$ actually appears in r, then any proper subset of $x_1, \ldots, x_q$ freely generate a free group.*

Note: Dedicated to Gilbert Baumslag, in memoriam.

Anthony E. Clement, Brooklyn College, City University of New York, 2900 Bedford Avenue, Brooklyn, NY 11210, USA, e-mail: aclement@brooklyn.cuny.edu

https://doi.org/10.1515/9783110638387-004

The lemma below is an immediate consequence of W. Magnus' breakdown in the proof of the theorem above.

Lemma 1.1 ([3]). *Let $G = \langle b, x, \dots, c; r = 1\rangle$ be a group with a single defining relation. Suppose that b occurs in r with exponent sum zero and that μ and ν are respectively the minimum and maximum subscripts of x occurring in r_0 (the rewrite of r with exponent sum zero). If $\mu < \nu$ and if x_μ and x_ν occur only once in r_0, then $N = gp_G(x, \dots, c)$ is free. Moreover, if G is a 2-generator group with generators b and x, then N is free of rank $\nu - \mu + 1$.*

The lemma below is a consequence of the lemma above.

Lemma 1.2. *Given a one-relator group on two generators say a and b, if a commutes with a^b, then the normal subgroup generated by b is free.*

The proof of the lemma below is done by a direct application of Lemma 1.1.

Lemma 1.3. *The Baumslag group*

$$G(1,1) = \langle a, b \mid b^{-1}a^{-1}bab^{-1}ab = a\rangle = \langle a, b \mid b^{-1}a^{-1}bab^{-1}aba^{-1} = 1\rangle$$

contains a free group rank 3.

Proof. Observe that the exponent sum of a in r, denoted, $\exp_r(a)$ is zero. Also, $\exp_r(b) = 0$. Let us use the fact that the first is true, that is, $\exp_r(a) = 0$. Put $b_i = a^{-i}ba^i$. Let $N = gp_G(b)$, the normal closure of b in G. Observe that $r = b^{-1}a^{-1}bab^{-1}aba^{-1}$ with $r_0 = b_0^{-1}b_1b_0^{-1}b_{-1}$. Notice that in r_0, $\mu = -1$ and $\nu = 1$, so $\mu < \nu$, and both $\mu = -1$ and $\nu = 1$ occur only once in r_0. Therefore, $N = gp_G(b)$ is free of rank $\nu - \mu + 1 = 1 - (-1) + 1 = 2 + 1 = 3$. □

The lemma below is a general form of the lemma immediately above.

Lemma 1.4. *The Baumslag group $G(m,n) = \langle a, b \mid b^{-1}a^{-1}ba^mb^{-1}ab = a^n \mid m \neq 0$, $n \neq 0 \in \mathbb{Z}$ where $m = n \geq 1$ for each n contains a free group rank $n + 2$.*

Proof. $G(n,n) = \langle a, b \mid b^{-1}a^{-1}ba^nb^{-1}ab = a^n\rangle = \langle a, b \mid b^{-1}a^{-1}ba^nb^{-1}aba^{-n} = 1\rangle$. Observe that the exponent sum of a in r, denoted, $\exp_r(a)$ is zero. Put $b_i = a^{-i}ba^i$. Let $N = gp_G(b)$. Observe that

$$\begin{aligned} r &= b^{-1}a^{-1}ba^nb^{-1}aba^{-n} \\ &= b^{-1}\underbrace{a^{-1}ba}\,\underbrace{a^{(n-1)}b^{-1}a^{(1-n)}}\,\underbrace{a^nba^{-n}} \\ &= b^{-1}\underbrace{a^{-1}ba}\,\underbrace{a^{-(1-n)}b^{-1}a^{(1-n)}}\,\underbrace{a^{-(-n)}ba^{-n}}. \end{aligned}$$

So $r_0 = b_0^{-1}b_1b_{(1-n)}b_{-n}$. Notice that in r_0, $\mu = -n$ and $\nu = 1$, so $\mu < \nu$, and both $\mu = -n$ and $\nu = 1$ occur only once in r_0. Therefore, $N = gp_G(b)$ is free of rank $\nu - \mu + 1 = 1 - (-n) + 1 = n + 2$. □

1.1 A subgroup of the group $G(1,2)$ as a certain generalized free product

With the use of the previous section's method, our analysis of the structure of $G(m,n)$ recognizes $B(m,n)$ as a subgroup. More precisely, we can state the result for $G(1,2)$ as follows.

Proposition 1.1 ([4]). *The Baumslag group* $G(1,2) = \langle a, b \mid b^{-1}a^{-1}bab^{-1}ab = a^2\rangle$ *contains a normal subgroup* N *that is an ascending union of generalized free products of isomorphic copies of the Baumslag–Solitar* $B(1,2) = \langle x, y \mid x^{-1}yx = y^2\rangle$ *with cyclic amalgam.*

Proof. We will use the "Magnus breakdown" to find a presentation of the subgroup N of $G(1,2)$. We will invoke the results of Magnus' proof of the *Freiheitssatz*. Since b occurs with exponent sum zero, put $a_i = b^{-i}ab^i$, $(i \in \mathbb{Z})$. Let $N = gp_{G(1,2)}(a) = gp(\dots, a_{-1}, a_0, a_1, a_2, \dots)$. Observe that $G(1,2)/N$ is infinite cyclic. Magnus incorporated the Reidemeister–Schreier method into his method. We first rewrite the relation of $G(1,2)$ in terms of the generators of N and "bump up" each term in the relation by 1 going from left to right according to the elements in the relation and then repeat the process. Notice that $\varrho(b^{-1}a^{-1}bab^{-1}ab = a^2)$ is $a_1^{-1}a_0a_1 = a_0^2$, where $\varrho(b^{-1}a^{-1}bab^{-1}ab = a^2)$ is the Reidemeister–Schreier rewrite of the relation $b^{-1}a^{-1}bab^{-1}ab = a^2$. So N has the following presentation:

$$N = \langle \dots, a_{-1}, a_0, a_1, \dots \mid a_0^{-1}a_{-1}a_0 = a_{-1}^2,\ a_1^{-1}a_0a_1 = a_0^2,\ a_2^{-1}a_1a_2 = a_1^2, \dots\rangle.$$

It follows from Magnus' work that $N_{-1,0} = \langle a_{-1}, a_0 \mid a_0^{-1}a_{-1}a_0 = a_{-1}^2\rangle$, $N_{0,1} = \langle a_0, a_1 \mid a_1^{-1}a_0a_1 = a_0^2\rangle, \dots, N_{i,i+1} = \langle a_i, a_{i+1} \mid a_{i+1}^{-1}a_ia_{i+1} = a_i^2\rangle$ and note that $N_{i,i+1} = \langle a_i, a_{i+1} \mid a_{i+1}^{-1}a_ia_{i+1} = a_i^2\rangle$ for each $i \in \mathbb{Z}$ is isomorphic to the Baumslag–Solitar group $B(1,2) = \langle x, y \mid x^{-1}yx = y^2\rangle$. Now, by Magnus' *Freiheitssatz*, a_{-1} and a_0 are both of infinite order in $N_{-1,0}$. Similarly, by the *Freiheitssatz* a_0 and a_1 are of infinite order in $N_{0,1}$. Thus, the generalized free product $N_{-1,1} = \{N_{-1,0} * N_{0,1} \mid a_0\}$ is well-defined: elements of finite order if any, in $N_{-1,1}$ must be conjugate to such elements in either $N_{-1,0}$ or $N_{0,1}$ [5], but neither of them have elements of finite order. Following a copy of this argument, we also have a well-defined torsion-free generalized free product $N_{-1,2} = \{N_{-1,1} * N_{1,2} \mid a_1\}$. Clearly, we can repeat the foregoing construction for all positive integers i, j, k and to obtain $N_{-i,j} = \{N_{-i,k} * N_{k,j} \mid a_k\}$. Now we note that the generalized free product $N_{-1,1}$ has the presentation $\langle a_{-1}, a_0, a_1 \mid a_0^{-1}a_{-1}a_0 = a_{-1}^2, a_1^{-1}a_0a_1 = a_0^2\rangle$. Similarly, $N_{-1,2} = \langle a_{-1}, a_0, a_1, a_2 \mid a_0^{-1}a_{-1}a_0 = a_{-1}^2, a_1^{-1}a_0a_1 = a_0^2, a_2^{-1}a_1a_2 = a_1^2\rangle$. Thus, the ascending union $N = \bigcup_{\substack{i>0 \\ j>0}}^{\infty} \uparrow N_{-i,j}$ has the presentation $\langle \dots, a_{-1}, a_0, a_1, \dots \mid a_0^{-1}a_{-1}a_0 = a_{-1}^2, a_1^{-1}a_0a_1 = a_0^2, a_2^{-1}a_1a_2 = a_1^2, \dots\rangle$ which coincides with that obtained through the Magnus breakdown and the Reidemeister–Schreier rewriting process. Following through with the details of the construction, it is clear that the copy of $B(1,2)$ in $N_{0,1}$, etc., is a subgroup of N, and the Baumslag group $G(1,2) = \langle a, b \mid b^{-1}a^{-1}bab^{-1}ab = a^2 \rangle$ contains N as a normal subgroup that is an ascending union of generalized free products of isomorphic copies of the Baumslag–Solitar $B(1,2) = \langle x, y \mid x^{-1}yx = y^2 \rangle$ with cyclic amalgam. □

Theorem 1.2. *The Baumslag group $G(m,n) = \langle a, b \mid b^{-1}a^{-1}ba^m b^{-1}ab = a^n \rangle$ for $n = m+1$ contains a normal subgroup N that is an ascending union of generalized free products of isomorphic copies of the Baumslag–Solitar group $B(m,n) = \langle x, y \mid x^{-1}y^m x = y^n \rangle$ for the same $n = m+1$ with cyclic amalgam.*

Proof. It is utterly clear that with the obvious adjustments with $(m, m+1)$ in place of $(1,2)$ in the proof for Proposition 1.1, generalizes for all such pair of nonzero integers m and n with $n = m+1$. □

The theorem above shows a unique structure of how some of the Baumslag–Solitar groups $B(m,n)$ fits inside the Baumslag groups $G(m,n)$. In general, the Baumslag groups $G(m,n)$ can be realized as HNN extensions of the Baumslag–Solitar groups $B(m,n)$ with stable letter b. With the use of Tietze transformations,

$$\begin{aligned} G(m,n) &= \langle a, b \mid b^{-1}a^{-1}ba^m b^{-1}ab = a^n \rangle \\ &= \langle t, a, b \mid b^{-1}a^{-1}ba^m b^{-1}ab = a^n, t = b^{-1}ab \rangle \\ &= \langle t, a, b \mid b^{-1}a^{-1}ba^m b^{-1}ab = a^n, t^{-1}a^m t = a^n, t = b^{-1}ab \rangle \\ &= \langle t, a, b \mid t^{-1}a^m t = a^n, t = b^{-1}ab \rangle \\ &= \langle B(m,n), b \mid t = b^{-1}ab \rangle. \end{aligned}$$

So every Baumslag group $G(m,n)$ can be realized as HNN extension of a Baumslag–Solitar group $B(m,n)$ for each m and n. However, we see that not every $G(m,n)$ can be realized as possessing a normal subgroup N that is an ascending union of generalized free products of isomorphic copies of $B(m,n)$ with cyclic amalgam.

Bibliography

[1] Baumslag G, Solitar D. Some Two-Generator One-Relator Non-Hopfian Groups. Bull Am Math Soc. 1962;68:199–201.

[2] Baumslag G, Solitar D. A Non-Cyclic One-Relator Group All of Whose Finite Quotients Are Cyclic. J Aust Math Soc. 1969;10:497–8.

[3] Baumslag G, Solitar D. Free Subgroups of Certain One-Relator Groups Defined by Positive Words. Math Proc Camb Philos Soc. 1983;93:247–51.

[4] Clement AE. On the Baumslag–Solitar Groups and Certain Generalized Free Products. Ph. D. Thesis, The Graduate Center, The City University of New York, New York 10016, October 2006.

[5] Neumann BH. An Essay on Free Products of Groups with Amalgamations. Philos Trans R Soc Lond, Ser A. 1954;246(919):503–54.

Anthony M. Gaglione and Dennis Spellman

Some model theory of the Heisenberg group: I

Unitriangular representations of models of a subtheory of its universal theory

Abstract: In this paper, all rings, R, are commutative, have a unity and are of characteristic zero. We denote by $U(R)$ the group of all 3×3 upper unitriangular matrices with entries in R. The Heisenberg group H is $U(\mathbb{Z})$. Here, we give a necessary and sufficient condition for a group to be a model of the universal theory of H. We also give sufficient conditions for a subgroup G of $U(R)$ to satisfy noncentral commutative transitivity (NZCT) and also for G to be a model of the universal theory of H. These conditions are not necessary. Nonetheless, our main result is that, in the special case where $G = U(R)$ where R is a locally residually-$\mathbb{Z}$ ring, G is a model of $\mathrm{Th}_{\forall}(H)$ if and only if G is a model of $\mathrm{diag}(H) \cup Q(H) \cup \{\mathrm{NZCT}\}$.

Keywords: Noncentral commutative transitivity, quasi-identity, locally residually-$\mathbb{Z}$, diagram

MSC 2010: Primary 05C38, 15A15, Secondary 05A15, 15A18

1 Preliminaries

We identify the set ω of nonnegative integers endowed with its natural order as the first limit ordinal ω hence the first infinite cardinal.

In this paper by a **ring**, we shall mean a commutative ring with identity $1 \neq 0$. All of the rings in this paper will be assumed to have characteristic zero, and hence be extensions of the ring $\mathbb{Z}$ of integers. Here, we require the identity 1 to be preserved in subrings as well as homomorphic images. Following Baumslag, Myasnikov, and Remeslennikov [1], if G is a group, then a G-**group** Γ is a group containing a distinguished copy of G as a subgroup. The meanings of G-subgroup and G-homomorphism are obvious.

If $(\Gamma_\lambda)_{\lambda \in \Lambda}$ is a nonempty family of G-groups, then the product $P = \prod_{\lambda \in \Lambda} \Gamma_\lambda$ becomes a G-group under the diagonal embedding $g \longmapsto (g, g, \ldots)$. Similarly, an ultraproduct of G-groups is again a G-group.

Note: Dedicated to the memory of G. Baumslag.

Anthony M. Gaglione, U.S. Naval Academy, 121 Blake Rd, Annapolis, MD 21402, USA, e-mail: agaglione@aol.com
Dennis Spellman, Crum Lynne, PA 19022, USA

https://doi.org/10.1515/9783110638387-005

Taking $G = \{1\}$ to be the trivial group, we see that a $\{1\}$-group is just a group, a $\{1\}$-subgroup is just a subgroup, and a $\{1\}$-homomorphism is just a homomorphism.

Definition 1.1. Let Γ and Γ_0 be G-groups

(i) Γ_0 ***G*-separates** Γ provided, given $y \in \Gamma\backslash\{1\}$ there is a G-homomorphism $\Gamma \to \Gamma_0$ which does not annihilate y.

(ii) Γ_0 ***G*-discriminates** Γ provided, given finitely many nontrivial elements $y_1, \dots, y_k \in \Gamma\backslash\{1\}$, there is a G-homomorphism $\Gamma \to \Gamma_0$ which does not annihilate any y_i, $i = 1, \dots, k$.

Of course, taking $G = \{1\}$, the notions of Γ_0 separating Γ and Γ_0 discriminating Γ are readily apparent.

1.1 Varieties of groups, discrimination, and universal theory

The **trivial variety** E of groups is the isomorphism class of the one element group. All other varieties of groups are **nontrivial.**

If $r > 0$ is a cardinal and V is a nontrivial variety of groups, then by $F_r(V)$ we mean a group free of rank r in V. (Here, we are following the notation of [10].) It is well known that such objects exist and are unique up to isomorphism; moreover, it is also well known that if r and s are cardinals with $0 < r < s$, then $F_r(V)$ and $F_s(V)$ are not isomorphic. Suppose V is a nontrivial variety of groups and $a_1, a_2, \dots, a_n, \dots$ is a set of free generators for $F_\omega(V)$. Then, for each integer $n, 0 < n < \omega$, the subgroup $\langle a_1, a_2, \dots, a_n \rangle$ is free of rank n in V. Let us say that V is **finitely discriminated** provided there is an integer $n, 0 < n < \omega$, such that $F_\omega(V) = \langle a_1, a_2, \dots \rangle$ is G-discriminated by $G = F_n(V) = \langle a_1, \dots, a_n \rangle$.

In that event, for all cardinals $r \geq n$, the $F_r(V)$ are universally equivalent with respect to the first-order language with equality $L_0[G]$ appropriate for group theory and containing constants naming the elements of G. (See [6].) Explicitly, $L_0[G]$ contains a binary operation symbol $\cdot$, a unary operation symbol $^{-1}$, and a constant symbol $\hat{g}$ for each element $g \in G$. We find it convenient to commit the abuses of dropping $\cdot$ in favor of juxtaposition and identifying $\hat{g}$ with g. If $c \geq 1$ is an integer and N_c is the variety of all groups nilpotent of class at most c, then it is known that, for $n = \max\{2, c-1\}$, $F_\omega(N_c)$ is G-discriminated by $G = F_n(N_c)$.

Such groups are **commutative transitive of level** $c - 1$ in the sense that the centralizer of any element lying outside of the $(c-1)$-st term of the upper central series is abelian (for a proof of this fact, see [4]). This property is captured by a universal sentence of $L_0[G]$. In particular, when $c = 2$, we call this property **noncentral commutative transitivity** and abbreviate it NZCT. It is captured by the universal sentence

$$\forall x, y, z, w((([y, w] \neq 1) \wedge ([x, y] = 1) \wedge ([y, z] = 1)) \to ([x, z] = 1)).$$

If G is any group, then by the **diagram of** G (abbreviated diag(G)) we mean the set of all atomic and negated atomic sentences of $L_0[G]$ true in G. One has that a group Γ is a G-group if and only if Γ satisfies diag(G). More precisely, a $L_0[G]$-structure Γ is a G-group if and only if it satisfies the group axioms and diag(G).

1.2 Unitriangular representations, quasivarieties of groups and rings, a question of A. G. Myasnikov and the Heisenberg group

Let R be a ring. We denote by $U(R)$ the group of all $3x3$ upper unitriangular matrices with entries in R. The group is nilpotent of class exactly 2.

Two such matrices

$$a = \begin{bmatrix} 1 & a_{12} & a_{13} \\ 0 & 1 & a_{23} \\ 0 & 0 & 1 \end{bmatrix} \quad \text{and} \quad b = \begin{bmatrix} 1 & b_{12} & b_{13} \\ 0 & 1 & b_{23} \\ 0 & 0 & 1 \end{bmatrix}$$

commute if and only if

$$\det\left(\begin{bmatrix} a_{12} & a_{23} \\ b_{12} & b_{23} \end{bmatrix}\right) = 0.$$

From the fact that any central element must commute with each of the matrices

$$a_1 = \begin{bmatrix} 1 & 0 & 0 \\ 0 & 1 & 1 \\ 0 & 0 & 1 \end{bmatrix} \quad \text{and} \quad a_2 = \begin{bmatrix} 1 & 1 & 0 \\ 0 & 1 & 0 \\ 0 & 0 & 1 \end{bmatrix}$$

one deduces that the center consists of all matrices of the form

$$z = \begin{bmatrix} 1 & 0 & z_{13} \\ 0 & 1 & 0 \\ 0 & 0 & 1 \end{bmatrix}.$$

It follows that if $\langle a_1, a_2 \rangle \leq G \leq U(R)$, then the center of G consists of all matrices of the form

$$z = \begin{bmatrix} 1 & 0 & z_{13} \\ 0 & 1 & 0 \\ 0 & 0 & 1 \end{bmatrix}.$$

The **Heisenberg group** H is $U(\mathbb{Z})$. It follows that if R is a ring of characteristic zero and G is a H-subgroup of $U(R)$, then the center of G consists of all matrices $z \in G$ of the form

$$z = \begin{bmatrix} 1 & 0 & z_{13} \\ 0 & 1 & 0 \\ 0 & 0 & 1 \end{bmatrix}.$$

H is free in the variety N_2 of all groups nilpotent of class at most 2 on the generators

$$a_1 = \begin{bmatrix} 1 & 0 & 0 \\ 0 & 1 & 1 \\ 0 & 0 & 1 \end{bmatrix} \quad \text{and} \quad a_2 = \begin{bmatrix} 1 & 1 & 0 \\ 0 & 1 & 0 \\ 0 & 0 & 1 \end{bmatrix}.$$

Lemma 1.2 (See [11]). *Let $H = U(\mathbb{Z})$ and let a_1 and a_2 be given as above. Then H is freely generated with respect to N_2 by $\{a_1, a_2\}$.*

Proof. Let

$$a_3 = [a_2, a_1] = a_2^{-1} a_1^{-1} a_2 a_1 = \begin{bmatrix} 1 & 0 & 1 \\ 0 & 1 & 0 \\ 0 & 0 & 1 \end{bmatrix}.$$

Now let $x \in H$ be arbitrary. Then a computation shows that

$$x = \begin{bmatrix} 1 & a & b \\ 0 & 1 & c \\ 0 & 0 & 1 \end{bmatrix} = a_1^c a_2^a a_3^b.$$

Suppose G is free in N_2 on the free generators $\{g_1, g_2\}$. Then every element g of G is uniquely expressed as

$$g = g_1^{e(1)} g_2^{e(2)} [g_2, g_1]^{e(2,1)}.$$

Since in the expression $a_1^{n(2,3)} a_2^{n(1,2)} a_3^{n(1,3)}$ the exponents are unique because they are the entries of

$$\begin{bmatrix} 1 & n(1,2) & n(1,3) \\ 0 & 1 & n(2,3) \\ 0 & 0 & 1 \end{bmatrix}.$$

We have the obvious isomorphism $H \to G$. □

Let L be a first-order language with equality appropriate for algebras of some fixed type. By a **quasi-identity** (See [7].) in L is meant as a universal sentence of L of the form

$$\forall \overline{x} \Big(\bigwedge_i (s_i(\overline{x}) = t_i(\overline{x})) \to (s(\overline{x}) = t(\overline{x})) \Big)$$

where the s_i, t_i, s, t are terms of L and $\overline{x}$ is a tuple of distinct variables. By an **identity** or **law** is meant as a universal sentence of L of the form $\forall \overline{x}(s(\overline{x}) = t(\overline{x}))$ where s and t are terms of L and $\overline{x}$ is a tuple of distinct variables.

Note that the above identity is equivalent to the following quasi-identity:

$$\forall \overline{x}, y((y = y) \to (s(\overline{x}) = t(\overline{x}))) \quad \text{of } L.$$

Thus, identities are considered special cases of quasi-identities.

In the case of rings, every quasi-identity is equivalent to one of the form

$$\forall \overline{x}\left(\bigwedge_i (f_i(\overline{x}) = 0) \rightarrow (g(\overline{x}) = 0)\right)$$

where the f_i and g are polynomials.

In the case of G-groups, every quasi-identity is equivalent to one of the form

$$\forall \overline{x}\left(\bigwedge_i (u_i(\overline{g}, \overline{x}) = 1) \rightarrow (w(\overline{g}, \overline{x}) = 1)\right)$$

where the u_i and w are words on the elements of G and the variables and their formal inverses.

By a **quasi-variety** (of algebras of some fixed type) is meant the model class of a set of quasi-identities and, of course, by a **variety** (of algebras of some fied type) is meant the model class of a set of laws. If G is a group, let $\mathrm{Th}_\forall(G)$ be the set of all universal sentences of $L_0[G]$ true in G. Vacuous quantifications are permitted; so, quantifier-free sentences are considered special cases of universal sentences. Let $c \geq 2$ be an integer and let $G = F_n(N_c)$ where $n = \max\{2, c-1\}$. A. G. Myasnikov posed the question of whether or not $\mathrm{Th}_\forall(G)$ is axiomatized by the diagram of G together with the set $Q(G)$ of all quasi-identities of $L_0[G]$ true in G and the assertion that the centralizer of any element $x \in G \backslash Z_{c-1}(G)$ outside the $(c-1)$-st term of the upper central series be abelian.

In the case $c = 2$, this becomes

$$\mathrm{diag}(H) \cup Q(H) \cup \{\mathrm{NZCT}\}$$

where $H = \langle a_1, a_2; [a_2, a_1, a_1] = [a_2, a_1, a_2] = 1 \rangle$ is the Heisenberg group. (It was shown in Gaglione, Jackson, and Spellman [5] that if $2 \leq r \leq \omega$ and $2 \leq c \leq 4$, then any fixed set of Hall basic commutators of weight $c+1$ on the free generators can be taken as a set of defining relators for $F_r(N_c)$. For the definition and properties of Hall basic commutators, see [10].)

Given a set of algebras of some fixed type, it was shown in Grätzer and Lakser [8] that the quasi-variety generated by the set consists of all algebras isomorphic to a subalgebra of a direct product of ultraproducts of algebras from the set. Moreover, we term a ring R **locally residually-$\mathbb{Z}$** provided that given any finitely generated subring $R_0 \leq R$ and any nonzero element $r \in R_0 \backslash \{0\}$, there is a retraction $\rho : R_0 \rightarrow \mathbb{Z}$ which does not annihilate r. From the above result of Grätzer and Lakser, we get the following.

Lemma 1.3. *Let G be a H-group. Then G embeds as a H-subgroup of $U(R)$ for some locally residually-$\mathbb{Z}$ ring R if and only if G is a model of*

$$\mathrm{diag}(H) \cup Q(H).$$

Proof. From the result of Grätzer and Lakser [8] mentioned above, we get that the quasi-variety generated by H consists of all groups isomorphic to subgroups of direct products of ultrapowers of H. If we, up to isomorphism, restrict ourselves to H-subgroups of direct products of ultrapowers of H, we get precisely the model class of $\operatorname{diag}(H) \cup Q(H)$. Moreover, the quasi-variety generated by the ring $\mathbb{Z}$ of integers, according to [8], consists precisely, up to isomorphism, of subrings of direct products of ultrapowers of $\mathbb{Z}$. Such rings R are locally residually-$\mathbb{Z}$ and conversely. We see by taking "the same" ultrapowers and direct products that any model G of $\operatorname{diag}(H) \cup Q(H)$ embeds as a H-subgroup of $U(R)$ for some locally residually-$\mathbb{Z}$ ring R.

Suppose conversely the group G admits such an embedding. Since G is a H-group, it is a model of $\operatorname{diag}(H)$. Let

$$\forall \overline{x}\Big(\bigwedge_i (u_i(\overline{h},\overline{x}) = 1) \to (w(\overline{h},\overline{x}) = 1)\Big)$$

be a quasi-identity of $L_0[H]$ true in H. Suppose the above were false in G. Then there is a tuple $\overline{g}$ from G such that $\bigwedge_i (u_i(\overline{h},\overline{g}) = 1)$ and $w(\overline{h},\overline{g}) \neq 1$. Suppose that

$$w(\overline{h},\overline{g}) = \begin{bmatrix} 1 & w_{12} & w_{13} \\ 0 & 1 & w_{23} \\ 0 & 0 & 1 \end{bmatrix}$$

where $(w_{12}, w_{13}, w_{23}) \in R^3 \backslash \{(0,0,0)\}$. Let R_0 be the subring of R generated by the entries of the $u_i(\overline{h},\overline{g})$ and $w(\overline{h},\overline{g})$. Then the $u_i(\overline{h},\overline{g})$ and $w(\overline{h},\overline{g})$ lie in $U(R_0)$; moreover, R_0 is residually-$\mathbb{Z}$. Let $w_{ij} \in \{w_{12}, w_{13}, w_{23}\} \backslash \{0\}$. Then there is a retraction $\rho : R_0 \to \mathbb{Z}$ such that $\rho(w_{ij}) \neq 0$.

But ρ induces a retraction $\overline{\rho} : G \cap U(R_0) \to H$ via

$$\begin{bmatrix} 1 & x_{12} & x_{13} \\ 0 & 1 & x_{23} \\ 0 & 0 & 1 \end{bmatrix} \mapsto \begin{bmatrix} 1 & \rho(x_{12}) & \rho(x_{13}) \\ 0 & 1 & \rho(x_{23}) \\ 0 & 0 & 1 \end{bmatrix}.$$

Now, in H, $\overline{\rho}(u_i(\overline{h},\overline{g})) = 1$ for all i but $\overline{\rho}(w(\overline{h},\overline{g})) \neq 1$. That contradicts the fact that the quasi-identity

$$\forall \overline{x}\Big(\bigwedge_i (u_i(\overline{h},\overline{x}) = 1) \to (w(\overline{h},\overline{x}) = 1)\Big)$$

holds in H. The contradiction shows that G is a model of $Q(H)$. □

So let G embed as an H-subgroup of $U(R)$ for some locally residually-$\mathbb{Z}$ ring R. Furthermore, we may assume R is generated by the entries of the elements of G since we can replace R with the subring so generated.

We say that the matrix representation satisfies the **weak property** provided for all

$$a = \begin{bmatrix} 1 & a_{12} & a_{13} \\ 0 & 1 & a_{23} \\ 0 & 0 & 1 \end{bmatrix} \in G$$

one has that either $a_{12}^2 + a_{23}^2$ is zero or it is not a zero divisor in R. We say that the matrix representation satisfies the **strong property** provided for all such $a \in G$ one has that either $a_{12}^2 + a_{13}^2 + a_{23}^2$ is zero or it is not a zero divisor in R. We shall presently see that the representation satisfying the weak property is a sufficient condition for G to satisfy NZCT and that the representation satisfying the strong property is a sufficient condition for G to be a model of $\mathrm{Th}_\forall(H)$. Neither condition is necessary. We shall show this in a later section. Call the representation **special** if, up to isomorphism, we have equality (i. e., G is not a proper H-subgroup of $U(R)$). If the representation is special, then G is a model of $\mathrm{Th}_\forall(H)$ if and only if G is a model of $\mathrm{diag}(H) \cup Q(H) \cup \{\mathrm{NZCT}\}$.

2 The main results

A residually-$\mathbb{Z}$ ring is ω-**residually**-Z provided, given finitely many nonzero elements of R, $r_i \in R\backslash\{0\}, i = 1, \ldots, k$, there is a retraction $\rho : R \to \mathbb{Z}$ which does not annihilate any of them. A characterization other than the above definition is given below. A residually-$\mathbb{Z}$ ring is ω-residually-$\mathbb{Z}$ if and only if R is an integral domain.

Suppose that R is ω-residually-$\mathbb{Z}$ and $a, b \in R\backslash\{0\}$. Then there is a retraction $\rho : R \to \mathbb{Z}$ such that $\rho(a) \neq 0$ and $\rho(b) \neq 0$. It follows that $\rho(ab) = \rho(a)\rho(b) \neq 0$ so $ab \neq 0$ in R and R is an integral domain. Conversely, if R is an integral domain and $r_1, \ldots, r_k \in R\backslash\{0\}$, then $r = r_1 r_2 \cdots r_k \neq 0$ so, since R is residually-$\mathbb{Z}$, there is a retraction $\rho : R \to \mathbb{Z}$ such that $\rho(r) = \rho(r_1)\rho(r_2)\cdots\rho(r_k) \neq 0$. It follows that $\rho(r_i) \neq 0$ for all $i = 1, \ldots, k$ and so R is ω-residually-$\mathbb{Z}$.

The locally ω-residually-$\mathbb{Z}$ rings are precisely the models of the universal theory of the ring $\mathbb{Z}$.

Let $\mathbb{A}$ and $\mathbb{B}$ be algebras of some fixed type with $\mathbb{A} \leq \mathbb{B}$. By a lemma in Bell and Slomson [2], $\mathbb{A}$ and $\mathbb{B}$ are universally equivalent in the language with constants from $\mathbb{A}$ if and only if there is a nonempty index set Λ and an ultrafilter D on Λ such that there is an injection

$$f : \mathbb{B} \to {}^*\mathbb{A} = \mathbb{A}^\Lambda/D$$

with the diagram below commutative

$$\begin{array}{ccc} & f & \\ \mathbb{B} & \rightarrow & {}^*\mathbb{A} \\ & \nwarrow & \uparrow \; d \\ & & \mathbb{A} \end{array}$$

where d is the canonical embedding and $\nwarrow$ is the inclusion map.

Theorem 2.1. *G is a model of* $\mathrm{Th}_\forall(H)$ *if and only if G has at least one representation $G \leq_H U(R)$ where the locally residually-$\mathbb{Z}$ ring R is an integral domain.*

Proof. Suppose $G \leq_H U(R)$ where the locally residually-$\mathbb{Z}$ ring R is an integral domain. Then R is locally ω-residually-$\mathbb{Z}$. Let R_0 be a finitely generated subring of R. Then R_0 is ω-residually-$\mathbb{Z}$. Let S be a finite nonempty subset of $(G \cap U(R_0))\backslash\{1\}$. Then there is a retraction $\rho : R_0 \to \mathbb{Z}$ which does not annihilate any of the nonzero (1, 2), (1, 3), and (2, 3) entries of any elements of S. In the obvious way, there is a retraction

$$\overline{\rho} : G \cap U(R_0) \to H$$

which does not annihilate any element of S. Thus, $G \cap U(R_0)$ is H-discriminated by H, and hence $G \cap U(R_0)$ is a model of $\mathrm{Th}_\forall(H)$. (For an argument to see why this is true, see the proof of Theorem 8.3.12 in [3].) Since G is the direct union of models of $\mathrm{Th}_\forall(H)$,

$$G = \varinjlim (G \cap U(R_0))$$

as the R_0 vary over the finitely generated subrings of R, G is a model of $\mathrm{Th}_\forall(H)$.

Now suppose we are given a model G of $\mathrm{Th}_\forall(H)$. By the lemma in Bell and Slomson alluded to before the statement of this theorem, there is a nonempty index set Λ and an ultrafilter D on Λ such that G embeds in the ultrapower ${}^*H = H^\Lambda/D$. Then G embeds in $U({}^*\mathbb{Z})$ where ${}^*\mathbb{Z} = \mathbb{Z}^\Lambda/D$ and is locally ω-residually-$\mathbb{Z}$, hence an integral domain. (${}^*\mathbb{Z}$ is the direct union over its finitely generated subrings ${}^*\mathbb{Z} = \varinjlim R_0$ and each R_0 is ω-residually-$\mathbb{Z}$ so an integral domain. A direct union of integral domains is again an integal domain.) □

Before we can prove the next result alluded to earlier about the weak property, we need some preliminaries. By the Hilbert basis theorem (see [9]), if R is a commutative Noetherian ring then so is the polynomial ring $R[x]$. An easy induction shows that if R is a commutative Noetherian ring then so is the polynomial ring $R[x_1, \ldots, x_n]$ for every positive integer n. Of course, the ring $\mathbb{Z}$ is a principal ideal domain and so is certainly Noetherian. Hence, $\mathbb{Z}[x_1, \ldots, x_n]$ is Noetherian for every positive integer n.

Theorem 2.2. *Let $R \neq \{0\}$ be a commutative ring. Then R is a model of the set of quasi-identities true in $\mathbb{Z}$ if and only if R is locally residually-$\mathbb{Z}$.*

Proof. Suppose first R is a model of the quasi-identities true in $\mathbb{Z}$. Let R_0 be a finitely generated subring of R. Say $r_1, \ldots, r_n$ generate R_0. Then R_0 is a homomorphic image of $\mathbb{Z}[x_1, \ldots, x_n]$ under the homomorphism determined by

$$x_i \mapsto r_i, \quad i = 1, \ldots, n.$$

Since $\mathbb{Z}[x_1, \ldots, x_n]$ is Noetherian, $\mathrm{Ker}(\mathbb{Z}[x_1, \ldots, x_n] \to R_0)$ is a finitely generated ideal. Say $f_1(x_1, \ldots, x_n), \ldots, f_k(x_1, \ldots, x_n)$ generate $\mathrm{Ker}(\mathbb{Z}[x_1, \ldots, x_n] \to R_0)$ as an ideal.

Now suppose to deduce a contradiction that R_0 is not residually-$\mathbb{Z}$. Suppose that $r \in R_0\backslash\{0\}$ is annihilated by every homomorphism $R_0 \to \mathbb{Z}$. (Such homomorphisms must be surjective since $1 \mapsto 1$.) Write r as $g(r_1,\ldots,r_n)$ where $g(x_1,\ldots,x_n) \in \mathbb{Z}[x_1,\ldots,x_n]$. Then the following quasi-identity must be true in $\mathbb{Z}$:

$$\forall x_1,\ldots,x_n\left(\left(\bigwedge_{i=1}^{k}(f_i(x_1,\ldots,x_n)=0)\right)\to(g(x_1,\ldots,x_n)=0)\right). \qquad (\circledast)$$

Then $\circledast$ must be true in R_0. But that is a contradiction since in R_0 $f_i(x_{r1},\ldots,r_n) = 0$, $i = 1,\ldots,k$, and $r = g(r_1,\ldots,r_n) \neq 0$. The contradiction shows R_0 is residually-$\mathbb{Z}$. Since R_0 was an arbitrary finitely generated subring of R, R is locally residually-$\mathbb{Z}$.

Now suppose R is locally residually-$\mathbb{Z}$. Then every finitely generated subring of R is isomorphic to a subdirect power of $\mathbb{Z}$. Explicitly, if R_0 is a finitely generated subring of R, then for all $r \in R_o\backslash\{0\}$ there is retraction $\rho_r : R_0 \to \mathbb{Z}$ such that $\rho_r(r) \neq 0$.

Consider the direct power $\mathbb{Z}^\Lambda$ where $\Lambda = R_0\backslash\{0\}$. We get a homomorphism

$$\rho : R_0 \to \mathbb{Z}^\Lambda$$

via

$$x \to \rho(x)$$

where $\rho(x)(r) = \rho_r(x)$ for all $r \in \Lambda$. Suppose $x \in \mathrm{Ker}(\rho)$. Then $\rho_r(x) = 0$ for all $r \in \Lambda$. But, if $x \neq 0$, then $x \in \Lambda$ and $\rho_x(x) \neq 0$ – a contradiction. Therefore, $x = 0$ and $\mathrm{Ker}(\rho) = \{0\}$; so, R_0 embeds in $\mathbb{Z}^\Lambda$.

Since quasi-identities are preserved in direct powers and subrings, R_0 is a model of the set of quasi-identities true in $\mathbb{Z}$. Now R is the direct union $R = \varinjlim R_0$ as the R_0 vary over the finitely generated subrings of R. Quasi-identities are preserved in direct unions; so, R is a model of the set of quasi-identities true in $\mathbb{Z}$. □

Theorem 2.3. *Let R be a locally residually-$\mathbb{Z}$ ring and let G be a H-subgroup of $U(R)$. If the representation satisfies the weak property, then NZCT holds in G.*

Proof. Each of the quasi-identities

$$\forall x_1,x_2((x_1^2+x_2^2=0)\to(x_i=0)), \quad i=1,2$$

holds in $\mathbb{Z}$. Hence, they hold in R by Theorem 2.2. Now assume to deduce a contradiction that $a,b,c \in G$ with b noncentral in G and each of a and c commute with b but a and c do not commute. Let R_0 be the subring of R generated by the entries of a, b, and c. Since R_0 is finitely generated, it is residually-$\mathbb{Z}$. Note that a, b, and c lie in $G \cap U(R_0)$. If $b_{12}^2 + b_{23}^2 = 0$, then $b_{12} = b_{23} = 0$. So $b = \begin{bmatrix} 1 & 0 & b_{13} \\ 0 & 1 & 0 \\ 0 & 0 & 1 \end{bmatrix}$ would be central in G contrary to hypothesis. The contradiction shows $b_{12}^2 + b_{23}^2 \neq 0$. Thus by the weak property, $b_{12}^2 + b_{23}^2$ is not a zero divisor.

Now a and c do not commute; hence

$$\det\left(\begin{bmatrix} a_{12} & a_{23} \\ c_{12} & c_{23} \end{bmatrix}\right) \neq 0.$$

Since $b_{12}^2 + b_{23}^2$ is not a zero divisor,

$$(b_{12}^2 + b_{23}^2)\det\left(\begin{bmatrix} a_{12} & a_{23} \\ c_{12} & c_{23} \end{bmatrix}\right) \neq 0.$$

Since R_0 is residually-$\mathbb{Z}$, there is a retraction $\rho : R_0 \to \mathbb{Z}$ which does not annihilate

$$(b_{12}^2 + b_{23}^2)\det\left(\begin{bmatrix} a_{12} & a_{23} \\ c_{12} & c_{23} \end{bmatrix}\right).$$

So, in $\mathbb{Z}$,

$$(\rho(b_{12})^2 + \rho(b_{23})^2)\det\left(\begin{bmatrix} \rho(a_{12}) & \rho(a_{23}) \\ \rho(c_{12}) & \rho(c_{23}) \end{bmatrix}\right) \neq 0.$$

ρ induces $\overline{\rho} : G \cap U(R_0) \to H$ via

$$\begin{bmatrix} 1 & g_{12} & g_{13} \\ 0 & 1 & g_{23} \\ 0 & 0 & 1 \end{bmatrix} \overset{\overline{\rho}}{\mapsto} \begin{bmatrix} 1 & \rho(g_{12}) & \rho(g_{13}) \\ 0 & 1 & \rho(g_{23}) \\ 0 & 0 & 1 \end{bmatrix}.$$

Now $\overline{\rho}(b)$ is not central in H since $\rho(b_{12})^2 + \rho(b_{23})^2 \neq 0$. $\overline{\rho}(a)$ commutes with $\overline{\rho}(b)$ and $\overline{\rho}(b)$ commutes with $\overline{\rho}(c)$. But $\overline{\rho}(a)$ and $\overline{\rho}(c)$ do not commute since $\det\left(\begin{bmatrix} \rho(a_{12}) & \rho(a_{23}) \\ \rho(c_{12}) & \rho(c_{23}) \end{bmatrix}\right) \neq 0$. That contradicts the fact that H satisfies NZCT. The contradiction shows that G satisfies NZCT. □

Theorem 2.4. *Let R be a locally residually-$\mathbb{Z}$ ring and let G be a H-subgroup of $U(R)$. If the representation satisfies the strong property, then G is a model of* $\mathrm{Th}_\forall(H)$.

Proof. Each of the quasi-identities

$$\forall x_1, x_2, x_3((x_1^2 + x_2^2 + x_3^2 = 0) \to (x_i = 0)), \quad i = 1, 2, 3$$

holds in $\mathbb{Z}$. Hence, they hold in R by Theorem 2.2. Let R_0 be a finitely generated subring of R. Hence, R_0 is residually-$\mathbb{Z}$. Let S be a finite nonempty subset of $(G \cap U(R_0))\backslash\{1\}$. Then, for each $g \in S$, we have

$$g_{12}^2 + g_{13}^2 + g_{23}^2 \neq 0.$$

By the strong property, for all $g \in S$, $g_{12}^2 + g_{13}^2 + g_{23}^2$ is not a zero divisor. Hence, $r = \prod_{g\in S}(g_{12}^2 + g_{13}^2 + g_{23}^2) \neq 0$. Since R_0 is residually-$\mathbb{Z}$ there is a retraction $\rho : R_0 \to \mathbb{Z}$ such that

$$\rho(r) = \prod_{g\in S}(\rho(g_{12})^2 + \rho(g_{13})^2 + \rho(g_{23})^2) \neq 0.$$

ρ induces a retraction $\overline{\rho} : G \cap U(R_0) \to H$ via

$$\begin{bmatrix} 1 & x_{12} & x_{13} \\ 0 & 1 & x_{23} \\ 0 & 0 & 1 \end{bmatrix} \overset{\overline{\rho}}{\mapsto} \begin{bmatrix} 1 & \rho(x_{12}) & \rho(x_{13}) \\ 0 & 1 & \rho(x_{23}) \\ 0 & 0 & 1 \end{bmatrix}.$$

Now, for all $g \in S$, $\overline{\rho}(g) \neq 1$ since $\rho(g_{12})^2 + \rho(g_{13})^2 + \rho(g_{23})^2 \neq 0$. Hence, $G \cap U(R_0)$ is H-discriminated by H and so $G \cap U(R_0)$ is a model of $\mathrm{Th}_\forall(H)$. Since G is the direct union of the models $G = \varinjlim(G \cap U(R_0))$ of $\mathrm{Th}_\forall(H)$ as R_0 vary over the finitely generated subrings of R. G itself must be a model of $\mathrm{Th}_\forall(H)$. □

Corollary 1. *Let R be a locally residually-$\mathbb{Z}$ ring and let G be a H-subgroup of $U(R)$. Suppose the representation satisfies the weak property and suppose that G contains no central element $z = \left[\begin{smallmatrix} 1 & 0 & z_{13} \\ 0 & 1 & 0 \\ 0 & 0 & 1 \end{smallmatrix}\right]$ where $z_{13} \neq 0$ is a zero divisor. Then the representation satisfies the strong property and, in particular, G is a model of* $\mathrm{Th}_\forall(H)$.

Proof. The following quasi-identities hold in $\mathbb{Z}$:

$$\forall x, y((x^2 y = 0) \to (xy = 0))$$
$$\forall x_1, x_2, x_3(((x_1^2 + x_2^2 + x_3^2) = 0) \to (x_1^2 + x_2^2 = 0))$$

and

$$\forall x((x^n = 0) \to (x = 0)) \quad \text{for every integer } n > 0.$$

So they hold in R by Theorem 2.2. Thus R contains no nonzero nilpotent elements. Suppose $a \in G$ and

$$a_{12}^2 + a_{13}^2 + a_{23}^2 \neq 0.$$

Case 1: $a_{12} = a_{23} = 0$.

Then $a = \left[\begin{smallmatrix} 1 & 0 & a_{13} \\ 0 & 1 & 0 \\ 0 & 0 & 1 \end{smallmatrix}\right]$ is central in G and so, if $a_{13} \neq 0$, then a_{13} is not a zero divisor.

Now if a_{13}^2 were a zero divisor, then there is a $b \neq 0$ such that $a_{13}^2 b = 0$. But that implies $a_{13} b = 0$ contrary to hypothesis since

$$a_{12}^2 + a_{13}^2 + a_{23}^2 = a_{13}^2 \neq 0 \quad \text{so } a_{13} \neq 0.$$

Case 2: $a_{12}^2 + a_{23}^2 \neq 0$.

Suppose there were $b \neq 0$ such that $(a_{12}^2 + a_{13}^2 + a_{23}^2)b = 0$. Then $(a_{12}^2 + a_{13}^2 + a_{23}^2)b^2 = 0$ and $(a_{12}b)^2 + (a_{13}b)^2 + (a_{23}b)^2 = 0$ from which $(a_{12}b)^2 + (a_{13}b)^2 = 0$ follows. Then $a_{12} = a_{23} = 0$ contrary to hypothesis in this case.

The contradiction settles Case 2 and so the representation satisfies the strong property. In particular, G is a model of $\mathrm{Th}_\forall(H)$. □

Theorem 2.5. *Let R be a locally residually-$\mathbb{Z}$ ring and let G be isomorphic to $U(R)$. If G satisfies NZCT, then the representation satisfies the weak property. Moreover, in that event R is an integral domain; hence, G is a model of* $\mathrm{Th}_\forall(H)$ *in that event.*

Proof. Suppose that the representation violates the weak property. Then there is $a \in G$ with $a_{12}^2 + a_{23}^2 \neq 0$ a zero divisor. Hence, there is $b_0 \neq 0$ such that $(a_{12}^2 + a_{23}^2)b_0 = 0$. Then $(a_{12}^2 + a_{23}^2)b_0^2 = 0$ so $(a_{12}b_0)^2 + (a_{23}b_0)^2 = 0$ and $a_{12}b_0 = a_{23}b_0 = 0$. Now let $b = \begin{bmatrix} 1 & b_0 & 0 \\ 0 & 1 & b_0 \\ 0 & 0 & 1 \end{bmatrix}$ and $c = \begin{bmatrix} 1 & -a_{23} & 0 \\ 0 & 1 & a_{12} \\ 1 & 0 & 0 \end{bmatrix}$. b is noncentral since $b_0 \neq 0$. a and b commute since $\det\left(\begin{bmatrix} a_{12} & a_{23} \\ b_0 & b_0 \end{bmatrix}\right) = a_{12}b_0 - a_{23}b_0 = 0 - 0 = 0$. b and c commute since $\det\left(\begin{bmatrix} b_0 & b_0 \\ -a_{23} & a_{12} \end{bmatrix}\right) = a_{12}b_0 + a_{23}b_0 = 0 + 0 = 0$. But a and c do not commute since $\det\left(\begin{bmatrix} a_{12} & a_{23} \\ -a_{23} & a_{12} \end{bmatrix}\right) = a_{12}^2 + a_{23}^2 \neq 0$. Thus violating the weak property implies violating NZCT. Taking the contrapositive, G satisfying NZCT implies the representation satisfies the weak property. Thus, in the case $G \cong U(R)$, G satisfying NZCT is equivalent to the representation satisfying the weak property.

So suppose that the representation does satisfy the weak Property (equivalently G satisfies NZCT).

Assume to deduce a contradiction that R contains a zero divisor $a_0 \neq 0$. Say $a_0b_0 = 0$ where $b_0 \neq 0$. Let $a = \begin{bmatrix} 1 & 0 & 0 \\ 0 & 1 & a_0 \\ 0 & 0 & 1 \end{bmatrix}$. Then $a_{12}^2 + a_{23}^2 = 0^2 + a_0^2 = a_0^2 \neq 0$. (Recall R contains no nonzero nilpotent elements.) But $a_{12}^2 + a_{23}^2 = a_0^2$ is a zero divisor since $a_0^2 b_0 = a_0(a_0b_0) = a_0 \cdot 0 = 0$. That contradicts the hypothesis that the representation satisfies the weak property. Therefore, if the representation satisfies the weak property, then R is an integral domain. In particular, in that event, G is a model of $\mathrm{Th}_\forall(H)$. □

In conclusion, we have the partial result that, if G is a model of $\mathrm{diag}(H) \cup Q(H) \cup \{\mathrm{NZCT}\}$ and G has a special representation (i. e., $G \cong U(R)$ for some locally residually-$\mathbb{Z}$ ring R. A priori R is locally residually-$\mathbb{Z}$ but then one deduces it must actually be locally ω-residually-$\mathbb{Z}$), then G must already be a model of $\mathrm{Th}_\forall(H)$.

3 Necessity of the conditions

Here, we give two examples. Example 1 below shows that the weak property is not necessary and Example 2 below shows that the strong property is not necessary.

Example 1. An example of $G \leq_H U(H)$ which satisfies NZCT but the representation violates the weak property is as follows. Let θ be an indeterminate over $\mathbb{Z}$. Let R be the subring of $\mathbb{Z}[\theta] \times \mathbb{Z}[\theta]$ generated by $(\theta, 0)$ and $(0, \theta)$.

Let $G = \langle H, a_3 \rangle$ where

$$a_3 = \begin{bmatrix} 1 & (\theta,0) & (0,\theta) \\ 0 & 1 & (\theta,0) \\ 0 & 0 & 1 \end{bmatrix}$$

(and we identify $f(\theta)$ with $(f(\theta).f(\theta))$ via the diagonal embedding). Then $(\theta,0)^2+(\theta,0)^2=(2\theta^2,0)$ is a nonzero divisor of zero in R; so, the representation violates the weak property. Now

$$a_1a_2=\begin{bmatrix}1&1&0\\0&1&1\\0&0&1\end{bmatrix}$$

and $\det\left(\left[\begin{smallmatrix}(\theta,0)&(\theta,0)\\1&1\end{smallmatrix}\right]\right)=(\theta,0)-(\theta,0)=0$ so a_3 commutes with a_1a_2.

A typical element $g\in G$ looks like

$$\begin{aligned}g&=a_1^{n(1)}a_2^{n(2)}a_3^{n(3)}[a_2,a_1]^{n(2,1)}[a_3,a_1]^{n(3,1)}[a_3,a_2]^{n(3,2)}\\&=\begin{bmatrix}1&(n(2)+n(3)\theta,n(2))&z_0\\0&1&(n(1)+n(3)\theta,n(1))\\0&0&1\end{bmatrix}\end{aligned}$$

where $z_0=n(2,1)+(0,n(3)\theta)+(\frac{n(3)(n(3)-1)}{2}\theta^2,0)+(n(3,1)\theta,0)-(n(3,2)\theta,0)=(n(2,1)+\frac{n(3)(n(3)-1)}{2}\theta^2+n(3,1)\theta-n(3,2)\theta,n(2,1)+n(3)\theta)$.

How could $g=1$? From $n(2,1)+n(3)\theta=0$, we need $n(3)=0$, and hence $n(2,1)=0$. Then from $n(2,1)+\frac{n(3)(n(3)-1)}{2}\theta^2+(n(3,1)-n(3,2))\theta=(n(3,1)-n(3,2))\theta=0$ or $n(3,1)=n(3,2)=n_0$. So the only possible relations are $[a_3,a_1]^{n_0}[a_3,a_2]^{n_0}=1$ or $[a_3,a_1a_2]^{n_0}=1$ which is a consequence of $[a_3,a_1a_2]=1$. So G is the centralizer extension (relative to N_2) of a_1a_2. We can find $N>0$ sufficiently large so that the retraction

$$\begin{cases}a_1\mapsto a_1\\a_2\mapsto a_2\\a_3\mapsto(a_1a_2)^N\end{cases}$$

does not annihilate finitely many a priori given nontrivial elements of G. Thus G is H-discriminated by H and so $G\models \mathrm{Th}_\forall(H)$. In particular, $G\models$ NZCT.

Example 2. An example of a model of $\mathrm{Th}_\forall(H)$ for which a representation violates the strong property is as follows. Let $R=\mathbb{Z}\times\mathbb{Z}$ and let

$$a_3=\begin{bmatrix}1&0&(1,0)\\0&1&0\\0&0&1\end{bmatrix}.$$

(We identify n with (n,n) via the diagonal embedding.) Let $G=\langle H,a_3\rangle$. Note $1-(1,0)=(0,1)$ so $(1,0)$ and $(0,1)$ both lie in R and are nonzero divisors of zero in R. But $0^2+(1,0)^2+0^2=(1,0)$ which is a nonzero divisor of zero in R; so, the representation violates the strong property. Given finitely many nontrivial elements of G, there is a $N>0$ sufficiently large so that the retraction $G\to H$ given by

$$\begin{cases}a_1\mapsto a_1\\a_2\mapsto a_2\\a_3\mapsto[a_2,a_1]^N\end{cases}$$

does not annihilate any of them. Hence, G is H-discriminated by H and so $G \models \mathrm{Th}_\forall(H)$.

In a sequel using the matrix representations, we shall characterize the 3-generator models of $\mathrm{diag}(H) \cup Q(H) \cup \{\mathrm{NZCT}\}$‘ and show they are all models of $\mathrm{Th}_\forall(H)$.

4 Question

Question (A. G. Myasnikov). Is $\mathrm{Th}_\forall(H)$ axiomatized by $\mathrm{diag}(H) \cup Q(H) \cup \{\mathrm{NZCT}\}$?

Bibliography

[1] Baumslag G, Myasnikov AG, Remeslennikov VN. Algebraic Geonetry over Groups I. J Algebra. 1999;219:16–79.
[2] Bell JL, Slomson AB. Models and Ultraproducts: An Introduction. Amsterdam: North-Holland; 1972. (Second revised printing).
[3] Fine B, Gaglione AM, Myasnikov AG, Rosenberger G, Spellman D. The Elementary Theory of Groups. Berlin: DeGruyter; 2014.
[4] Fine B, Gaglione AM, Myasnikov AG, Spellman D. Discriminating Groups. J Group Theory. 2001;4:463–74.
[5] Gaglione AM, Jackson D, Spellman D. Basic Commutators as Relators. J Group Theory. 2002;5:351–63.
[6] Gaglione AM, Spellman D. The Persistence of Universal Formulae in Free Algebras. Bull Aust Math Soc. 1987;36(1):11–7.
[7] Grätzer G. Universal Algebra. New Jersey: Van Nostrand; 1968.
[8] Grätzer G, Lakser H. A Note on the Implication Class Generated by a Class of Structures. Can Math Bull. 1973;16:603–5.
[9] Hungerford TW. Algebra. New York: Holt, Rinehart, and Winston, Inc.; 1974.
[10] Neumann H. Varieties of Groups. New York: Springer; 1967.
[11] Weinstein M. Examples of Groups. 2nd ed. New Jersey: Polygonal Publishing House; 2006.

Anthony M. Gaglione and Dennis Spellman

Some model theory of the Heisenberg group: II

The three generator case

Abstract: The Heisenberg group H is the group of all 3×3 upper unitriangular matrices with entries in the ring $\mathbb{Z}$ of integers. We may paraphrase an instance of a question of A. G. Myasnikov as follows: Is the universal theory of H in the language of H, here denoted $\mathrm{Th}_\forall(H)$, axiomatizable by the set $\mathrm{diag}(H)$ of atomic and negated atomic sentences true in H together with the set $Q(H)$ of quasi-identities true in H and the single additional axiom NZCT (or noncentral commutative transitivity) asserting that the centralizer of every noncentral element be abelian? In this paper, we characterize the 3-generator models of $\mathrm{diag}(H) \cup Q(H) \cup \{\mathrm{NZCT}\}$ and show they are all models of $\mathrm{Th}_\forall(H)$.

Keywords: Noncentral commutative transitivity, quasi-identity, locally residually-$\mathbb{Z}$, diagram

MSC 2010: Primary 05C38, 15A15, Secondary 05A15, 15A18

1 Introduction

This paper is a sequel to [2] and we heavily rely on that paper for preliminaries. Nonetheless, in order to keep the manuscript relatively self-contained, we explain our notation in this section.

Let R b a commutative ring with unity having characteristic zero so that R may be considered an extension of the ring $\mathbb{Z}$ of integers. $U(R)$ is the group of all 3×3 upper unitriangular matrices with entries from R. The **Heisenberg group** H is $U(\mathbb{Z})$. It is free in the variety N_2 of all groups nilpotent of class 2 on the generators

$$a_1 = \begin{bmatrix} 1 & 0 & 0 \\ 0 & 1 & 1 \\ 0 & 0 & 1 \end{bmatrix} \quad \text{and} \quad a_2 = \begin{bmatrix} 1 & 1 & 0 \\ 0 & 1 & 0 \\ 0 & 0 & 1 \end{bmatrix}.$$

$L_0[H]$ shall be the first-order language with equality containing constants naming the elements of H as well as a binary operation symbol $\cdot$ (suppressed in favor of juxtaposition) and a unary operation symbol $^{-1}$. Furthermore, we commit the convenient abuse of identifying the name of an element with the element itself.

Anthony M. Gaglione, U.S. Naval Academy, 121 Blake Rd, Annapolis, MD 21402, USA, e-mail: agaglione@aol.com
Dennis Spellman, Crum Lynne, PA 19022, USA

https://doi.org/10.1515/9783110638387-006

$\mathrm{Th}_\forall(H)$ shall be the set of all universal sentences of $L_0[H]$ true in H. $\mathrm{diag}(H)$, the **diagram** of H, shall be the set of all atomic and negated atomic sentences of $L_0[H]$ true in H. $Q(H)$ shall be the set of all quasi-identities of $L_0[H]$ true in H. Note that, up to logical equivalence, the group axioms lie in $Q(H)$. Models of the group axioms and $\mathrm{diag}(H)$ are precisely the H-groups of Baumslag, Myasnikov, and Remeslennikov [1]. Equivalently, a H-group is a group containing a distinguished copy of H as a subgroup. Finally, NZCT (or **noncentral commutative transitivity**) is expressed by the universal sentence

$$\forall x, y, z, w(([y,w] \neq 1) \wedge ([x,y] = 1) \wedge ([y,z] = 1)) \to ([x,z] = 1))$$

asserting that the centralizer of every noncentral element be abelian. We may paraphrase an instance of a question of A. G. Myasnikov as follows: Is $\mathrm{Th}_\forall(H)$ axiomatizable by $\mathrm{diag}(H) \cup Q(H) \cup \{\mathrm{NZCT}\}$?

It was shown in [2] that every model of $\mathrm{diag}(H) \cup Q(H)$ embeds as a H-subgroup (the embedding preserving H) of $U(R)$ for some locally residually-$\mathbb{Z}$ ring R (and conversely). Here, R is locally residually-$\mathbb{Z}$ provided, for every finitely generated unital subring $R_0 \leq R$ and every $r \in R_0\backslash\{0\}$ there is a retraction $\rho : R_0 \to \mathbb{Z}$ which does not annihilate r. A sufficient condition for a H-group G to be a model of $\mathrm{Th}_\forall(H)$ is that G be H-discriminated by H. That is, given finitely many nontrivial elements of G there is a retraction $\rho : G \to H$ which does not annihilate any of them. Suppose G is free in N_2 of rank $r \geq 3$. (So G may be considered a H-group.) It is known that G is H-discriminated by H; hence, G is a model of $\mathrm{Th}_\forall(H)$. If $h \in H$, let $C_H(h)$ be its centralizer in H. Writing $x = a_1^{x(1)} a_2^{x(2)} [a_2, a_1]^{x(2,1)}$ for an arbitrary element $x \in H$, we have the following.

Lemma 1.1. *Let $h \in H\backslash Z(H)$. Let G be the group with relative presentation*

$$\langle \overline{a}_1, \overline{a}_2, a_3; [a_3, \overline{a}_1^{c(1)} \overline{a}_2^{c(2)} [\overline{a}_2, \overline{a}_1]^{c(2,1)}] = 1 \,\forall c \in C_H(h)\rangle_{N_2}.$$

Then the assignment

$$\begin{cases} \overline{a}_1 \mapsto a_1 \\ \overline{a}_2 \mapsto a_2 \end{cases}$$

extends to an isomorphism from $\langle \overline{a}_1, \overline{a}_2\rangle$ onto H. Identifying $\overline{a}_i$ with a_i, $i = 1, 2$ we view G as a H-group and may rewrite G as $\langle a_1, a_2, a_3; [a_3, c] = 1 \,\forall c \in C_H(h)\rangle_{N_2}$. Furthermore, for every integer $N > 0$, the assignment

$$\begin{cases} a_1 \mapsto a_1 \\ a_2 \mapsto a_2 \\ a_3 \mapsto h^N \end{cases}$$

extends to a retraction $G \to H$.

Proof. We claim that, for any $y \in C_H(h)$, the assignment

$$\begin{cases} \overline{a}_1 \mapsto a_1 \\ \overline{a}_2 \mapsto a_2 \\ \overline{a}_3 \mapsto y \end{cases}$$

extends to a well-defined surjective homomorphism $\rho_y : G \to H$. This is so since the relations are preserved and the generators a_1 and a_2 lie in the image of ρ_y. Now suppose $\overline{x} = \overline{a}_1^{x(1)}\overline{a}_2^{x(2)}[\overline{a}_2, \overline{a}_1]^{x(2,1)}$ where $x(1)^2 + x(2)^2 + x(2,1)^2 \neq 0$. Then $\rho_y(\overline{x}) = a_1^{x(1)}a_2^{x(2)}[a_2, a_1]^{x(2,1)} \neq 1$ in H. It follows that $Ker(\rho_y) \cap \langle \overline{a}_1, \overline{a}_2 \rangle = \{1\}$ and so the restriction $\rho_y \mid_{\langle \overline{a}_1, \overline{a}_2 \rangle}$ is an isomorphism from $\langle \overline{a}_1, \overline{a}_2 \rangle$ onto H. So G is a H-group and we may identify $\overline{a}_i$ with a_i, $i = 1, 2$, and $\overline{a}_1^{c(1)}\overline{a}_2^{c(2)}[\overline{a}_2, \overline{a}_1]^{c(2,1)}$ with c for all $c \in C_H(h)$. Furthermore, since, if $N > 0$ is an integer, $y = h^N$ lies in $C_H(h)$ and so the assignment

$$\begin{cases} a_1 \mapsto a_1 \\ a_2 \mapsto a_2 \\ a_3 \mapsto h^N \end{cases}$$

extends to a well-defined retraction $G \to H$. □

If $h \in H \backslash Z(H)$, let us call the group G with relative presentation,

$$\langle a_1, a_2, a_3;\ [a_3, c] = 1\, \forall c \in C_H(h) \rangle_{N_2},$$

the **free rank one extension of $C_H(h)$ relative to N_2.**

Given finitely many nontrivial elements of G, there is an integer $N > 0$ sufficiently large so that the retraction $\rho : G \to H$ determined by

$$\begin{cases} a_1 \mapsto a_1 \\ a_2 \mapsto a_2 \\ a_3 \mapsto h^N \end{cases}$$

does not annihilate any of them. Hence, G is H-discriminated by H and so G is a model of $\mathrm{Th}_\forall(H)$. In this paper, we shall prove the following.

Theorem 1.2 (Three generator theorem). *Let G be a rank 3 model of* $\mathrm{diag}(H) \cup Q(H) \cup \{\mathrm{NZCT}\}$. *Then, up to isomorphism, one of the following three cases must occur:*

(1) $G = F_3(N_2)$
(2) $G = \langle a_1, a_2, t;\ [t, c] = 1\, \forall c \in C_H(h) \rangle_{N_2}$
(3) $G = H \times \langle z; \rangle$

In particular, every rank 3 model of $\mathrm{diag}(H) \cup Q(H) \cup \{\mathrm{NZCT}\}$ *must already be a model of* $\mathrm{Th}_\forall(H)$.

2 The main theorem

It was observed in [2] that, for any H-subgroup G of $U(R)$, the two elements $a = \left[\begin{smallmatrix} 1 & a_{12} & a_{13} \\ 0 & 1 & a_{23} \\ 0 & 0 & 1 \end{smallmatrix}\right]$ and $b = \left[\begin{smallmatrix} 1 & b_{12} & b_{13} \\ 0 & 1 & b_{23} \\ 0 & 0 & 1 \end{smallmatrix}\right]$ of G commute if and only if $\det\left(\left[\begin{smallmatrix} a_{12} & a_{23} \\ b_{12} & b_{23} \end{smallmatrix}\right]\right) = 0$ and, moreover, the center of G consists precisely of all elements of G of the form $z = \left[\begin{smallmatrix} 1 & 0 & z_{13} \\ 0 & 1 & 0 \\ 0 & 0 & 1 \end{smallmatrix}\right]$.

Proof of the three generator theorem. Let $R \supsetneqq \mathbb{Z}$ be locally residually-$\mathbb{Z}$. Let $(r_{12}, r_{13}, r_{23})R^3\setminus\mathbb{Z}^3$ and let $a_3 = \left[\begin{smallmatrix} 1 & r_{12} & r_{13} \\ 0 & 1 & r_{23} \\ 0 & 0 & 1 \end{smallmatrix}\right]$. Let $G = \langle H, a_3\rangle$.

Case (I): $r_{12}^2 + r_{23}^2 \notin \mathbb{Z}$ (In particular, $r_{12}^2 + r_{23}^2 \neq 0$.)

Subcase (I.A) No nontrivial $\mathbb{Z}$-linear combination of r_{12} and r_{23} lies in $\mathbb{Z}$. (In particular, r_{12} and r_{23} are linearly independent over $\mathbb{Z}$.)

A typical element $g \in G$ looks like

$$g = a_1^{n(1)} a_2^{n(2)} a_3^{n(3)} [a_2, a_1]^{n(2,1)} [a_3, a_1]^{n(3,1)} [a_3, a_2]^{n(3,2)}$$
$$= \begin{bmatrix} 1 & n(2) + n(3)r_{12} & z_0 \\ 0 & 1 & n(1) + n(3)r_{23} \\ 0 & 0 & 1 \end{bmatrix}$$

where $z_0 = n(3)r_{13} + \frac{n(3)(n(3)-1)}{2} r_{12}r_{23} + n(2,1) + n(3,1)r_{12} - n(3,2)r_{23}$.

How could $g = 1$? We must have, in particular, $n(2) + n(3)r_{12} = 0 = n(1) + n(3)r_{23}$. So $n(3)r_{12} = -n(2)$ and $n(3)r_{23} = -n(1)$. Then $n(3)n(1)r_{12} = -n(1)n(2) = n(3)n(2)r_{23}$ and $n(3)(n(1)r_{12} - n(2)r_{23}) = 0$. If $n(3) \neq 0$, then $n(1)r_{12} - n(2)r_{23} = 0$ and, by linear independence, $n(1) = 0 = n(2)$.

Suppose $n(3) = 0$. Then from $n(2)+n(3)r_{12} = 0 = n(1)+n(3)r_{23}$ we get $n(2) = 0 = n(1)$. So, in all cases, $n(1) = 0 = n(2)$. Then from $n(2) + n(3)r_{12} = 0 = n(1) + n(3)r_{23}$, we get $n(3)r_{12} = 0 = n(3)r_{23}$. If $n(3) \neq 0$, we would have $r_{12} = 0 = r_{23}$ which contradicts the linear independence of r_{12} and r_{23}. So $n(3) = 0$. Hence, $z_0 = n(3)r_{13} + \frac{n(3)(n(3)-1)}{2} r_{12}r_{23} + n(2,1) + n(3,1)r_{12} - n(3,2)r_{23} = n(2,1) + n(3,1)r_{12} - n(3,2)r_{23}$ must be zero. But then $n(3,1)r_{12} - n(3,2)r_{23} = -n(2,1) \in \mathbb{Z}$. But then by hypothesis $n(3,1) = 0 = n(3,2)$. So z_0 reduces to $n(2,1)$ and we must have $n(2,1) = 0$ also. It follows that G can satisfy no nontrivial relation on $\{a_1, a_2, a_3\}$ relative to N_2 and so G is free on $\{a_1, a_2, a_3\}$ relative to N_2.

Subcase (I.B) There are $(m_1, m_2, m_0) \in \mathbb{Z}^3$ with $(m_1, m_2) \neq (0,0)$ such that $m_1r_{12} + m_2r_{23} = m_0$. Let R_0 be the subring of R generated by r_{12} and r_{23}. Then R_0 is residually-$\mathbb{Z}$ and $r_{12}^2 + r_{23}^2 \in R_0\setminus\{0\}$. Thus, there is a retraction $\rho : R_0 \to \mathbb{Z}$ which does not annihilate $r_{12}^2 + r_{23}^2$. Applying ρ to the equation $m_1r_{12} + m_2r_{23} = m_0$, we get $m_1\rho(r_{12}) + m_2\rho(r_{23}) = m_0$.

Let

$$h = \begin{bmatrix} 1 & -m_2 & 0 \\ 0 & 1 & m_1 \\ 0 & 0 & 1 \end{bmatrix} = a_1^{m_1} a_2^{-m_2}.$$

Note that, since $(m_1, m_2) \in \mathbb{Z}^2 \backslash \{(0.0)\}$, $h \in H \backslash Z(H)$.

Now G is also generated by a_1, a_2 and

$$b = a_1^{-\rho(r_{23})} a_2^{-\rho(r_{12})} a_3 = \begin{bmatrix} 1 & r_{12} - \rho(r_{12}) & r_{13} \\ 0 & 1 & r_{23} - \rho(r_{23}) \\ 0 & 0 & 1 \end{bmatrix}.$$

We claim that b commutes with h. To see that, observe

$$\begin{aligned} &\det\left(\begin{bmatrix} r_{12} - \rho(r_{12}) & r_{23} - \rho(r_{23}) \\ -m_2 & m_1 \end{bmatrix}\right) \\ &= m_1(r_{12} - \rho(r_{12})) + m_2(r_{23} - \rho(r_{23})) \\ &= (m_1 r_{12} + m_2 r_{23}) - (m_1 \rho(r_{12}) + m_2 \rho(r_{23})) \\ &= m_0 - m_0 = 0. \end{aligned}$$

Now let $s_{12} = r_{12} - \rho(r_{12})$ and $s_{23} = r_{23} - \rho(r_{23})$ so that we can write b as

$$\begin{bmatrix} 1 & s_{12} & r_{13} \\ 0 & 1 & s_{23} \\ 0 & 0 & 1 \end{bmatrix}.$$

Since ρ is a retraction, we have $\rho^2 = \rho$. Now $\rho(s_{12}) = \rho(r_{12} - \rho(r_{12})) = \rho(r_{12}) - \rho(r_{12}) = 0$. Similarly, $\rho(s_{23}) = 0$. A typical element $g \in G$ looks like

$$\begin{aligned} g &= a_1^{n(1)} a_2^{n(2)} b^{n(3)} [a_2, a_1]^{n(2,1)} [b, a_1]^{n(3,1)} [b, a_2]^{n(3,2)} \\ &= \begin{bmatrix} 1 & n(2) + n(3)s_{12} & z_0 \\ 0 & 1 & n(1) + n(3)s_{23} \\ 0 & 0 & 1 \end{bmatrix} \end{aligned}$$

where $z_0 = n(3)r_{13} + \frac{n(3)(n(3)-1)}{2} s_{12}s_{23} + n(2,1) + n(3,1)s_{12} - n(3,2)s_{23}$.

How could $g = 1$? We would need, in particular, $n(2) + n(3)s_{12} = 0 = n(1) + n(3)s_{23}$. Applying ρ, we conclude $n(2) = 0 = n(1)$. So $n(2) + n(3)s_{12} = 0 = n(1) + n(3)s_{23}$ becomes $n(3)s_{12} = 0 = n(3)s_{23}$.

Could $s_{12} = 0 = s_{23}$? If that were true, then $r_{12} = \rho(r_{12}) \in \mathbb{Z}$ and $r_{23} = \rho(r_{23}) \in \mathbb{Z}$. So $r_{12}^2 + r_{23}^2 \in \mathbb{Z}$, contrary to hypothesis.

So at least one of s_{12} or s_{23} is nonzero and from $n(3)s_{12} = 0 = n(3)s_{23}$ we deduce $n(3) = 0$.

So $z_0 = n(3)r_{13} + \frac{n(3)(n(3)-1)}{2}s_{12}s_{23} + n(2,1) + n(3,1)s_{12} - n(3,2)s_{23} = n(2,1) + n(3,1)s_{12} - n(3,2)s_{23} = 0$. Applying ρ to this, we get $n(2,1) = 0$. So $z_0 = n(3,1)s_{12} - n(3,2)s_{23} = 0$.

So the only possible relations are equivalent to $[b, a_1]^{n(3,1)}[b, a_1]^{n(3,2)} = 1$ or $[b, a_1^{n(3,1)}a_2^{n(3,2)}] = 1$.

Since we are assuming that G satisfies NZCT and $b \notin Z(G)$, we deduce $a_1^{n(3,1)}a_2^{n(3,2)}$ must commute with $h = a_1^{m_1}a_2^{-m_2} \in H\backslash Z(H)$ and G is just the free rank one centralizer extension

$$\langle a_1, a_2, b;\ [b, c] = 1\ \forall c \in C_H(a_1^{m_1}a_2^{-m_2})\rangle_{N_2}$$

relative to N_2.

Case (II): $r_{12}^2 + r_{23}^2 \in \mathbb{Z}$.

Subcase (II.A): $(r_{12}, r_{23}) \in \mathbb{Z}^2$.

Then G is also generated by a_1, a_2, and $b = a_1^{-r_{23}}a_2^{-r_{12}}a_3 = \begin{bmatrix} 1 & 0 & r_{13} \\ 0 & 1 & 0 \\ 0 & 0 & 1 \end{bmatrix}$. Since $(r_{12}, r_{13}, r_{23}) \in R^3\backslash\mathbb{Z}^3$ and $(r_{12}, r_{23}) \in \mathbb{Z}^2$, we must have $r_{13} \notin \mathbb{Z}$. Thus, G is isomorphic to $H \times \langle z; \rangle$.

Subcase (II.B): $(r_{12}, r_{23}) \notin \mathbb{Z}^2$.

Thus at least one of r_{12} or r_{23} is nonintegral. We may assume $r_{12} \notin \mathbb{Z}$. The case $r_{23} \notin \mathbb{Z}$ is similar.

G is also generated by a_1, a_2, and

$$b = a_2a_3 = \begin{bmatrix} 1 & r_{12}+1 & r_{13} \\ 0 & 1 & r_{23} \\ 0 & 0 & 1 \end{bmatrix}.$$

We claim that $(r_{12}+1)^2 + r_{23}^2 \notin \mathbb{Z}$. $(r_{12}+1)^2 + r_{23}^2 = 1 + r_{12}^2 + r_{23}^2 + 2r_{12}$. Now $1 + r_{12}^2 + r_{23}^2 \in \mathbb{Z}$. So if $(r_{12}+1)^2 + r_{23}^2$ were integral, we would have to have $2r_{12} \in \mathbb{Z}$. Say $2r_{12} = k$ where $k \in \mathbb{Z}$. Since we are assuming $r_{12} \notin \mathbb{Z}$, we certainly have $r_{12} \neq 0$. Let R_0 be the subring of R generated by r_{12}. Then there is a retraction $\rho : R_0 \to \mathbb{Z}$ which does not annihilate r_{12}. Applying ρ to $2r_{12} = k$, we get $2\rho(r_{12}) = k$ in $\mathbb{Z}$ so k is even. Thus from $2r_{12} = k$ we get $r_{12} = \frac{k}{2} \in \mathbb{Z}$ contrary to hypothesis. The contradiction shows that $(r_{12}+1)^2 + r_{23}^2 \notin \mathbb{Z}$ and we are back in Case (I). □

Bibliography

[1] Baumslag G, Myasnikov AG, Remeslennikov VN. Algebraic Geonetry over Groups I. J Algebra. 1999;219:16–79.

[2] Gaglione AM, Spellman D. Some Model Theory of the Heisenberg Group: I. Preprint.

Benjamin Fine, Anja I. S. Moldenhauer, and Gerhard Rosenberger

On vector-valued Hecke forms

Abstract: We discuss a general theory of vector-valued Hecke forms associated to finite-dimensional representations ρ of the Hecke group $\overline{G}(\lambda)$.

Keywords: Hecke groups, vector-valued Hecke forms, dimensions, Poincaré series and Eisenstein series

MSC 2010: Primary 11F03, 11F12, 30F35, Secondary 11F30, 20H10, 30M50

1 Introduction

Knapp and Mason, in [7] and [8], initiated a theory of vector-valued modular forms associated to a finite-dimensional representation ρ of the (homogenous) modular group $\mathrm{SL}(2,\mathbb{Z})$. In this paper, we extend their approaches for the (homogenous) Hecke group $\overline{G}(\lambda), \lambda = 2\cos(\frac{\pi}{q})$, $q \geq 3$, which is generated by the two matrices

$$\overline{T} = \begin{pmatrix} 0 & -1 \\ 1 & 0 \end{pmatrix} \quad \text{and} \quad \overline{S}_\lambda = \begin{pmatrix} 1 & \lambda \\ 0 & 1 \end{pmatrix}. \tag{1}$$

$\overline{G}(\lambda)$ acts discontinuously on the upper half-plane and is a natural extension of the modular group. In this paper, we introduce in an analogous manner vector-valued Hecke forms associated to a finite-dimensional representation ρ of $\overline{G}(\lambda)$ and establish analogs of some basic results from the theory of vector-valued modular forms. It would be of great interest to extend the described results and to consider some important questions. Is, for large enough k, the space of entire vector-valued Hecke forms of weight k spanned by polynomial combination of Poincaré series or even by Eisenstein series. Does Hecke's estimate $a_n(j) = O(n^{k-1})$ continue to apply to the Fourier coefficients $a_n(j)$ of the component functions f_j of vectors-valued Hecke forms for large enough weight k? These two questions have a positive answer in the case of vector-valued modular forms (see [8]).

To answer the question, it should be very possible in the case $q = 4$ and 6, that is, $\lambda = \sqrt{2}$ and $\lambda = \sqrt{3}$, respectively, because here we may describe the elements of $\overline{G}(\lambda)$ explicitly, just as in the case of the modular group $\mathrm{SL}(2,\mathbb{Z})$, because $\overline{G}(\sqrt{2})$ and $\overline{G}(\sqrt{3})$ are commensurable (up to conjugacy) to the modular group.

Note: Dedicated to Benjamin Fine and Tony Gaglione on their 70th birthdays.

Benjamin Fine, Department of Mathematics, Fairfield University, 1073 North Benson Road, Fairfield, CT 06430, USA, e-mail: fine@fairfield.edu
Anja I. S. Moldenhauer, Heidberg 22, 22301 Hamburg, Germany, e-mail: a.moldenhauer@gmx.net
Gerhard Rosenberger, Fachbereich Mathematik, Universität Hamburg, Bundesstrasse 55, 20146 Hamburg, Germany, e-mail: gerhard.rosenberger@math.uni-hamburg.de

https://doi.org/10.1515/9783110638387-007

The last two authors dedicate the paper to Benjamin Fine and Tony Gaglione on their 70th birthdays.

We thank Andrea Hennekemper to leave parts of her thesis [5] to us.

2 Hecke groups $G(\lambda)$

Definition 2.1. Let $\lambda = 2\cos(\frac{\pi}{q})$, $q \in \mathbb{N}$, $q \geq 3$. The Hecke group $G(\lambda)$ is the group generated by the linear fraction transformation

$$\begin{aligned} T &: \mathbb{H} \to \mathbb{H}, \quad \tau \mapsto -\frac{1}{\tau}, \quad \text{and} \\ S_\lambda &: \mathbb{H} \to \mathbb{H}, \quad \tau \mapsto \tau + \lambda. \end{aligned}$$

Here, $\mathbb{H}$ is the upper half-plane $\{\tau \in \mathbb{C} \mid \operatorname{Im}(\tau) > 0\}$.

Remark 2.2.

1. Let $U_\lambda = TS_\lambda$. U_λ has order q. Group theoretically G_λ is the free product $\langle T\rangle \star \langle U_\lambda\rangle$ of two cyclic groups of order 2 and q, respectively.
2. G_λ acts discontinuously on the upper half-plane $\mathbb{H}$; it is a subgroup of $\operatorname{Aut}(\mathbb{H})$, the group of the biholomorphic maps $\mathbb{H} \to \mathbb{H}$. A fundamental domain for G_λ is given by
$$F_\lambda = \left\{\tau \in \mathbb{H} \mid |\tau| > 1,\ |\operatorname{Re}(\tau)| < \frac{\lambda}{2}\right\}.$$
The details for this may be found in the book [3].
3. Let $\phi : \mathrm{SL}(2,\mathbb{R}) \to \operatorname{Aut}(\mathbb{H})$, $M \mapsto \phi_M$ where
$$\phi_M(\tau) := \frac{a\tau + b}{c\tau + d} \quad \text{if } M = \begin{pmatrix} a & b \\ c & d \end{pmatrix}.$$
ϕ is a group epimorphism with $\ker(\phi) = \{E_2, -E_2\}$, $E_2 \left(\begin{smallmatrix} 1 & 0 \\ 0 & 1 \end{smallmatrix}\right)$. We call
$$\overline{G}(\lambda) = \phi^{-1}(G(\lambda)) = \left\langle \overline{T} = \begin{pmatrix} 0 & -1 \\ 1 & 0 \end{pmatrix}, \overline{S_\lambda} = \begin{pmatrix} 1 & \lambda \\ 0 & 1 \end{pmatrix}\right\rangle$$
the *Hecke matrix group for* λ.
If there is no misunderstanding, we do not distinguish between $V \in G(\lambda)$ and $\phi^{-1}(V)$. If it is necessary to distinguish, we also write $\overline{V} \in \overline{G}(\lambda)$ if $V(\tau) = \frac{a\tau+b}{c\tau+d}$ to denote one of the matrices
$$\begin{pmatrix} a & b \\ c & d \end{pmatrix} \quad \text{and} \quad -\begin{pmatrix} a & b \\ c & d \end{pmatrix}.$$
If $\overline{V} = \left(\begin{smallmatrix} a & b \\ c & d \end{smallmatrix}\right) \in \overline{G}(\lambda)$, then we also write $\overline{V}(\tau) := \frac{a\tau+b}{c\tau+d}$. This is well-defined.
In the following section, we also just write V for the matrix.
4. If $\lambda = 1$, that is, $q = 3$, then $G(1) = \Gamma$; the modular group $\Gamma \cong \mathrm{PSL}(2,\mathbb{Z})$.

Definition 2.3. If $V \in G(\lambda)$, then we define by

$$L_\lambda(V) := L(V) := \min\{\ell \in \mathbb{N} \mid V = W_1 \cdots W_\ell \text{ with } W_j \in \{T\} \cup \{S_\lambda^k \mid k \in \mathbb{Z}\}\}$$

the *Eichler length of V in* $G(\lambda)$.

Theorem 2.4. *For each* λ, *there exist* $k_1, k_2 \in \mathbb{R}$, $k_1, k_2 > 0$, *such that*

$$L(V) \leq k_1 \ln(a^2 + b^2 + c^2 + d^2) + k_2$$

for each $V \in G(\lambda)$ *with* $\overline{V} = \left(\begin{smallmatrix} a & b \\ c & d \end{smallmatrix}\right)$.

Theorem 2.4 is given by Eichler [2]. His proof contains a technical error. A corrected proof is given by Hinz [6].

Remark 2.5. In what follows, we need the little observation that

$$M = \pm \begin{pmatrix} 1 & n\lambda \\ 0 & 1 \end{pmatrix} \quad \text{for some } n \in \mathbb{Z},$$

if

$$M = \begin{pmatrix} a & b \\ 0 & d \end{pmatrix} \in \overline{G}(\lambda).$$

Lemma 2.6. *Let* $\rho : \overline{G}(\lambda) \to \mathrm{GL}(p, \mathbb{C})$, $p \in \mathbb{N}$, *be a group homomorphism.*

If $\|\rho(\overline{T})\|_\infty$, $\|\rho(\overline{S}_\lambda^n)\| \leq c_1$ *for all* $n \in \mathbb{Z}$ *with* $c_1 \in \mathbb{R}$, $c_1 \geq 1$, *then there exist* c_2, $C \in \mathbb{R}$, $c_2, C > 0$, *such that*

$$p\|\rho(V)\|_\infty \leq c_2(a^2 + b^2 + c^2 + d^2)^{C \ln(pc_1)}$$

for all $V = \left(\begin{smallmatrix} a & b \\ c & d \end{smallmatrix}\right) \in \overline{G}(\lambda)$.

Here, $\|\cdot\|_\infty$ *denotes the matrix maximum norm.*

Proof. If $V \in \overline{G}(\lambda)$, then

$$V = \pm W_1 \cdots W_L, \quad W_j \in \{\overline{T}\} \cup \{\overline{S}^k \mid k \in \mathbb{Z}\}.$$

By Theorem 2.4, there exist $k_1, k_2 \in \mathbb{R}$, $k_1, k_2 > 0$, such that

$$L \leq k_1 \ln(a^2 + b^2 + c^2 + d^2) + k_2$$

which leads to

$$\rho(V) = \rho(W_1) \cdots \rho(W_L) = \left(\sum_{m_1, \ldots, m_{L-1}=1}^{p} \rho_{jm_1}(W_1)\rho_{m_1m_2}(W_2) \cdots \rho_{m_{L-1}n}(W_L) \right)_{j,n=1,\ldots,p}.$$

Therefore,

$$\|\rho(V)\|_\infty \le p^{L-1} c_1^L,$$

and we get

$$\begin{aligned} p\|\rho(V)\|_\infty \le (pc_1)^L = e^{L\ln(pc_1)} &\le e^{(k_1 \ln(a^2+b^2+c^2+d^2)+k_2)\ln(pc_1)} \\ &\le c_2(a^2+b^2+c^2+d^2)^{C\ln(pc_1)}. \end{aligned}$$

□

Remark 2.7. In group theory, given a subgroup H of a group G, a right transversal is a set containing one element from each right coset of H in G. These cosets are mutually disjoint, that is, form a partition of G.

Theorem 2.8. *There exists a right transversal $\mathcal{R}$ of $\langle S_\lambda \rangle$ in $G(\lambda)$ and a $k \in \mathbb{R}$, $k > 0$, such that*

$$a^2 + b^2 + c^2 + d^2 \le k(c^2 + d^2)$$

for all $R \in \mathcal{R}$ with $\overline{R} = \begin{pmatrix} a & b \\ c & d \end{pmatrix}$.

Proof. From each right coset $\langle S_\lambda \rangle R$ of $\langle S_\lambda \rangle$ in $G(\lambda)$, we choose a representative R such that

$$-\frac{\lambda}{2} \le \operatorname{Re}(R(2i)) \le \frac{\lambda}{2}.$$

This is possible by slotting a suitable translation from $\langle S_\lambda \rangle$ in ahead.

These R's form a desired right transversal $\mathcal{R}$: Certainly,

$$0 < \operatorname{Im}(R(2i)) = \frac{2}{|2ci+d|^2} = \frac{2}{4c^2+d^2}.$$

Since $c \neq 0$ or $d \neq 0$, we get $|\operatorname{Im}(R(2i))| \le 2$. Altogether, we obtain

$$\frac{4a^2+b^2}{4c^2+d^2} = \left|\frac{2ai+b}{2ci+d}\right|^2 = |R(2i)|^2 \le 4 + \frac{\lambda^2}{4} \le 5,$$

that gives

$$a^2 + b^2 \le 4a^2 + b^2 \le 5(4c^2 + d^2) \le 20(c^2 + d^2).$$

Hence, we may chose $k = 21$. □

Remark 2.9.

1. If $\mathcal{R}$ is a right transversal of $\langle S_\lambda \rangle$ in $G(\lambda)$, then $\overline{\mathcal{R}} := \phi^{-1}(\mathcal{R})$ is a right transversal of $\langle \overline{S}_\lambda \rangle$ in $\overline{G(\lambda)}$ because $-E_2 \notin \langle \overline{S}_\lambda \rangle$.
   Therefore, Theorem 2.8 holds analogously for $\overline{G}(\lambda)$.

2. If $\overline{\mathcal{R}}$ is a right transversal of $\langle \overline{S}_\lambda \rangle$ in $\overline{G(\lambda)}$, then we denote by $\overline{\mathcal{R}}^\star$ the set of all $M \in \overline{R}$ with $M = \left(\begin{smallmatrix} a & b \\ c & d \end{smallmatrix}\right)$ and $c \neq 0$.
   Since $-E_2 \notin \langle \overline{S}_\lambda \rangle$, we may in the following always assume that
$$\overline{\mathcal{R}} = \overline{\mathcal{R}}^\star \cup \{E_2, -E_2\}.$$
3. If $\overline{\mathcal{R}}$ is a right transversal of $\langle \overline{S}_\lambda \rangle$ in $\overline{G}(\lambda)$ and $V \in \overline{G}(\lambda)$, then also $\overline{\mathcal{R}}V$ is one.

Theorem 2.10. *Let $k \in \mathbb{R}$, $k > 2$, and $\overline{\mathcal{R}}$ a right transversal of $\langle \overline{S}_\lambda \rangle$ in $\overline{G}(\lambda)$. Then:*
1. *$\overline{\mathcal{R}}$ is countable, and the series*
$$\sum_{M \in \overline{\mathcal{R}}} \frac{1}{|c\tau + d|^k}, \quad M = \begin{pmatrix} a & b \\ c & d \end{pmatrix},$$
   converges for all $\tau \in \mathbb{H}$.
2. *In each vertical strip,*
$$V_k := \{\tau = x + iy \in \mathbb{H} \mid |x| \leq \epsilon^{-1},\ y \geq \epsilon\},$$
   $\epsilon > 0$, the series converges uniformly.
3. *The series converges uniformly on compact sets in $\mathbb{H}$.*
4. $$\lim_{\tau \to i\infty} \sum_{M \in \overline{\mathcal{R}}^\star} \frac{1}{|c\tau + d|^k} = 0, \quad M = \begin{pmatrix} a & b \\ c & d \end{pmatrix}, \quad c \neq 0.$$

   A proof can be found on pages 68–72 of the book [9].

3 Vector-valued Hecke forms

From now on, we use the following notation. If $M \in \mathrm{SL}(2, \mathbb{R})$, $M = \left(\begin{smallmatrix} a & b \\ c & d \end{smallmatrix}\right)$, then we write
$$M : \tau := c\tau + d.$$
Certainly, $(MN) : \tau = (M : N\tau)(N : \tau)$.

Definition 3.1. Let $k \in \mathbb{R}$. A map
$$v : \overline{G}(\lambda) \to \{z \in \mathbb{C} \mid |z| = 1\}, \quad V \mapsto v(V)$$
is called a k-multiplicator system for $\overline{G}(\lambda)$ if the following hold:
1. $$v(V_1V_2)((V_1V_2) : \tau)^k = v(V_1)v(V_2)(V_1 : V_2\tau)^k(V_2 : \tau)^k$$
   for all $V_1, V_2 \in \overline{G}(\lambda)$, $\tau \in \mathbb{H}$,
2. $v(-E_2) = (-1)^{-k}$.

Remark 3.2. If $V_1 = V_2 = E_2$, then we see that $v(E_2) = 1$.

Definition 3.3. Let $k \in \mathbb{R}$, v a k-multiplicator system for $\overline{G}(\lambda)$.

1. The slash operator $|_k V$ is defined by

$$f|_k V(\tau) := f|_k^v V(\tau) := v(V)^{-1}(V : \tau)^{-k} f(V\tau)$$

for $f \in O(\mathbb{H})$, $V \in \overline{G}(\lambda)$, $\tau \in \mathbb{H}$.

2. If $f_1, \ldots, f_p \in O(\mathbb{H})$ and $F := \begin{pmatrix} f_1 \\ \vdots \\ f_p \end{pmatrix}$, then we define the slash operator by

$$F|_k V(\tau) := F|_k^\tau V(\tau) := \begin{pmatrix} f_1|_k V(\tau) \\ \vdots \\ f_p|_k V(\tau) \end{pmatrix}$$

for $V \in \overline{G}(\lambda)$, $\tau \in \mathbb{H}$.

Here, $O(\mathbb{H})$ denotes—as usual—the $\mathbb{C}$-vector space of the holomorphic functions in $\mathbb{H}$.

Theorem 3.4. *Let $k \in \mathbb{R}$ and v be a k-multiplicator system for $\overline{G}(\lambda)$. Then we have:*

1. *If $c_1, c_2 \in \mathbb{C}$, $f_1, f_2 \in O(\mathbb{H})$ and $V \in \overline{G}(\lambda)$, $\tau \in \mathbb{H}$, then*

$$(c_1 f_1 + c_2 f_2)|_k^v V(\tau) = c_1 f_1|_k^v V(\tau) + c_2 f_2|_k^v V(\tau).$$

2. *If $f \in O(\mathbb{H})$ and $V_1, V_2 \in \overline{G}(\lambda)$, $\tau \in \mathbb{H}$, then*

$$f|_k^v V_1 V_2(\tau) = (f|_k^v V_1)|_k^v V_2(\tau).$$

The proof of the statements is a straightforward calculation from the definitions.

Recall that a complex-linear representation of degree p of a group G is a group homomorphism $\rho : G \to \mathrm{GL}(p, \mathbb{C})$.

Definition 3.5. Let $k \in \mathbb{R}$, $p \in \mathbb{N}$, v be a k-multiplicator system for $\overline{G}(\lambda)$ and $\rho : \overline{G}(\lambda) \to \mathrm{GL}(p, \mathbb{C})$ be a complex linear representation of degree p with

$$\rho(-E_2) =: E_p = \begin{pmatrix} 1 & & 0 \\ & \ddots & \\ 0 & & 1 \end{pmatrix} \in \mathrm{GL}(p, \mathbb{C}).$$

Let further be $f_1, f_2, \ldots, f_p \in O(\mathbb{H})$ and $F = \begin{pmatrix} f_1 \\ \vdots \\ f_p \end{pmatrix}$.

Then (F, ρ), more concrete (F, v, ρ), is called a vector-valued Hecke form of weight k for $\overline{G}(\lambda)$ if and only if the following hold:

1. $F|_k^v V(\tau) = \rho(V)F(\tau)$, for all $V \in \overline{G}(\lambda)$ and $\tau \in \mathbb{H}$.
2. Each component function f_j has an expansion

$$f_j(\tau) = \sum_{n \geq m_j} a_n(j) e^{2\pi i \frac{n}{\lambda N_j} \tau}$$

with $m_j \in \mathbb{Z}$, $N_j \in \mathbb{N}$, $\tau \in \mathbb{H}$, which converges absolutely-uniformly on compact subsets of $\mathbb{H}$ and which converges meromorphically at infinity.

Agreements:
1. In the following, we always assume that $\rho(-E_2) = E_p$ for a complex linear representation $\rho : \overline{G}(\lambda) \to \mathrm{GL}(p, \mathbb{C})$.
   If we talk about a representation ρ, then we automatically assume that $\rho(-E_2) = E_p$.
2. In the following, we fix $k \in \mathbb{R}$ and the k-multiplicator system v.
   Further, let $\kappa \in [0, 1]$ so that

$$v(\overline{S}_\lambda) = e^{2\pi i \kappa}.$$

Remark 3.6.
1. We often take $q := e^{2\pi i \tau}$. Then we write $f_j(\tau) = \sum_{n \geq m_j} a_n(j) q^{\frac{n}{\lambda N_j}}$ for the expansion, called now the q-expansion.
2. Using standard arguments in complex analysis, we get the following.
   The component functions f_j are holomorphic in $\mathbb{H}$, they have a pole at most at infinity $i\infty$, they are periodic with period λN_j, and the coefficients $a_n(j)$ in the q-expansion are uniquely determined.
   Since there is a $n_0 \in \mathbb{Z}$ with $a_n = 0$ for $n < n_0$, then for each $\epsilon \in \mathbb{R}$, $\epsilon > 0$, there exists a $C \in \mathbb{R}$, $C > 0$, with

$$|f(\tau)| \leq Ce^{\frac{-2\pi n_0 \operatorname{Im}(\tau)}{\lambda N_j}}$$

   for $\tau \in \mathbb{H}$, $\operatorname{Im}(\tau) > \epsilon$ (for concrete calculations of these facts see, for instance, the book [3]).
3. The numbers N_j are certainly not uniquely determined by the f_j. For instance, if $\ell \in \mathbb{N}$ then we define

$$\tilde{a}_n = \begin{cases} a_m, & \text{if } n = \ell m,\ m \in \mathbb{Z},\ n \geq m_j, \\ 0, & \text{otherwise.} \end{cases}$$

   Then

$$f_j(\tau) = \tilde{f}_j(\tau) = \sum_{n \geq \ell m_j} \tilde{a}_n(j) e^{\frac{2\pi i n \tau}{\lambda \ell N_j}}.$$

On the other side, if $\ell \in \mathbb{N}$, $\ell \mid N_j$ and $a_n(j) = 0$ for $n \in \mathbb{Z}$, $\ell \nmid n$, then we define $\tilde{a}_n(j) := a_{\ell n}(j)$ for $n \in \mathbb{Z}$, $n \geq \lceil \frac{m_j}{\ell} \rceil$, and have

$$f_j(\tau) = \tilde{f}_j(\tau) = \sum_{n \geq \lceil \frac{m_j}{\ell} \rceil} \tilde{a}_n(j) e^{2\pi i \frac{n}{\lambda(\frac{N_j}{\ell})} \tau}.$$

4. Also each linear combination of the $f_1, \ldots, f_p$ has a q-expansion $\sum_{n \geq m} a_n q^{\frac{n}{\lambda N_j}}$. If $N := \ell cm(N_1, \ldots, N_p)$, then all $f_1, \ldots, f_p$ are periodic with period $N\lambda$, and each linear combination of the $f_1, \ldots, f_p$ has therefore period $N\lambda$. Hence we get the following result.

Theorem 3.7. *Let $\mathcal{F}(k, v, \rho)$ be the set of all vector-valued Hecke forms (F, v, ρ) of weight k. Then $\mathcal{F}(k, v, \rho)$ is a $\mathbb{C}$-vector space.*

Definition 3.8.
1. A vector-valued Hecke form of weight k is called *entire* if in the q-expansions of the component functions f_j all $a_n(j) = 0$ for $n < 0$, $j = 1, \ldots, p$. The set of all entire vector-valued Hecke forms (F, v, ρ) of weight k is denoted by $\mathcal{M}(k, v, \rho)$.
2. An entire vector-valued Hecke form of weight k is called a *cusp form* if in the q-expansions of the component functions f_j all $a_n(j) = 0$ for $n \leq 0$, $j = 1, \ldots, p$. The set of cusp forms (F, v, ρ) of weight k is denoted by $\mathcal{S}(k, v, \rho)$.
   Certainly, $\mathcal{M}(k, v, \rho)$ is a subspace of $\mathcal{F}(k, v, \rho)$, and $\mathcal{S}(k, v, \rho)$ is a subspace of $\mathcal{M}(k, v, \rho)$.

Theorem 3.9. *Let $\left(F = \begin{pmatrix} f_1 \\ \vdots \\ f_p \end{pmatrix}, \rho\right)$ and $\left(\tilde{F} = \begin{pmatrix} \tilde{f}_1 \\ \vdots \\ \tilde{f}_{\tilde{p}} \end{pmatrix}, \tilde{\rho}\right)$ be two vector-valued Hecke forms of weight k.*

Then $(F \bigoplus \tilde{F}, \rho \bigoplus \tilde{\rho})$, where

$$F \bigoplus \tilde{F} := \begin{pmatrix} f_1 \\ \vdots \\ f_p \\ \tilde{f}_1 \\ \vdots \\ \tilde{f}_{\tilde{p}} \end{pmatrix}, \quad \rho \bigoplus \tilde{\rho} := \begin{pmatrix} \rho & 0 \\ 0 & \tilde{\rho} \end{pmatrix}$$

is a vector-valued Hecke form of weight k.

Definition 3.10.
1. We call a q-expansion

$$f_j = \sum_{n \geq m_j} a_n(j) q^{\frac{n}{\lambda N_j}}$$

elementary if N_j is elementary, that is, if f_j is constant, then $N_j = 1$, and if f_j is not constant with smallest positive period ω_j, then $\lambda N_j = k_j\omega_j$ with $k \in \mathbb{N}$ and $\gcd(N_j, k_j) = 1$. By Remark 3.6, part 3 from above, it is clear that each f_j has exactly one elementary q-expansion.

2. Let $\left(\begin{pmatrix} f_1 \\ \vdots \\ f_p \end{pmatrix}, \rho\right)$ be a vector-valued Hecke form of weight k with elementary N_j's. Then $N := \operatorname{lcm}(N_1, \dots, N_p)$ is called its *level*.

Remark 3.11. Let $\left(\begin{pmatrix} f_1 \\ \vdots \\ f_p \end{pmatrix}, \rho\right)$ be a vector-valued Hecke form of weight k, and let $\tilde{f}_j$ be the elementary q-expansion of f_j. Then the level of $\left(\begin{pmatrix} f_1 \\ \vdots \\ f_p \end{pmatrix}, \rho\right)$ is defined to be the level of $\left(\begin{pmatrix} \tilde{f}_1 \\ \vdots \\ \tilde{f}_p \end{pmatrix}, \rho\right)$.

Theorem 3.12. *Let $\left(\begin{pmatrix} f_1 \\ \vdots \\ f_p \end{pmatrix}, \rho\right)$ be a vector-valued Hecke form of weight k. Assume that it has a level N and that the component functions are linearly independent. Then*

$$\rho(\overline{S}_\lambda^N) = \rho(\overline{S}_\lambda)^N = e^{-2\pi i N\kappa} E_p.$$

Proof. For $j = 1, \dots, p$ and $\tau \in \mathbb{H}$, we have

$$\begin{aligned} f_j|_k \overline{S}_\lambda^N(\tau) &= v(\overline{S}_\lambda^N)^{-1} f_j(\overline{S}_\lambda^N(\tau)) = e^{-2\pi i N\kappa} f_j(\overline{S}_\lambda^N(\tau)) \\ &= e^{-2\pi i N\kappa} \sum_{n\geq 0} a_n(j) e^{2\pi i \frac{n}{\lambda N_j}(\tau + N\lambda)} \\ &= e^{-2\pi i N\kappa} \sum_{n\geq 0} a_n(j) e^{2\pi i \frac{n}{\lambda N_j}\tau} \\ &= e^{-2\pi i N\kappa} f_j(\tau) \quad \text{since } e^{2\pi i n \frac{N}{N_j}} = 1. \end{aligned}$$

Hence,

$$e^{-2\pi i N\kappa} f_j(\tau) = \sum_{k=1}^{p} \rho_{jk}(\overline{S}_\lambda^N) f_k.$$

Now the statement follows because the $f_1, \dots, f_p$ are linearly independent. □

Definition 3.13. Two vector-valued Hecke forms $\left(\begin{pmatrix} f_1 \\ \vdots \\ f_p \end{pmatrix}, \rho\right)$ and $\left(\begin{pmatrix} \tilde{f}_1 \\ \vdots \\ \tilde{f}_{\tilde{p}} \end{pmatrix}, \tilde{\rho}\right)$ are called *equivalent* (or more concrete f-equivalent) if

$$\langle f_1, \dots, f_p \rangle = \langle \tilde{f}_1, \dots, \tilde{f}_p \rangle$$

in the $\mathbb{C}$-vector space $O(\mathbb{H})$.

Theorem 3.14. *Equivalent vector-valued Hecke forms of weight k have the same level.*

Proof. Let $\left(\begin{pmatrix} f_1 \\ \vdots \\ f_p \end{pmatrix}, \rho\right), \left(\begin{pmatrix} \tilde{f}_1 \\ \vdots \\ \tilde{f}_{\tilde{p}} \end{pmatrix}\right)$ be equivalent vector-valued Hecke forms of weight k, and level S and $\tilde{S}$, respectively.

We may assume that no f_j is constant.

For $j = 1, \dots, \tilde{p}$ we have that $\tilde{f}_j$, as linear combination of $f_1, \dots, f_p$, is periodic with period S which gives $\tilde{S} \leq S$. By symmetry, $S \leq \tilde{S}$. □

Theorem 3.15. *Let (F, ρ) be a vector-valued Hecke form of weight k, and let $U \in \mathrm{GL}(p, \mathbb{C})$.*

We define $U\rho U^{-1} : \overline{G}(\lambda) \to \mathrm{GL}(p, \mathbb{C})$ by $U\rho U^{-1}(V) := U\rho(V)U^{-1}$. Then $(UF, U\rho U^{-1})$ is a vector-valued Hecke form of weight k which is equivalent to (F, ρ).

Proof. Let $V \in G(\lambda)$, $\tau \in \mathbb{H}$. By Theorem 3.4, we get

$$(UF)|_k V(\tau) = U(F|_k V(\tau)) = U\rho(V)F = U\rho(V)U^{-1}(UF).$$ □

Theorem 3.16. *Let $\left(F = \begin{pmatrix} f_1 \\ \vdots \\ f_p \end{pmatrix}, \rho\right)$ be a vector-valued Hecke form of weight k such that not all component functions f_j are constant 0. Then there exist a vector-valued Hecke form*

$$\left(\tilde{F} = \begin{pmatrix} \tilde{f}_1 \\ \vdots \\ \tilde{f}_{\tilde{p}} \end{pmatrix}, \tilde{\rho}\right)$$

which is equivalent to (F, ρ) and for which the component functions $\tilde{f}_1, \dots, \tilde{f}_{\tilde{p}}$ are linearly independent in $\mathcal{O}(\mathbb{H})$.

Proof. Assume that the component functions $f_1, \dots, f_p$ are linearly dependent. Without loss of generality, let $(f_1, \dots, f_r)$, $r < p$, be a basis of $\langle f_1, \dots, f_p \rangle$. Let K be the $(p - r) \times r$-matrix such that

$$\begin{pmatrix} f_{r+1} \\ \vdots \\ f_p \end{pmatrix} = K \begin{pmatrix} f_1 \\ \vdots \\ f_r \end{pmatrix}.$$

We define $U := \begin{pmatrix} E_r & 0 \\ K & -E_{p-r} \end{pmatrix} \in \mathrm{GL}(p, \mathbb{C})$. Then $UF = \begin{pmatrix} f_1 \\ \vdots \\ f_r \\ 0 \\ \vdots \\ 0 \end{pmatrix}$, and $(UF, U\rho U^{-1})$ is equivalent to (F, ρ) by Theorem 3.15. We consider $U\rho U^{-1}(V)$ as a block matrix

$$\begin{pmatrix} \alpha(V) & \beta(V) \\ \gamma(V) & \delta(V) \end{pmatrix}$$

with $\alpha(V)$ a $(r \times r)$-matrix. Especially, there are linear combinations of the form

$$0 = \gamma(V)_{11} f_1 + \cdots + \gamma(V)_{\ell r} f_r + \delta(V)_{11} \cdot 0 + \cdots + \delta(V)_{\ell(p-r)} \cdot 0$$

which belong to the rows $\ell = r+1, \ldots, p$. We get $\gamma(V) = 0$ because $f_1, \ldots, f_r$ are linearly independent.

Since the column vectors of $U\rho U^{-1}(V)$ are linearly independent, we have that $\alpha(V) \in \mathrm{GL}(r, \mathbb{C})$. Since

$$f_\ell|_k V = \alpha(V)_{11} f_1 + \cdots + \alpha(V)_{\ell r} f_r + \beta(V)_{\ell 1} \cdot 0 + \cdots + \beta(V)_{\ell(p-r)} \cdot 0$$

for $\ell = 1, \ldots, r$ we get that $\left(\begin{pmatrix} f_1 \\ \vdots \\ f_r \end{pmatrix}, \alpha\right)$ is a vector-valued Hecke form of the desired form. □

Theorem 3.17. *Let $\left(\begin{pmatrix} f_1 \\ \vdots \\ f_p \end{pmatrix}, \rho\right)$ be a vector-valued Hecke form of weight k. Let not all component functions f_j be constant 0.*

Then there exists a vector-valued Hecke form $\left(\begin{pmatrix} \tilde{f}_1 \\ \vdots \\ \tilde{f}_{\tilde{p}} \end{pmatrix}, \tilde{\rho}\right)$ of weight k, which is equivalent to $\left(\begin{pmatrix} f_1 \\ \vdots \\ f_p \end{pmatrix}, \rho\right)$, such that:

1. *The component functions $\tilde{f}_1, \ldots, \tilde{f}_{\tilde{p}}$ are linearly independent in $O(\mathbb{H})$.*
2. $\tilde{\rho}(\tilde{S}_\lambda) = e^{-2\pi i\kappa} \begin{pmatrix} e^{2\pi i\varphi_1} & & 0 \\ & \ddots & \\ 0 & & e^{2\pi i\varphi_{\tilde{p}}} \end{pmatrix}$ *with $\varphi_1, \ldots, \varphi_{\tilde{p}} \in (0,1] \cap \mathbb{Q}$.*
3. *$\tilde{f}_j$ has a q-expansion*

$$q^{\frac{\varphi_j}{\lambda}} \sum_{n \geq m_j} a_n(j) q^{\frac{n}{\lambda}}.$$

Proof. By Theorem 3.16, we may assume that $f_1, \ldots, f_p$ are linearly independent.

Let N be the level of our vector-valued Hecke form.

Then

$$\left(e^{2\pi i\kappa} \rho(\overline{S}_\lambda)\right)^N = E_p$$

by Theorem 3.12.

We choose an $U \in \mathrm{GL}(p, \mathbb{C})$ such that

$$U\rho(\overline{S}_\lambda)U^{-1} = e^{-2\pi i\kappa} \begin{pmatrix} e^{2\pi i\varphi_1} & & 0 \\ & \ddots & \\ 0 & & e^{2\pi i\varphi_p} \end{pmatrix}$$

with $\varphi_1, \dots, \varphi_p \in (0,1] \cap \mathbb{Q}$. The transition to $\left(U \begin{pmatrix} f_1 \\ \vdots \\ f_p \end{pmatrix}, U\rho U^{-1} \right)$ gives an equivalent Hecke form which satisfies 1 and 2. The point 3 now follows easily from straightforward calculations just using the definition and the construction in Remark 3.6 part 3. For more details, see [5]. □

Definition 3.18. A vector-valued Hecke form $\left(\begin{pmatrix} f_1 \\ \vdots \\ f_p \end{pmatrix}, \rho \right)$ of weight k is called *reduced* if it satisfies the condition 1, 2, and 3 from Theorem 3.17.

We close this section with two results concerning the construction of vector-valued Hecke forms and the growth of the coefficients $a_n(j)$ of the q-expansions.

Theorem 3.19. *Let $\mathcal{U}$ be a finite dimensional subspace of $O(\mathbb{H})$ with the following properties:*

(i) *Each element of $\mathcal{U}$ has a q-expansion of the form*
$$\sum_{n \geq m} a_n q^{\frac{n}{\lambda N}}.$$

(ii) *$\mathcal{U}$ is closed with respect to the slash operators $|_k V$.*

If $f_1, \dots, f_p$ is a generating system of $\mathcal{U}$, then there exists a representation $\rho : \overline{G}(\lambda) \to \mathrm{GL}(p, \mathbb{C})$ such that $\left(\begin{pmatrix} f_1 \\ \vdots \\ f_p \end{pmatrix}, \rho \right)$ is a vector-valued Hecke form of weight k.

Proof. We may assume that $\mathcal{U} \neq \{0\}$.

Let $(f_1, \dots, f_p)$ be a basis of $\mathcal{U}$. Then also $(f_1|_k, \dots, f_p|_k)$ is a basis of $\mathcal{U}$. We define $\rho_{jm}(V) \in \mathbb{C}$ by

$$f_j|_k V(\tau) = \sum_{m=1}^{p} \rho_{jm}(V) f_m(\tau)$$

and get a representation

$$\rho : \overline{G}(\lambda) \to \mathrm{GL}(p, \mathbb{C}) \quad \text{via } \rho(V) := (\rho_{jm})_{j,m=1,\dots,p} \in \mathrm{GL}(p, \mathbb{C}).$$

The necessary property $\rho(V_1)\rho(V_2) = \rho(V_1 V_2)$ for $V_1, V_2 \in \overline{G}(\lambda)$ follows directly from the properties of the slash operator.

Now, let be $(f_1, \dots, f_p)$ be not a basis of $\mathcal{U}$. We may assume that $f_1, \dots, f_r$ form a basis of $\mathcal{U}$. We choose now a representation $\rho : \overline{G}(\lambda) \to \mathrm{GL}(r, \mathbb{C})$ such that $\left(\begin{pmatrix} f_1 \\ \vdots \\ f_r \end{pmatrix}, \rho \right)$ is a vector-valued Hecke form of weight k. Then $\left(\begin{pmatrix} f_1 \\ \vdots \\ f_r \\ 0 \\ \vdots \\ 0 \end{pmatrix}, \begin{pmatrix} \rho & 0 \\ 0 & E_{p-r} \end{pmatrix} \right)$ is a vector-valued Hecke form of weight k by Theorem 3.9. The result now follows from Theorem 3.16. □

Theorem 3.20.

(a) *Let* $\left(\begin{pmatrix} f_1 \\ \vdots \\ f_r \end{pmatrix}, \rho\right)$ *be an entire vector-valued Hecke form of weight* k. *Then there exists a constant* $\gamma \in \mathbb{R}$, $\gamma > 0$, *such that the coefficients* $a_n(j)$ *of* f_j *satisfy the growth condition* $a_n(j) = O(n^{k+2\gamma})$ *for* $n \to \infty$ *and for all* $j = 1, \ldots, p$.

(b) *Let* $\left(\begin{pmatrix} f_1 \\ \vdots \\ f_r \end{pmatrix}, \rho\right)$ *be a vector-valued Hecke cusp form of weight* k. *Then there exists a constant* $\gamma \in \mathbb{R}$, $\gamma > 0$, *such that the coefficients* $a_n(j)$ *of* f_j *satisfy the growth condition* $a_n(j) = O(n^{\frac{k}{2}+\gamma})$ *for* $n \to \infty$ *and for all* $j = 1, \ldots, p$.

Remark 3.21. Here, we use the big O-notation which means that there exists a $n_0 \in \mathbb{N}$ and a constant C such that $|a_n(j)| \leq Cn^{k+2\alpha}$ and $|a_n(j) \leq Cn^{\frac{k}{2}+\alpha}|$, respectively, for all $n > n_0$.

The proof uses Theorem 2.8 and heavily the fact that each f_j has a convergent q-expansion holomorphic at infinity to calculate the common growth for the f_j. The details can be found in [7] for the modular group and in [4] for the general Hecke group.

We notice that the arguments are similar to the arguments in the proof of Theorem 4.1 in Section 4.

4 Dimensions

From now on for the rest of the paper, we assume that all representations $\rho : \overline{G}(\lambda) \to \mathrm{GL}(p, \mathbb{C})$ and all vector-valued Hecke forms $\left(F = \begin{pmatrix} f_1 \\ \vdots \\ f_p \end{pmatrix}, \rho\right)$ satisfy the conditions:

1.
$$\rho(\overline{S}_\lambda) = e^{-2\pi i\kappa} \begin{pmatrix} e^{2\pi i\varphi_1} & & 0 \\ & \ddots & \\ 0 & & e^{2\pi i\varphi_p} \end{pmatrix}$$
with $\varphi_1, \ldots, \varphi_p \in (0, 1] \cap \mathbb{Q}$, $\varphi_j = \frac{c_j}{d_j}$, $c_j, d_j \in \mathbb{N}$, $c_j \leq d_j$ and

2. each f_j has a q-expansion
$$q^{\frac{\varphi_j}{\lambda}} \sum_{n \geq m_j} a_n(j) q^{\frac{n}{\lambda}} = \sum_{n \geq m_j} a_n(j) q^{\frac{d_j n + c_j}{d_j \lambda}}$$
with $m_j = 0$ if $f_j = 0$ and $a_{m_j}(j) \neq 0$ otherwise.

By Theorem 3.17, this is not an essential restriction.

We define $\Lambda(F) := (m_1, \ldots, m_p) \in \mathbb{Z}^p$ for (F, ρ). We provide $\mathbb{Z}^p$ with the partial order:

If $\Lambda = (\lambda_1, \ldots, \lambda_p)$, $\Omega = (\omega_1, \ldots, \omega_p)$, then $\Lambda \le \Omega$ if and only if

$$\lambda_j \le \omega_j \quad \text{for } j = 1, \ldots, p.$$

In this context, let

$$\mathbf{0} := (0, \ldots, 0) \quad \text{and} \quad \mathbf{1} := (1, \ldots, 1),$$
$$G := \{j \mid 1 \le j \le p, \varphi_j = 1\}$$

and $\tilde{G} = (j_1, \ldots, j_p)$ with $j_i = -1$ if $i \in G$ and $j_i = 0$ otherwise.

By 1, we have

$$\|\rho(\overline{S}_\lambda^n)\|_\infty = \|\rho^n(\overline{S}_\lambda)\|_\infty = 1,$$

and we may use $c_1 = \max(\|\rho(T)\|_\infty, 1)$ to satisfy the premise in Lemma 2.6. With some C as in Lemma 2.6, we define

$$\alpha := \alpha(\rho) := C \ln(p \max(\|\rho(T)\|_\infty, 1)).$$

Theorem 4.1. *There exists a constant $\hat{C} \in \mathbb{R}$, $\hat{C} > 0$, such that*

$$|f_j(\tau)| = \begin{cases} \hat{C} y^{-2\alpha-k}, & \text{for } k \ge -2\alpha, \\ \hat{C} y^{-\alpha-\frac{k}{2}}, & \text{for } k < -2\alpha, \end{cases}$$

for all $\left(\begin{pmatrix} f_1 \\ \vdots \\ f_p \end{pmatrix}, \rho\right) \in \mathcal{M}(k, v, \rho)$, $j = 1, \ldots, p$, $\tau = x + iy \in \mathbb{H}$, $y = \operatorname{Im}(\tau) \le 1$.

Proof. For $j = 1, \ldots, p$ and $\tau = x + iy \in \mathbb{H}$, we define $g_j : \mathbb{H} \to \mathbb{R} \setminus (-\infty, 0)$, $\tau \mapsto y^\sigma |f_j(\tau)|$ where $\sigma := \alpha + \frac{k}{2}$. Now $\operatorname{Im}(\overline{S}_\lambda(\tau)) = \operatorname{Im}(\tau)$ and

$$f_j(\overline{S}_\lambda(\tau)) = e^{2\pi i \frac{\varphi_j}{\lambda} \tau} e^{2\pi i \varphi_j} \sum_{n \ge 0} a_n(j) e^{2\pi i \frac{n}{\lambda} \tau} e^{2\pi i n}.$$

Now let $V = \left(\begin{smallmatrix} a & b \\ c & d \end{smallmatrix}\right) \in \overline{G}(\lambda)$ be arbitrary. Then

$$g_j(V(\tau)) = \left(\frac{y}{|c\tau + d|^2}\right)^\sigma |f_j(V(\tau))|$$

because $\operatorname{Im}(V(\tau)) = \frac{\operatorname{Im}(\tau)}{|c\tau+d|^2}$. Now

$$g_j(V(\tau)) = y^\sigma |c\tau + d|^{k-2\sigma} |f_j|_k V(\tau)| = y^\sigma |c\tau + d|^{k-2\sigma} \left| \sum_{\ell=1}^{p} \rho_{j\ell}(V) f_\ell(\tau) \right|$$

and, therefore,

$$g_j(\tau) \le |c\tau + d|^{k-2\sigma} \sum_{\ell=1}^{p} |\rho_{j\ell}(V)| g_\ell(\tau).$$

Using Remark 3.6 part 2, we get

$$g_j(\tau) = y^\sigma |f_j(\tau)| \le C_1 y^\sigma e^{-2\pi \frac{0}{d_j\lambda}} = C_1 y^\sigma$$

for some $C_1 \in \mathbb{R}$, $C_1 > 0$. Hence, by Lemma 2.6, there exists a constant $C_2 \in \mathbb{R}$, $C_2 > 0$, such that

$$g_j(V(\tau)) \le C_2 |c\tau + d|^{k-2\sigma} (a^2 + b^2 + c^2 + d^2)^\alpha y^\sigma$$

for $f = 1, \dots, p$, $\tau \in \overline{F}_\lambda = \{\tau \in \mathbb{H} \mid |\tau| \ge 1, |\operatorname{Re}(\tau)| \le \frac{\lambda}{2}\}$, $V \in \overline{G}(\lambda)$.

Now, let $\mathcal{R}$ be a right transversal of $\langle \overline{S}_\lambda \rangle$ in $\overline{G}(\lambda)$. Then $V = \pm \overline{S}_\lambda^n R$ for some $R \in \mathcal{R}$ and $n \in \mathbb{Z}$. We have $g_j(V(\tau)) = g_j(R(\tau))$.

We want to use Theorem 2.8. We just write

$$R = \begin{pmatrix} a & b \\ c & d \end{pmatrix} \in \overline{G}(\lambda).$$

This is okay because

$$g_j(V(\tau)) = g_j(R(\tau)).$$

Then $a^2 + b^2 + c^2 + d^2 \le K(c^2 + d^2)$ for some $K \in \mathbb{R}$, $K > 0$. This gives

$$g_j(V(\tau)) = g_j(R(\tau)) \le C_3 |c\tau + d|^{k-2\sigma} (c^2 + d^2)^\alpha y^\sigma,$$

for some real C_3 with $C_3 > 0$.

Straightforward calculations give the following fact:

$$\frac{y^2}{1 + |z|^2}(c^2 + d^2) \le |cz + d|^2 \le 2(|z|^2 + y^{-2})(c^2 + d^2)$$

for $c, d \in \mathbb{R}$ and $z = x + iy \in \mathbb{C}$. We use this fact to get

$$g_j(V(\tau)) \le C_3 |c\tau + d|^{k+2(\alpha-\sigma)} \left(\frac{1 + |\tau|^2}{y^2} \right)^\alpha y^\sigma.$$

Hence,

$$g_j(V(\tau)) \le C_3 \left(\frac{1 + |\tau|^2}{y^2} \right)^\alpha y^\sigma$$

by the definition of σ.

If $\tau \in \overline{F}_\lambda$, then $|y|^2 \le \frac{\lambda^2}{4} + y^2$ and $y^2 \ge 1 - \frac{\lambda^2}{4}$, which means that $\frac{1+|\tau|^2}{y^2}$ is bounded for $\tau \in \overline{F}_\lambda$.

Hence we get $g_j(V(\tau)) \le C_4 y^\sigma$ for some $C_4 \in \mathbb{R}$, $C_4 > 0$, for all $j = 1, \dots, p$, $V \in \overline{G}_\lambda$ and $\tau \in \overline{F}_\lambda$.

For $\tau = x + iy \in \mathbb{H}$, we choose a $V \in \overline{G}(\lambda)$ and a $\tilde{\tau} \in \overline{F}_\lambda$ such that $\tau = V(\tilde{\tau})$. Then we have a $C_5 \in \mathbb{R}$, $C_5 > 0$, such that

$$\begin{aligned} |f_j(\tau)| &= y^{-\alpha-\frac{k}{2}} g_j(V(\tilde{\tau})) \le y^{-\alpha-\frac{k}{2}} C_4 (\mathrm{Im}(\tilde{\tau}))^{\alpha+\frac{k}{2}} \\ &\le C_4 \left(\frac{1}{y}\right)^{\alpha+\frac{k}{2}} \left(y + \frac{1}{y}\right)^{\alpha+\frac{k}{2}} \\ &\le C_4 \left(1 + \frac{1}{y^2}\right)^{\alpha+\frac{k}{2}} \\ &\le C_5 y^{-2\alpha-k} \end{aligned}$$

for all $j = 1, \dots, p$, $\tau \in \mathbb{H}$. $y = \mathrm{Im}(\tau) \le 1$, $k \ge -2\alpha$.

Here, we used that $y \le \frac{1}{\mathrm{Im}(V(\tilde{\tau}))}$ if $V \in \overline{G}(\lambda) \setminus \langle \overline{S}_\lambda \rangle$ and $y = \mathrm{Im}(V(\tilde{\tau}))$ if $V \in \langle \overline{S}_\lambda \rangle$, which we get straightforward by induction on the Eichler length of V (as an element of $G(\lambda)$).

If $k < -2\alpha$, then $(\mathrm{Im}(\tilde{\tau}))^{\alpha+\frac{k}{2}}$ is bounded above because $\mathrm{Im}(\tilde{\tau}) \ge \sqrt{1 - \frac{\lambda^2}{4}}$. This proves the statement for $k < -2\alpha$. □

Corollary 4.2.

$$\dim(\mathcal{M}(k, v, \rho)) = 0 \quad \textit{for } k < -2\alpha.$$

Proof. Use Theorem 4.1 for small positive $y = \mathrm{Im}(\tau)$. □

Definition 4.3. Let $\Lambda \in \mathbb{Z}^p$. Then we define

$$\mathcal{M}(k, v, \rho, \Lambda) := \{F \in \mathcal{F}(k, v, \rho) \mid \Lambda(F) \ge \Lambda\}.$$

With this definition, we may write $\mathcal{M}(k, v, \rho) = \mathcal{M}(k, v, \rho, \tilde{G})$ and $\mathcal{S}(k, v, \rho) = \mathcal{M}(k, v, \rho, \mathbf{0})$.

Lemma 4.4. *Let $\lambda = \lambda_q = 2\cos(\frac{\pi}{q})$, $q \in \mathbb{N}$, $q \ge 3$. There exists a Hecke form*

$$\hat{f}_\infty \in \mathcal{M}(4q, v \equiv 1, \rho \equiv E_1)$$

with the following properties:
(1) *$\hat{f}_\infty$ has in $i\infty$ a $(q-2)$-fold zero.*
(2) *$\hat{f}_\infty$ has no zero in $\mathbb{H}$.*

Proof. This follows from Theorem 5.5 in [1], if we take there $\hat{f}_\infty := f_\infty^{q-2}$. Alternatively, we may use [3]. □

Lemma 4.5. *The allocation*

$$\begin{pmatrix} f_1 \\ \vdots \\ f_p \end{pmatrix} \to \begin{pmatrix} f_1 \cdot \hat{f}_\infty \\ \vdots \\ f_p \cdot \hat{f}_\infty \end{pmatrix}$$

defines a vector space isomorphism $\mathcal{M}(k, v, \rho, \Lambda) \to \mathcal{M}(k + 4q, v, \rho, \Lambda + (q - 2)\mathbf{1})$.

Proof. The proof follows from Lemma 2.6 and the respective definitions. □

Theorem 4.6.
1. $\dim(\mathcal{M}(k, v, \rho, \Lambda)) < \infty$ *for all* $\Lambda \in \mathbb{Z}^p$.
2. *If* $k \geq -2\alpha$, *then* $\dim(\mathcal{M}(k, v, \rho)) \leq p(q - 2)(\frac{k+2\alpha}{4q} + 1)$.

Proof. We choose $\ell \in \mathbb{Z}$ such that

$$\frac{k + 2\alpha}{4q} \leq \ell \leq \frac{k + 2\alpha}{4q} + 1.$$

Then $k - 4q\ell < -2\alpha$.

(1) We choose $\ell_0 \in \mathbb{Z}$, $\ell_0 < 0$, so, that the coefficients in the q-expansions of all elements from $\mathcal{M}(k - 4q\ell, v, \rho, \Lambda - (q - 2)\ell 1)$ for n-exponents of q with $n < \ell_0$ are equal 0.
Then we define a linear map

$$\mathcal{M}(k - 4q\ell, v, \rho, \Lambda - (q - 2)\ell\mathbf{1}) \to \mathbb{C}^{p|\ell_0|}$$

in such a way, that, in a given ordering, we allocate to each (F, ρ) a vector of all coefficients for n-exponents of q with $\ell_0 \leq n \leq -1$.
By Corollary 4.2 and because $k - 4q\ell < -2\alpha$, we get that the kernel of this linear map is trivial. Hence, the dimension of

$$\mathcal{M}(k - 4q\ell, v, \rho, \Lambda - (q - 2)\ell\mathbf{1})$$

is equal to the dimension of the image of the map, and hence, at most $p|\ell_0|$. By Lemma 4.5, we have that $\mathcal{M}(k-4q\ell, v, \rho, \Lambda-(q-2)\ell\mathbf{1})$ is isomorphic to $M(k, v, \rho, \Lambda)$.

(2) This is the special case $\Lambda = \tilde{G}$. We may chose $\ell_0 = -\frac{q-2}{\ell}$ and get

$$\begin{aligned}\dim(\mathcal{M}(k, v, \rho)) &= \dim(\mathcal{M}(k - 4q\ell, v, \rho, \tilde{G} - (q - 2)\ell\mathbf{1})) \leq p(q - 2)\ell \\ &\leq p(q - 2)\left(\frac{k + 2\alpha}{4q} + 1\right).\end{aligned}$$

□

5 Poncaré series and Eisenstein series

In this section, we briefly discuss certain concrete vector-valued Hecke forms.

In the following, let $\vec{e}_1, \ldots, \vec{e}_p$ be the p unit vectors in $\mathbb{C}^p$, and let δ_{jr} be the Kronecker-delta symbols, that is, $\delta_{jr} = 1$ if $j = r$ and $\delta_{jr} = 0$ if $j \neq r$.

Definition 5.1 (Vector-valued Poincaré series). We fix integers μ and r with $1 \le r \le p$. Let $\mathcal{R}$ be a right transversal for $\langle \overline{S}_\lambda \rangle$ in $\overline{G}(\lambda)$. The Poincaré series $P(\tau)$ is defined as

$$P(\tau) = P(\tau; k, v, \rho; \mu, r) := \frac{1}{2} \sum_{M \in \mathcal{R}} \frac{e^{2\pi i \frac{\mu+\varphi_r}{\lambda} M(\tau)}}{v(M)(M:\tau)^k} (\rho(M))^{-1} \vec{e}_r.$$

Here, the φ_r are as in Section 4. The theory of Poincaré series for vector-valued Hecke forms runs parallel to the classical theory (see [10]).

Theorem 5.2. *If $k > 2 + 2\alpha$, then $P \in \mathcal{F}(k, v, \rho)$, and the component functions*

$$p_j(\tau) = \frac{1}{2} \sum_{M \in \mathcal{R}} \frac{e^{2\pi i \frac{\mu+\varphi_r}{\lambda} M(\tau)}}{v(M)(M:\tau)^k} (\rho(M))^{-1}_{jr},$$

have a q-expansion of the form

$$p_j(\tau) = \delta_{jr} q^{\frac{\mu+\varphi_r}{\lambda}} + q^{\frac{\varphi_j}{\lambda}} \sum_{n=0}^{\infty} a_n(j) q^{\frac{n}{\lambda}}.$$

The proof is rather technical but complicated. It just uses standard techniques in complex analysis. The details can be found in [8] for the case of the modular group and in [5] for the general case of Hecke groups.

Agreement:
From now on, let $k > 2 + 2\alpha$.

Definition 5.3 (Vector-valued Eisenstein series). The Eisenstein series G_r is defined as

$$G_r(\tau) := P(\tau; k, v, \rho; -1, r)$$

for $r \in G = \{j \mid 1 \le j \le p, \varphi_j = 1\}$.

From Theorem 5.2, we automatically get the following.

Theorem 5.4. *If $r \in G$, then $G_r \in \mathcal{M}(k, v, \rho)$ and the component functions*

$$(g_r)_j(\tau) = \frac{1}{2} \sum_{M \in \mathcal{R}} \frac{(\rho(M))^{-1}_{jr}}{v(M)|M:\tau|^k}$$

have a q-expansion of the form

$$(g_r)_j(\tau) = \delta_{jr} + q^{\frac{\varphi_j}{\lambda}} \sum_{n=0}^{\infty} a_n(j) q^{\frac{n}{\lambda}}.$$

Definition 5.5.

$$E(k, v, \rho) := \langle \{G_r \mid r \in G\} \rangle \subset \mathcal{M}(k, v, \rho).$$

Theorem 5.6.
(i) $\dim(E(k,\nu,\rho)) = |G|$.
(ii) $\mathcal{M}(k,\nu,\rho) = E(k,\nu,\rho) \bigoplus \mathcal{S}(k,\nu,\rho)$.

Proof.
(i) From the definition of the Eisenstein series, we have

$$G_r(\tau) = \vec{e}_r + \begin{pmatrix} a_0(1)q^{\frac{\varphi_1}{\lambda}} \\ \vdots \\ a_0(p)q^{\frac{\varphi_p}{\lambda}} \end{pmatrix} q^{\frac{0}{\lambda}} + \cdots .$$

This shows that the G_r are linearly independent.
(ii) First of all, we have

$$E(k,\nu,\rho) \cap \mathcal{S}(k,\nu,\rho) = \{0\}.$$

Let $F \in \mathcal{M}(k,\nu,\rho)$, $F \notin S(k,\nu,\rho)$. Let

$$R = \{1 \le r \le p \mid \text{The } q\text{-expansions of } f_r \text{ begins with } c_r q^{\frac{0}{\lambda}}, c_r \ne 0\}.$$

Then $R \subset G$ by definition, and then

$$\Big(F - \sum_{r\in\mathcal{R}} c_r G_r\Big) \in \mathcal{S}(k,\nu,\rho).$$

□

Theorem 5.7. $\dim(\mathcal{M}(k,\nu,\rho)) \ge \dim(\mathcal{S}(k,\nu,\rho)) \ge p(q-2)(\frac{k-(2+2\alpha)}{4q} - 1)$.

Proof. We choose $\ell \in \mathbb{N}$ such that

$$\frac{k-(2+2\alpha)}{4q} - 1 \le \ell < \frac{k-(2+2\alpha)}{4q}.$$

Then $k - 4q\ell > 2 + 2\alpha$. By Lemma 4.5, we have

$$\mathcal{M}(k,\nu,\rho) = \mathcal{M}(k,\nu,\rho,G) \cong \mathcal{M}(k-4q\ell,\nu,\rho,G-(q-2)\ell\mathbf{1}).$$

We may apply Theorem 5.2 for $\mathcal{M}(k-4q\ell,\nu,\rho,G-(q-2)\ell\mathbf{1})$. Since $\mu = -1,\dots,-(q-2)\ell-1$ and $r = 1,\dots,p$, we get $p(q-2)\ell$ linearly independent, nonentire Hecke forms, and in addition the elements from $E(k-4q\ell,\nu,\rho)$. Hence

$$\begin{aligned}\dim(\mathcal{M}(k,\nu,\rho)) &= \dim(\mathcal{M}(k-4q\ell,\nu,\rho,G-(q-2)\ell\mathbf{1})) \ge p(q-2)\ell + |G| \\ &\ge p(q-2)\Big(\frac{k-(2+2\alpha)}{4q} - 1\Big) + |G|\end{aligned}$$

which gives the asserted inequality.

□

Bibliography

[1] Berndt BC, Knopp MI. Hecke's theory of modular forms and Dirichlet series. Singapore: World Scientific; 2008.
[2] Eichler M. Grenzkreisgruppen und kettenbruchartige Algorithmen. Acta Arith. 1965;11:169–80.
[3] Hecke E. Lectures on Dirichlet-Series. Schoenberg B, Maak W, editors. Modular Functions and Quadratic Forms. Gottingen: Vandenhoeck & Ruprecht; 1983.
[4] Hennekemper A. Vektorwertige Heckeformen und ihre Fouriekoeffizienten. Dortmund: Diplomarbeit; 2006.
[5] Hennekemper A. Über die Dimesion von Vektorräumen ganzer vektorvertiger Heckeformen. Dortmund: Dissertation; 2010.
[6] Hinz K. Kettenbruchartige Algorithmen bei Hecke-Gruppen. Hamburg: Staatsarbeit, Univ. Hamburg; 1975.
[7] Knapp M, Mason G. On vector-valued modular forms and their Fourier coefficients. Acta Arith. 2003;110:117–24.
[8] Knapp M, Mason G. Vector-valued Modular Forms and Poincaré Series. Ill J Math. 2004;48:1345–66.
[9] Lehner J. A short course in automorphic functions. New York: Holt. Rinehart and Winston; 1966.
[10] Lehner J. Discontinuous groups and automorphic functions. Math. Surveys. vol. VIII. Providence: AMS; 1964.

Rita Gitik

Two algorithms in group theory

Abstract: We present a new algorithm deciding if the intersection of a quasiconvex subgroup of a negatively curved group with a conjugate is finite. We also give a short proof of decidability of the membership problem for quasiconvex subgroups of finitely generated groups with decidable word problem.

Keywords: Group, algorithm, conjugate, membership problem

MSC 2010: Primary 20F10, Secondary 20F65, 20F67

1 Introduction

Let H be a subgroup of a group G and let g be an element of G. The subgroup $g^{-1}Hg$ is called the conjugate of H by g. We would like to check if the intersection $H \cap g^{-1}Hg$ is finite. This is a very old problem, closely connected to the study of the behavior of different lifts of subspaces of topological spaces in covering spaces.

However in general, the question if a subgroup of a group is finite is undecidable. That was shown by Adian in 1957 [1], and independently, by Rabin in 1958 [35]. The modern statement of the Adian–Rabin theorem utilizes a notion of a Markov property for finitely presented groups.

Definition 1. A property M of finitely presented groups is called Markov if it is preserved under group isomorphisms and the following holds:
(1) There exists a finitely presented group K with property M.
(2) There exists a finitely presented group H which cannot be embedded in any finitely presented group with property M.

The Adian–Rabin theorem can be stated as follows.

Theorem 1. *Let M be a Markov property of finitely presented groups and let G be a finitely presented group. It is undecidable whether or not G has property M.*

Note that being a finite group is a Markov property. The group K can be chosen to be the trivial group and the group H can be chosen to be infinite cyclic group.

We would like to mention several families of subgroups which have well understood intersections with their conjugates.

Acknowledgement: The author would like to thank Peter Scott for helpful conversations.

Rita Gitik, Department of Mathematics, University of Michigan, Ann Arbor, MI 48109, USA, e-mail: ritagtk@umich.edu

https://doi.org/10.1515/9783110638387-008

A subgroup H is normal in G if $H = g^{-1}Hg$ for any $g \in G$. The study of normal subgroups goes back to the origins of group theory. The concept was introduced by Évariste Galois at the beginning of the 19th century, who called normal subgroups “invariant subgroups,” [20]. G. Baumslag, Boone, and B. Newmann showed in 1959 that being normal is an undecidable property for a subgroup [6].

A subgroup H is malnormal in G if for any $g \in G$ such that $g \notin H$ the intersection $H \cap g^{-1}Hg$ is trivial. This concept for infinite groups was introduced by B. Baumslag in 1968 [5]. However, malnormality was investigated in finite groups at the end of the 19th century by Ferdinand Georg Frobenius [19]. A proper nontrivial malnormal subgroup of a finite group is called a Frobenius complement or a Frobenius subgroup. Bridson and Wise showed in 2001 that malnormality of a finitely generated subgroup in a negatively curved group is undecidable [12]; however, the author proved in 2016 [23], that malnormality is decidable for a torsion-free quasi-convex subgroup of a negatively curved group.

Malnormality of a subgroup has been generalized in different ways. One of them, namely the height, introduced by the author in 1995 [24], has been used by Agol in 2013 in his proof of Thurston's conjecture that hyperbolic 3-manifolds are virtual bundles (over a circle with fiber a surface) [37], [3], and [4].

A subgroup H of G is almost malnormal in G if for any $g \in G$ such that $g \notin H$ the intersection $H \cap g^{-1}Hg$ is finite. Using a result of Rips from 1982 [36], Bridson and Wise showed in 2001 [12], that almost malnormality of a finitely generated subgroup of a negatively curved group is undecidable. However, the author showed in 2016 [23], that almost malnormality is decidable for quasi-convex subgroups of negatively curved groups. An informative paper on malnormality and almost malnormality was published by de la Harpe and Weber in 2014 [27].

Most subgroups are neither normal nor malnormal, so the study of the intersection pattern of conjugates of a subgroup is a challenging problem. We restrict ourselves to the special case of H being quasi-convex and G being negatively curved.

2 Notation and definitions

Let X be a set and let $X^* = \{x, x^{-1} \mid x \in X\}$, where for $x \in X$ we define $(x^{-1})^{-1} = x$. Let G be a group generated by the set X. As usual, we identify a word in X^* with the corresponding element in G.

Let Cayley(G) be the Cayley graph of G with respect to the generating set X. The set of vertices of Cayley(G) is G, the set of edges of Cayley(G) is $G \times X^*$, and the edge (g, x) joins the vertex g to gx. The Cayley graph was first considered by Cayley in 1878 [13].

A path p in Cayley(G) is a sequence of edges of the following form: $p = (g, x_1)(gx_1, x_2) \cdots (gx_1x_2 \cdots x_{n-1}, x_n)$. The length of the path p is the number of edges forming it. A geodesic between two vertices in Cayley(G) is a shortest path in Cayley(G) connecting these vertices.

A group G is δ-negatively curved if any side of any geodesic triangle in Cayley(G) belongs to the δ-neighborhood of the union of the two other sides. Negatively curved groups were introduced by Gromov in 1987 [26]. Negatively curved groups are also called word hyperbolic and Gromov hyperbolic groups.

Note that Cayley(G) can be effectively constructed if and only if the word problem in G is decidable. The word problem asks if there exists an algorithm to decide if any word in the alphabet X^* represents the trivial element of $G = \langle X|R\rangle$. The word problem was introduced by Dehn in 1911 [16]. It was shown by Novikov in 1955 [33], and independently, by Boone in 1958 [9], that the word problem in groups in undecidable. However, it follows from the work of Greendlinger in 1960 [25], that the word problem is decidable in negatively curved groups.

A subgroup H of G is K-quasi-convex in G if any geodesic in Cayley(G) with endpoints in H belongs to the K-neighborhood of H. Quasi-convex subgroups were introduced by Gromov in 1987 [26]. Some important properties of quasi-convex subgroups were presented by the author in 1997 [22].

Remark 1. G. Baumslag, C. F. Miller III, and Short showed in 1992 that given a finite presentation $\langle X|R\rangle$ for a group G, it is undecidable if G is negatively curved [7]. However, if G is known to be negatively curved, a negative curvature constant δ can be determined. That was demonstrated by Epstein and Holt in 2000 [17], and independently, by Papasoglu in 1996 [34].

Bridson and Wise showed in 2001 that the property of being quasi-convex is undecidable for a finitely generated subgroup of a torsion-free negatively curved group [12]. However, I. Kapovich showed in 1996 that if a subgroup of a negatively curved group is known to be quasi-convex, then a quasi-convexity constant K for such a subgroup can be computed [28].

3 An algorithm deciding if the intersection of a quasi-convex subgroup of a negatively curved group with a conjugate is finite

Input: a finite presentation $\langle X|R\rangle$ for a negatively curved group G, a finite generating set for a quasi-convex subgroup H of G, and an element $g \in G$ which is not in H.

Output: a finite group isomorphic to $H \cap g^{-1}Hg$ or a statement that the intersection is infinite.

(1) Find a constant δ (not necessarily minimal) of negative curvature of G. This can be done using the results of Epstein and Holt [17], or of Papasoglu [34].
(2) N. Brady in 2000 [10], showed that the orders of finite subgroups of a δ-negatively curved group G generated by a finite set X are bounded by a constant $C = (2|X|)^{2\delta+1} + 1$. A similar result was obtained independently by Bogopolskii and Gerasimov in 1996 [8].

Make a list L of all finite groups with fewer than C elements. This can be done, for example, by considering the multiplication tables of group elements. Note that all finite groups are negatively curved because their Cayley graphs have finite diameters.

(3) The author showed in 1996 [21], that conjugation, in general, does not preserve quasi-convexity. However, the author proved in 1997 [23], that a conjugate of a quasi-convex subgroup of a negatively curved group is quasi-convex. It follows that $g^{-1}Hg$ is a quasi-convex subgroup of G. Gromov proved in 1987 [26], that the intersection of two quasi-convex subgroups of a negatively curved group is quasi-convex. Hence, the subgroup $H \cap g^{-1}Hg$ is quasi-convex in G. Bridson and Haefliger proved in 1999 that a quasi-convex subgroup of a negatively curved group is negatively curved [11], page 462. Therefore, the group $H \cap g^{-1}Hg$ is negatively curved.
The isomorphism problem in groups asks if there exists an algorithm to decide if any two presentations define isomorphic groups. The isomorphism problem was introduced by Dehn is 1911 [16]. It was shown by Adian in 1957 [2], and independently by Rabin in 1958 [35], that the isomorphism problem in groups is undecidable. However, Dahmani and Guirardel showed in 2011 that the isomorphism problem for negatively curved groups is decidable [15]. Therefore, we can determine whether $H \cap g^{-1}Hg$ is finite by checking if it is isomorphic to an element of L.
If positive, output the intersection.
If negative, output the statement that the intersection is infinite.

4 The membership problem

We will need additional notation. Denote the equality of two words in X^* by $\equiv$. The length of the word w is the number of symbols from X^* forming w. Denote the length of the word w by $|w|$.

The membership problem for a subgroup H of $G = \langle X|R\rangle$ asks if there exists an algorithm which for any word w in the alphabet X^* decides whether or not w represents an element of H. The membership problem is also called the generalized word problem. If G has a decidable membership problem for the trivial subgroup, then G has a decidable word problem. As the word problem, in general, is undecidable, [33] and [9], the membership problem, in general, is undecidable.

The solution of the word problem in negatively curved groups follows from the result proved by Greendlinger in 1960 [25], for certain small cancellations groups. A good exposition of this result was given by Lyndon and Schupp in 1977 [30], page 249. That result was generalized to negatively curved groups by Gromov in [26].

Theorem 2. *Let G be a negatively curved group. There exists a finite presentation $G = \langle X|R\rangle$, called Dehn's presentation, with the following property. If w is a non-trivial freely* *reduced word in* X^* *such that* $w = 1_G$, *then there exists a relator* $r \in R$ *and an initial subword* v *of* r *with* $|v| > \frac{1}{2}|r|$ *such that* v *is a subword of* w.

A good exposition of Dehn's presentation was given by Bridson and Haefliger in 1999 [11], page 450.

The solution of the word problem for a negatively curved group G is given by the following procedure, called Dehn's algorithm. A good exposition of Dehn's algorithm can be found in [11], page 449 and in [30], page 246.

Start by choosing Dehn's presentation $\langle X|R\rangle$ for G. Let w be a nontrivial freely reduced word in X^*.

For any relator $r \in R$, *check if there exists an initial subword* v *of* r *with* $|v| > \frac{1}{2}|r|$ *such that* v *is a subword of* w. *If no, then* $w \neq 1_G$. *If yes, let* $r \equiv vu$, *replace the word* v *in* w *by the word* u^{-1} *and freely reduce. Denote the resulting freely reduced word by* w_1. *As the words* v *and* u^{-1} *represent the same element in* G, *it follows that the words* w *and* w_1 *represent the same element in* G. *As* $|v| > |u|$, *it follows that* $|w| > |w_1$. *Repeat the procedure with the word* w_1.

Dehn's algorithm terminates in at most $|w|$ steps. If it terminates with the trivial word, then $w = 1_G$.

It was shown by Rips in 1982 [36], that the membership problem is undecidable for arbitrary subgroups of negatively curved groups. A good exposition of that result can be found in [7]. However, the membership problem for quasi-convex subgroups of negatively curved groups is decidable which was shown, for example, by the author in 1995 [21], and independently, by Farb in 1994 [18], and by I. Kapovich in 1995 [28]. The author in 2016 [23], and independently, Kharlampovich, Miasnikov, and Weil in 2017 [29], gave new proofs of that fact.

We present a short solution of the membership problem for quasi-convex subgroups of finitely generated groups with decidable word problem. Our solution utilizes the concept of a weakly Nielsen generating set of a subgroup, introduced by the author in 1995 [21].

Nielsen generating sets are an important tool in the study of free groups. They originated in work of Nielsen in 1921 [32]. The main characteristic of Nielsen generating sets is a small amount of cancellations between all the members. A good description of Nielsen generating sets and their applications in free groups was given by Magnus, Karrass, and Solitar in 1966 [31], page 128.

Definition 2. Let H be a subgroup of a group $G = \langle X|R\rangle$. We say that a finite generating set $S = \{s \equiv l_i n_i r_i \mid 1 \leq i \leq m, n_i \neq 1\}$ of H is Nielsen if the cancellations in any freely reduced product of elements of S do not affect the words n_i.

A theorem of Nielsen proven in 1921 [32], states that any finitely generated subgroup of a free group has a Nielsen generating set.

The existence of generating sets with similar strong noncancellation properties in nonfree groups was a topic of extensive research which showed that Nielsen generating sets are very rare. See, for example, a paper of Collins and Zieschang from 1988 [14]. However, a modified version of the Nielsen generating set turned out to be very useful.

Definition 3. Let H be a subgroup of the group $G = \langle X|R\rangle$. We say that a finite generating set S of H is weakly Nielsen if for any $h \in H$ and for any shortest word w in X^* representing h in G there exist finitely many elements $s_i \in S$ and decompositions $s_i \equiv l_i n_i r_i$ (which depends on w), with $n_i \neq 1$ such that $h = s_1 \cdots s_m$ and $w \equiv l_1 n_1 n_2 \cdots n_m r_m$.

We need an additional definition.

Definition 4. The label of the path $p = (g, x_1)(gx_1, x_2) \cdots (gx_1x_2 \cdots x_{n-1}, x_n)$ in Cayley(G) is the word $\text{Lab}(p) \equiv x_1 \cdots x_n$.

The inverse of a path p is denoted $\bar{p}$.

The author proved in 1996 that quasi-convex subgroups of finitely generated groups have weakly Nielsen generating sets [21]. Below is a new proof of that fact.

Lemma 1. *Let H be a K-quasi-convex subgroup of a group G, generated by a finite set X. Then H has a weakly Nielsen generating set consisting of elements of G of length at most $2K + 1$.*

Proof. Consider $h \in H$. Let w be a shortest word in X^* representing h in G, and let γ be a geodesic in Cayley(G) beginning at 1_G with $\text{Lab}(\gamma) \equiv w$. Let $v_0 = 1_G, v_1, \ldots, v_n = h$ be the vertices of γ listed in order, and let $e_1, e_2, \ldots, e_n$ be the edges of γ listed in order, so the length of γ is n. As H is K-quasi-convex in G, for any vertex v_i of γ there exists a path t_i not longer than K, which joins v_i to an element of h. Note that some t_i might be empty. Then

$$h = (\text{Lab}(\overline{t_0})\,\text{Lab}(e_1)\,\text{Lab}(t_1))(\text{Lab}(\overline{t_1})\,\text{Lab}(e_2)\,\text{Lab}(t_2)) \cdots (\text{Lab}(\overline{t_n})\,\text{Lab}(e_n)\,\text{Lab}(t_{n+1})).$$

Let $l_i \equiv \text{Lab}(\overline{t_{i-1}})$, $n_i \equiv \text{Lab}(e_i)$, $r_i \equiv \text{Lab}(t_i)$, and $s_i \equiv l_i n_i r_i$.

Let S be the union of all s_i, constructed for all $h \in H$. By construction, each s_i is no longer than $2K + 1$, hence as the set X is finite, the set S is also finite. By construction, S is a weakly Nielsen generating set of H. □

Theorem 3. *Let G be a group generated by a finite set X and let H be a K-quasi-convex subgroup of G. If G has a decidable word problem then the membership problem for H in G is decidable.*

Proof. Let g be an element of G. As G has a decidable word problem, we can find a shortest word w in X^* representing g in G. Let the length of w be n. Lemma 1 states that H has a weakly Nielsen generating set S. If g belongs to H then, by definition of a weakly Nielsen generating set, g can be written as a product of at most n elements of S. As S is finite, we can generate all the products of at most n elements of S. As G has a decidable word problem, we can check if any of these products is equal to g. □

Bibliography

[1] Adian SI. Algorithmic Unsolvability of Problems of Recognition of Certain Properties of Groups. Dokl Akad Nauk SSSR. 1955;103:533–5.
[2] Adian SI. The Unsolvability of Certain Algorithmic Problems in the Theory of Groups. Tr Mosk Mat Obŝ. 1957;6:231–98.
[3] Agol I. The virtual Haken conjecture. With an appendix by Agol, Daniel Groves, and Jason Manning. Doc Math. 2013;18:1045–87.
[4] Agol I, Groves D, Manning J. Residual finiteness, QCERF and fillings of hyperbolic groups. Geom Topol. 2009;13:1043–73.
[5] Baumslag B. Generalized Free Products Whose Two-Generator Subgroups Are Free. J Lond Math Soc. 1968;43:601–6.
[6] Baumslag G, Boone WW, Neumann BH. Some Unsolvable Problems About Elements and Subgroups of Groups. Math Scand. 1959;7:191–201.
[7] Baumslag G, Miller CF III, Short H. Unsolvable Problems about Small Cancellation and Word Hyperbolic Groups. Bull Lond Math Soc. 1994;26:97–101.
[8] Bogopolskii OV, Gerasimov VN. Finite Subgroups of Hyperbolic Groups. Algebra Log. 1996;34:343–5.
[9] Boone WW. The Word Problem. Proc Natl Acad Sci. 1958;44:1061–5.
[10] Brady N. Finite Subgroups of Hyperbolic Groups. IJAC. 2000;10:399–406.
[11] Bridson MR, Haefliger A. Metric Spaces of Non-Negative Curvature. A Series of Comprehensive Studies in Mathematics. vol. 319. Berlin, Heidelberg: Springer; 1999.
[12] Bridson MR, Wise DT. Malnormality is Undecidable in Hyperbolic Groups. Isr J Math. 2001;124:313–6.
[13] Cayley A. Desiderata and Suggestions: No. 2. The Theory of Groups: Graphical Representations. Am J Math. 1878;1:174–6.
[14] Collins DJ, Zieschang H. On the Nielsen Method in Free Products with Amalgamated Subgroups. Math Z. 1988;197:97–118.
[15] Dahmani F, Guirardel V. The Isomorphism Problem for All Hyperbolic Groups. Geom Funct Anal. 2011;21:223–300.
[16] Dehn M. Über Unendliche Diskontinuierliche Gruppen. Math Ann. 1911;71:116–44.
[17] Epstein DBA, Holt DF. Efficient Computations in Word-Hyperbolic Groups. In: Computational and Geometric Aspects of Modern Algebra. London Math Society Lecture Note Ser. vol. 275. 2000. p. 66–77.
[18] Farb B. The Extrinsic Geometry of Subgroups and Generalized Word Problem. Proc LMS. 1994;68:577–93.
[19] Frobenius G. Über Auflösbare Gruppen. IV. Berl Ber. 1901. 1216–1230.
[20] Galois É. Analyse d'un Mémoire sur la Résolution Algébrique des Équations. Bull Sci Math. 1830;13:413–27.
[21] Gitik R. Nielsen Generating Sets and Quasiconvexity of Subgroups. J Pure Appl Algebra. 1996;112:287–92.
[22] Gitik R. On Quasiconvex Subgroups of Negatively Curved Groups. J Pure Appl Algebra. 1997;119:155–69.
[23] Gitik R. On Intersection of Conjugate Subgroups. IJAC. 2017;27:403–20.
[24] Gitik R, Mitra M, Rips E, Sageev M. Width of Subgroups. Trans Am Math Soc. 1998;350:321–9.
[25] Greendlinger M. Dehn's Algorithm for the Word Problem. Commun Pure Appl Math. 1960;13:67–83.
[26] Gromov M. Hyperbolic Groups. In: Gersten SM, editor. Essays in Group Theory. MSRI series. vol. 8. New York: Springer; 1987. p. 75–263.

[27] de la Harpe P, Weber C. Malnormal Subgroups and Frobenius Groups: Basics and Examples. Confluentes Math. 2014;6:65–76.
[28] Kapovich I. Detecting Quasiconvexity: Algorithmic Aspects. In: Geometric and Computational Perspectives on Infinite Groups. DIMACS Ser Discrete Math Theoret Comput Sci. vol. 25. 1996. p. 91–9.
[29] Kharlampovich O, Miasnikov A, Weil P. Stallings Graphs for Quasi-Convex Subgroups. J Algebra. 2017;488:442–83.
[30] Lyndon RC, Schupp PE. Combinatorial Group Theory. Berlin, Heidelberg, New York: Springer; 1977.
[31] Magnus W, Karrass A, Solitar D. Combinatorial Group Theory. New York, London, Sydney: Interscience Publishers; 1966.
[32] Nielsen J. Om Regning med ikke Kommutative Faktoren og dens Anvendelse i Gruppeteorien. Math Tidsskr. 1921;B:77–94.
[33] Novikov PS. On the Algorithmic Unsolvability of the Word Problem in Group Theory. Proc Steklov Inst Math. 1955;44:1–143.
[34] Papasoglu P. An Algorithm Deciding Hyperbolicity. In: Geometric and Computational Perspectives on Infinite Groups. DIMACS Ser Discrete Math Theoret Comput Sci. vol. 25. 1996. p. 193–200.
[35] Rabin MO. Recursive Unsolvability of Group Theoretic Problems. Ann Math. 1958;67:172–94.
[36] Rips E. Subgroups of Small Cancellation Groups. Bull LMS. 1982;14:45–7.
[37] Thurston W. Geometry and Topology of 3-Manifolds. Lecture Notes. Princeton: Princeton University; 1977.

Rita Gitik and Eliyahu Rips

On products of closed subsets in free groups

Abstract: We present examples of closed subsets of a free group such that their product is not closed in the profinite topology. We discuss how to characterize a subset of a free group which is closed in the profinite topology and its product with any finitely generated subgroup of a free group is also closed in the profinite topology.

Keywords: Topological group, free group, profinite topology, closed set

MSC 2010: Primary 20E05, Secondary 20E26, 22A05

1 Introduction

The products of closed subsets of topological groups have been studied for a long time. The following well-known fact is an important result in the general theory of topological groups.

Theorem 1 (cf. [8], Theorem 4.4). *Let G be a topological group, let A be a compact subset of G, and let B be a closed subset of G. Then the set AB is closed.*

It has been known for a long time that if the set A is closed but not compact, then the set AB might fail to be closed.

Example 1. Let $G = \mathbf{R}$, $A = \mathbf{Z}$, and $B = \pi \cdot \mathbf{Z}$. Both A and B are closed subsets of G, but the set $A + B$ is a proper dense subset of $\mathbf{R}$, so it is not closed in G.

Note that in Example 1 both A and B are subgroups of G, so the product of closed subgroups is not closed in an arbitrary topological group. The situation is different in the profinite topology, which is defined by proclaiming all subgroups of finite index of G and their cosets to be basic open sets.

Denote the profinite topology on a group G by $\mathrm{PT}(G)$. An open set in $\mathrm{PT}(G)$ is a (possibly infinite) union of cosets of various subgroups of finite index and the closed sets in $\mathrm{PT}(G)$ are the complements of such unions in G.

A well-known theorem of M. Hall [6] states (in different language) that any finitely generated subgroup of a free group F is closed in $\mathrm{PT}(F)$. This result has been generalized by many researchers.

Acknowledgement: The first author would like to thank the Institute of Mathematics of the Hebrew University for generous support.

Rita Gitik, Department of Mathematics, University of Michigan, Ann Arbor, MI 48109, USA, e-mail: ritagtk@umich.edu
Eliyahu Rips, Institute of Mathematics, Hebrew University, Jerusalem, 91904, Israel, e-mail: eliyahu.rips@mail.huji.ac.il

https://doi.org/10.1515/9783110638387-009

The authors proved in [3] and [4] that the product of two finitely generated subgroups of a free group F is closed in $\mathrm{PT}(F)$. A different proof of that fact was published by G. A. Niblo, cf. [10]. Note that a finitely generated subgroup of a free group need not be compact, so this result cannot be deduced from Theorem 1.

A more general result saying that for any finitely generated subgroups $H_1, \dots, H_n$ of a free group F the product $H_1 H_2 \cdots H_{n-1} H_n$ is closed in $\mathrm{PT}(F)$ was obtained by L. Ribes and P. A. Zalesskii in [11], by K. Henckell, S. T. Margolis, J. E. Pin, and J. Rhodes in [7], and by B. Steinberg in [12].

T. Coulbois in [1] proved that property RZ_n is closed under free products, where a group G is said to have property RZ_n if for any n finitely generated subgroups $H_1, \dots H_n$ of G, the product $H_1 H_2 \cdots H_{n-1} H_n$ is closed in $\mathrm{PT}(G)$.

The first author showed in [2] that if H and K are quasi-convex subgroups of a negatively curved LERF group G and H is malnormal in G then the double coset KH is closed in $\mathrm{PT}(G)$. A. Minasyan generalized that result in [9], showing that a product of any finite number of quasi-convex subgroups of a negatively curved GFERF group G is closed in $\mathrm{PT}(G)$. (A negatively curved group G is GFERF if all its quasi-convex subgroups are closed in $\mathrm{PT}(G)$.)

The aforementioned results lead to the following question: is it true that for any finitely generated subgroup H of a free group F and for any subset S of F, which is closed in $\mathrm{PT}(F)$, the product SH is closed in $\mathrm{PT}(F)$.

As a finitely generated subgroup of F need not be compact in $\mathrm{PT}(F)$, we cannot apply Theorem 1 to resolve this question. Not surprisingly, the answer to this question is negative, in general, as shown in Example 2, Example 3, and Example 4 in Section 2 of this paper.

So we would ask a more restricted question: can we characterize subsets S of F, which are closed in $\mathrm{PT}(F)$, such that for any finitely generated subgroup H of F the product SH is closed in $\mathrm{PT}(F)$. The following special case might be considered first.

Question 1. Consider a subset S of a free group F such that S is closed in $\mathrm{PT}(F)$. Assume that for any element c of F the set $\langle c \rangle \cdot S$ is closed in $\mathrm{PT}(F)$. Is it true that for any elements a and b of F the set $\langle a, b \rangle \cdot F$ is closed in $\mathrm{PT}(F)$?

Remark 1. Note that for any subset S of F which is closed in $\mathrm{PT}(F)$ there exists a finitely generated subgroup H_S of F such that SH_S is closed in $\mathrm{PT}(F)$. Indeed, take $H_S = \langle 1 \rangle$.

2 The examples

Let F be the free group of rank two generated by elements a and b.

The following example describes an infinitely generated normal subgroup N of F which is closed in $\mathrm{PT}(F)$ such that the double coset $\langle a^2 \rangle N$ is not closed in $\mathrm{PT}(F)$.

Example 2. Consider the Higman's group $G = \langle a, b | b^{-1}ab = a^2\rangle$. As G is metabelian and linear, it is residually finite, however, the cyclic subgroup $\langle a^2\rangle$ is not separable from the element $a \in G$ in PT(G). Indeed, $a = b^n a^{2^n} b^{-n}$ (which can be shown by induction) and a finite index subgroup of G should contain an element $b^n \in G$ for some positive n.

Let N be the kernel of the quotient map from $F = \langle a, b\rangle$ to G. As G is residually finite and N is the preimage of the trivial subgroup of G in F, it follows that N is closed in PT(F). The subgroup $\langle a^2\rangle$ of F is closed in PT(F) because it is finitely generated. However, the double coset $\langle a^2\rangle N$ is not closed in PT(F) because it is not separable from the element $a \in F$.

Indeed, let F_0 be any finite index subgroup of F such that $\langle a^2\rangle N \subset F_0$. Let $G_0 < G$ be the projection of F_0 to G. As $a^2 \in G_0$ and G_0 has a finite index in G, it follows that $a \in G_0$, hence there exists an element $n_0 \in N$ such that $an_0 \in F_0$. However, $a^2 n \in N$ for all $n \in N$, hence $(n_0^{-1}a^{-1})(a^2 n) \in F_0$ for all $n \in N$. As N is normal in N, there exists $n_1 \in N$ such that $n_0^{-1}a = an_1$. Then $an_1 n \in F_0$ for all $n \in N$, hence $aN \subset F_0$. It follows that $a \in F_0$.

The following example describes a set S which is closed in PT(F) such that its product with a free factor of F is not closed in PT(F).

Example 3 (cf. [5]). Let $F = \langle a, b\rangle$ be the free group of rank two. Consider an infinite sequence $A = \{a, a^{2!}, a^{3!}, \ldots, a^{k!}, \ldots\} \subset F$. Note that A converges to 1_F. Indeed, let M be a normal subgroup of finite index m in F. If $k \geq m$, then $a^{k!}$ is contained in M. Hence any open neighborhood of 1_F in F contains all, but finitely many elements of A, therefore, A converges to 1_F. Note that $1_F \notin A$, so A is not closed in PT(F).

Let $m_k, k \geq 1$ be integers such that $m_k \to m_0 \in \hat{Z} \setminus Z$ in PT($\hat{Z}$), where $\hat{Z}$ is the completion of Z in PT(Z). Then $a^k b^{m_k} \to a^0 b^{m_0} \in \hat{F} \setminus F$, where $\hat{F}$ is the completion of F in PT(F). Hence the sequence $a^k b^{m_k}$ has no other limit points. In particular, it has no limit points in F. Therefore, for every $w \in F$ with $w \neq a^k b^{m_k}$ for all $k \geq 1$, there exists an open neighborhood U of w such that $a^k b^{m_k} \notin U$, for all $k \geq 1$. It follows that the set $S = \{ab^{m_1}, a^{2!}b^{m_2}, \ldots, a^{k!}b^{m_k}, \ldots\}$ is closed in PT(F).

Note that $1_F \notin S\langle b\rangle$, however $A \subseteq S\langle b\rangle$, so $1_F \in \bar{A} \subseteq \overline{S\langle b\rangle}$. We conclude that $S\langle b\rangle$ is not closed in PT(F).

Example 3 motivates the following question.

Question 2. Is it possible to impose some restrictions on a set S which is closed in PT(F), such that the product of S with a free factor of F would be closed in PT(F)?

The following example demonstrates that such restrictions on S should be severe.

Example 4 (cf. [5]). Let F be a finitely generated free group on free generators $K \cup L$ such that $F = \langle K\rangle * \langle L\rangle$. There exists a discrete subset S of F which is closed in PT(F) such that $S\langle K\rangle$ is not closed in PT(F) and the last syllable of all elements of S is in $\langle L\rangle$.

Bibliography

[1] Coulbois T. Free Product, Profinite Topology and Finitely Generated Subgroups. Int J Algebra Comput. 2001;11:171–84.
[2] Gitik R. On the Profinite Topology on Negatively Curved Groups. J Algebra. 1999;219:80–6.
[3] Gitik R, Rips E. On Separability Properties of Groups. Int J Algebra Comput. 1995;5:703–17.
[4] Gitik R, Rips E. On Double Cosets in Free Groups. IOSR J Math. 2016;12(6 version IV):14–5.
[5] Gitik R, Rips E. On Closed Subsets of Free Groups. IOSR J Math. 2017;13(6 version III):45–7.
[6] Hall M Jr. Coset Representations in Free Groups. Trans Am Math Soc. 1949;67:431–51.
[7] Henckell K, Margolis ST, Pin JE, Rhodes J. Ash's Type II Theorem, Profinite Topology, and Malcev Products I. Int J Algebra Comput. 1991;1:411–36.
[8] Hewitt E, Ross KA. Abstract Harmonic Analyses I. Berlin-Heidelberg: Springer; 1963.
[9] Minasyan A. Separable Subsets of GFERF Negatively Curved Groups. J Algebra. 2006;304:1090–100.
[10] Niblo GA. Separability Properties of Free Groups and Surface Groups. J Pure Appl Algebra. 1992;78:77–84.
[11] Ribes L, Zalesskii PA. On the Profinite Topology on a Free Group I. Bull LMS. 1993;25:37–43.
[12] Steinberg B. Inverse Automata and Profinite Topologies on a Free Group. J Pure Appl Algebra. 2002;167:341–59.

Ron Hirshon

Misbehaved direct products

Abstract: The question whether or not there exists a finitely presented group with

$$E \approx E \times B, \quad B \neq 1 \tag{1}$$

has been an elusive one. A form of this question was posed by Peter Hilton in the 1950s in connection with some topological problems related to the work of Lusternik and Schnirelman (private communication). The question was also raised in Hirshon (J Algebra 1986;99:232–238), Hirshon and Meyer (Bull Aust Math Soc 1992;43:513–520), and Hirshon (J Algebra 1994;167(2):284–290) with some pessimistic predictions for an easy resolution. In this paper, we give a general method for constructing groups as in Higman (J Lond Math Soc 1951;26:59–61).

Keywords: Direct product, finitely presented groups

MSC 2010: 20J05, 20F10, 20E18, 20F65

1 Introduction

We note that (1) implies

$$E \approx E \times B^n.$$

Here, B^n designates the nth Cartesian direct product of B. If G is a nontrivial finite group, the minimal number of generators of G^n goes to infinity with n [9]. Hence any group B that arises in our considerations cannot have any nontrivial finite images. Moreover, the B that arises in our construction does not seem to be related in any apparent way to standard constructions. At the present time, it is an open question whether our methods yield or can yield a nontrivial B. We leave this battle for a future time, recalling the immensity of solving the word problem for groups with even a single defining relation [7].

2 Preliminaries

Notation: If G is a finitely presented group, $G = \langle a_1, a_2, \ldots, a_n; R_1, R_2, \ldots, R_t\rangle$ with the R_i relations in the a_j, we write $G \approx F/R$ with F the free group on $x_1, x_2, \ldots, x_n$ and R the normal subgroup of F generated by the corresponding words $R_i(x_j)$.

Ron Hirshon, Mathematics Department, College of Staten Island, Staten Island, NY 10314, USA, e-mail: rhirshon@aol.com

https://doi.org/10.1515/9783110638387-010

Lemma. *Let*

$$A = \langle a_1, a_2, \ldots, a_n; R_1, R_2, \ldots, R_t\rangle \quad and \quad B = \langle b_1, b_2, \ldots, b_n; S_1, S_2, \ldots, S_s\rangle.$$

In $G = A \times B$, let $e_i = a_i b_i$ for all i. Let E be the subgroup of G generated by the e_i. Then $E \approx F/R \cap S$. The map $x_i R \cap S \to e_i$ induces an isomorphism of $F/R \cap S$ onto E.

3 The idea

Let A and B be as in the lemma, with A a non-Hopfian finitely presented group A such that A is isomorphic to F/R under the map $a_i \to x_i R$. Let K be a proper normal subgroup of F, properly contained in R with $F/K \approx F/R$. Choose a set of generators of A, $u_1, u_2, \ldots, u_n$, such that A is isomorphic to F/K under the map $u_i \to x_i K$. Moreover, choose the e_i so that they generate G and choose S so that in F, K is contained in S. Consequently, $A \times B \approx F/R \cap S$ and the map θ given by $u_i\theta = e_i$ induces a homomorphism of A onto $A \times B$. Let $A_1 = A\theta^{-1}$ and for $i > 0$ let $A_{i+1} = A_i\theta^{-1}$. The A_i form a descending chain of normal subgroups and if $\tilde{A}$ is the intersection of the A_i, then $\tilde{A}\theta = \tilde{A}$ and $\tilde{A}$ contains the kernel of θ. Hence, $A/\tilde{A} \approx G/\tilde{A} \approx (A/\tilde{A}) \times B$. Consequently, it suffices to find A and B as above with $B \neq 1$, B finitely presented and $\tilde{A}$ the normal closure of a finite number of elements. It will be useful to note in the sequel that $\tilde{A}$ may be characterized as consisting precisely of those elements of A with $a\theta^n$ in A for all a in A and all positive integers n. We call $\tilde{A}$ the upper kernel of θ. (See [4].)

4 A possible solution

Let A_1 be the non-Hopfian group (see [1])

$$A_1 = \langle z.s, t; z^{-1}sz = t^{-r}st^r = s^h\rangle.$$

Let

$$A = \langle z.s, t; z^{-1}sz = t^{-r}st^r = s^h, z^i sz^{-i} = t^n st^{-n}, i = 1, 2, \ldots, k\rangle.$$

Then $A = A_1$ and may be obtained from A_1 by considering y the surjective endomorphism of A_1 induced by $zy = z$, $sy = s^h$, $ty = t$ and then considering y^k.

If in A, $u = z^k sz^{-k}$, an isomorphism of A onto A_1 is induced by the map α given by $u\alpha = s$, $z\alpha = z$, $t\alpha = t$. Let $k > 2$ and let B have presentation

$$B = \langle Z, S.T; W_1, W_2, W_3, W_4, W_5, W_6\rangle,$$

where $W_1 = Z(ZSZ^{-1}T^rS^{-1}T^{-r})^{h_1}$, $W_2 = S(Z^2SZ^{-2}T^{2r}S^{-1}T^{-2r})^{h_2}$, $W_3 = T(Z^3SZ^{-3}T^{3r}S^{-1} \times T^{-3r})^{h_3}$, $W_4 = [Z^{2k}, T]$, $W_5 = Z^{-1}SZT^{-r}S^{-1}T^r$, $W_6 = Z^{-1}SZS^{-h}$. Here, h_1, h_2, and h_3 are arbitrary positive integers. Now we consider the map θ of A into $A \times B$ induced by $z\theta = e_1 = zZ$, $u\theta = e_2 = sS$, $t\theta = e_3 = tT$. Then $w_1(e_1, e_2, e_3) = w_1(z, s, t)w_1(Z, S, T) = z$. Similarly, $w_2(e_1, e_2, e_3) = s$ and $w_3(e_1, e_2, e_3) = t$. Hence θ maps A onto $A \times B$. We now argue that the upper kernel of θ is the normal closure of a finite number of elements. This will complete what we set out to do.

First, let $W = W(Z, S, T)$ be a word in B with

$$W(Z, S, T) = W(Z, Z^{-k}SZ^{k}, T) = 1. \tag{2}$$

Then we claim that

$$W(z, u, t)\theta^{n} = W(z, z^{-nk}uz^{nk}, t). \tag{3}$$

This is true for $n = 1$ **and assuming** (3) inductively, then

$$W(z, u, t)\theta^{n+1} = W(z, z^{-nk}sz^{nk}, t)W(Z, Z^{-nk}SZ^{nk}, T). \tag{4}$$

Now if we conjugate both sides of (2) by Z^{2rk}. we have

$$W(Z, Z^{-2rk}SZ^{2rk}, T) = W(Z, Z^{-2rk-k}SZ^{2rk+k}, T) = 1. \tag{5}$$

That is,

$$W(Z, Z^{-2rk}SZ^{2rk}, T) = W(Z, Z^{-(2r+1)k}SZ^{2r+1)k}, T) = 1. \tag{6}$$

Hence, $W(Z, Z^{-nk}SZ^{nk}, T) = 1$ for all n so that (4) says

$$W(z, u, t)\theta^{n+1} = W(z, z^{-nk}sz^{nk}, t) = W(z, z^{-(n+1)k}uz^{(n+1)k}, t). \tag{7}$$

On the other hand, if $W(Z, S, T)$ is an arbitrary word with $W(z, u, t)$ in $\tilde{A}$, it is easy to check that (2) holds. Hence, $\tilde{A}$ may be described simply as the set of words $W(z, u, t)$ such that (2) holds. Now we show that $\tilde{A}$ is the normal closure of a finite number of elements in A. In order to see this, let B_1 be an isomorphic replica of B on the corresponding generators, Z_1, S_1, T_1. In $B \times B_1$, let

$$Z^{*} = ZZ_1, \quad S^{*} = SZ_1^{-k}S_1Z_1^{-k}, \quad T^{*} = TT_1. \tag{8}$$

It suffices to show that Z^{*}, S^{*}, T^{*} generate $B \times B_1$ because if this is the case, a set of defining relations for $B \times B_1$ on the generators (8) will be a finite set of words $W(Z^{*}, S^{*}, T^{*})$ such that (2) holds. A set of defining relations for $B \times B_1$ on generators Z^{*}, S^{*}, T^{*} induces a set of elements in A whose normal closure is $\tilde{A}$. To see that Z^{*}, S^{*}, T^{*} generate $B \times B_1$, note as before zZ, sS, tT generate $A \times B$. Hence an arbitrary element of B can be written as a word W in zZ, sS, tT,

$$W(zZ, sS, tT) = W(Z, S, T) \quad W(z, s, t) = 1. \tag{9}$$

Now apply θ to $W(z,s,t) = 1$ to obtain

$$W(z,s,t)\theta = W(z,z^{-k}uz^k,t)\theta = W(z,z^{-k}sz^k,t)W(Z,Z^{-k}SZ^k,T) = 1.$$

That is, an arbitrary element of B has the form $W(Z,S,T)$ with $W(Z,Z^{-k}SZ^k,T) = 1$. So this observation shows that B is in the subgroup generated by Z^*, S^*, T^* so that Z^*, S^*, T^* generate $B \times B_1$.

5 Some final remarks

In an attempt to elucidate the previous section, we emphasize that the integer k is not completely arbitrary in the exposition but comes from the presentation of A obtained from considering the surjective endomorphism γ^k of A. In an original attempt at the above idea, instead of using the defining relation $[Z^{2k},T]$, in B we used the defining relation $[Z^k,T]$ in B. In this case, David Meier very cleverly showed that the resulting group, $A/\tilde{A}$, collapsed. We thank him for that cogent observation. Also, thanks to Ken Ross for a lot of editorial help including the making of a LaTeX file for this paper.

Bibliography

[1] Higman G. A finitely related group with an isomorphic proper factor group. J Lond Math Soc. 1951;26:59–61.
[2] Hilton P. Private communication.
[3] Hirshon R. Finitely presented groups L with $L \approx L \times M$, $M \neq 1$, M finitely presented. J Algebra. 1986;99:232–8.
[4] Hirshon R, Meyer D. Groups with a quotient that contains the original group as a direct factor. Bull Aust Math Soc. 1992;43:513–20.
[5] Hirshon R. Some properties of groups which allow homomorphisms onto their direct square. J Algebra. 1994;167(2):284–90.
[6] Jones J. Direct products and the Hopf property. J Aust Math Soc. 1974;17:174–96.
[7] Magnus K. Solitar Presentations of Groups in Terms of Generators and Relations. second revised ed. New York: Dover Publications; 1976. p. 274. Theorem 4.14.
[8] Meier D. Non hopfian groups. J Lond Math Soc. 1982;26:265–70.
[9] Wiegold J. Growth of finite groups. J Aust Math Soc. 1974;27:133–43.

Neha Hooda

A survey on Albert algebras and groups of type F_4

Abstract: The aim of this paper is to provide a detailed account of the recent developments in the area of Albert algebras over fields of characteristics different from 2 and 3. We document the new developments in understanding the role of the three cohomological invariants in determining the isomorphism classes of Albert algebras.

Keywords: Automorphisms, Albert algebras, structure group, invariants Mod-2, octonions, F_4

MSC 2010: Primary 20G15, Secondary 17C30

1 Introduction

This paper is focused on the study of Albert algebras and exceptional algebraic groups of type F_4 and is motivated by the connections between group theory, geometry, and quadratic forms. Algebraic groups are important in many major areas of mathematics and mathematical physics. We find them in diverse roles, notably as groups of automorphisms of geometric structures, as symmetries of differential systems, or as basic tools in the theory of automorphic forms. In this paper, we discuss the developments in providing an answer to the question raised by Jean-Pierre Serre on the classification of Albert algebras ([17], p. 205), asking whether the three cohomological invariants (see Section 4 for details), determine the isomorphism classes of Albert algebras? We discuss an interesting approach of looking at the Serre's problem by studying the invariants of Albert algebras via subgroups embeddings into their automorphism groups. The research work discussed in the paper involves results on factorising of the mod 3 invariant of Albert algebras, the Skolem–Noether-type theorems for Albert algebras, and essential dimension of Albert algebras.

Acknowledgement: I thank the organizing committee of the conference on The Elementary Theory of Groups and Groups Rings and Related Topics for giving me the opportunity to talk about my research work. I thank Professor Ben Fine and Professor Paul Baginski for their encouragement. I thank Professor Maneesh Thakur for introducing me to Albert algebras and many fruitful discussions I shared with him.

Neha Hooda, Department of Mathematics, Fairfield University, 1073 N Benson Road, Fairfield, CT 06824, USA, e-mail: nhooda@fairfield.edu

https://doi.org/10.1515/9783110638387-011

2 Jordan algebras

We will assume throughout the paper that the characteristic of the field is different from 2 and 3.

Definition. A Jordan algebra J over k is a commutative finite dimensional, not necessary associative, k-algebra with identity element e together with a nondegenerate quadratic form Q on A such that the condition $Q(x^2) = Q(x)^2$ if $\langle x, e\rangle = 0$, $\langle xy, z\rangle = \langle x, yz\rangle$, $Q(e) = \frac{3}{2}$ are satisfied.

Q is called the norm of J. The relation $x^2(xy) = x(x^2y)$ holds for a Jordan algebra. For any associative algebra A, the product

$$a.b = \frac{1}{2}(ab + ba)$$

gives A the structure of a Jordan algebra, which we write A^+. If B is an associative algebra with involution τ, the set $\mathrm{Sym}(B, \tau)$ of symmetric elements is a Jordan subalgebra of B^+, denoted by $H(B, \tau)$.

Theorem 2.1 ([9], Corollary 40.15). *Let B and B' be K-central simple with involutions τ and τ' of degree 3. Any isomorphism $H(B,\tau) \xrightarrow{\sim} H(B',\tau')$ of Jordan algebras extend to a unique isomorphism $(B,\tau) \xrightarrow{\sim} (B',\tau')$ of K-algebras with involutions.*

In particular, if $H(B,\tau)$ and $H(B',\tau')$ are isomorphic Jordan algebras then (B,τ) and (B',τ') are isomorphic as K-algebras with involution.

Definition. A Jordan algebra A is called exceptional if there does not exists an injective homomorphism $A \to D^+$ for any associative algebra D.

We are in particular interested in exceptional simple Jordan algebras or Albert algebras of dimension 27 as their automorphism groups are the groups of type F_4. Let us start by defining the structure of such algebras. Let C be a octonion algebra over k. For fixed $\gamma_i \in k^*$, let $A = H_3(C,\Gamma)$ be the set of $\Gamma = (\gamma_1, \gamma_2, \gamma_3)$-hermition 3×3 matrices

$$x = \begin{pmatrix} \xi_1 & c_3 & \gamma_1^{-1}\gamma_3\overline{c_2} \\ \gamma_2^{-1}\gamma_1\overline{c_3} & \xi_2 & c_1 \\ c_2 & \gamma_3^{-1}\gamma_2\overline{c_1} & \xi_3 \end{pmatrix}$$

with $\xi_i \in k$ and $c_i \in C (i = 1,2,3)$; here, $^- : C \to C$ defined by $\overline{x} =< x, e > e - x$ denotes conjugation. The product is defined as $xy = \frac{1}{2}(x.y + y.x)$, where the dot indicates the standard matrix product. Together with the usual addition of matrices and multiplication by elements of k, it makes A into a commutative, nonassociative k-algebra with the 3×3 identity matrix e as the identity element, called an reduced Albert algebra. The octonion algebra C associated with A is called the coordinate algebra of A.

Theorem 2.2 ([21], Theorem 1). *Let J be a Jordan algebra of degree 3 over k. Then J is either a division algebra or J is isomorphic to $H_3(C,\Gamma)$, where C is a composition algebra over K and $\Gamma = \mathrm{diag}(\gamma_1,\gamma_2,\gamma_3) \in \mathrm{GL}_3(k)$ is a diagonal matrix.*

Definition. An algebra A is called an Albert algebra if $k' \otimes_k A$ is isomorphic to a matrix algebra $H_3(C',\Gamma)$ for some field extension k' of k and some octonion algebra C' over k'.

We call a field extension F/k a **reducing field** of an Albert algebra A over k if the extended algebra $A \otimes_k F$ over F is reduced.

Theorem 2.3 ([18], Section 7). *Let A be an Albert algebra over a field k. Then there exists an unique octonion algebra C/k such that for any reducing field extension k' of k, $C \otimes_k k'$ is the coordinate algebra of the reduced algebra $A \otimes_k k'$. The unique octonion algebra C associated to A is called* Oct(A).

Isotopy and the structure group of Jordan algebras

Let J be a Jordan algebra over k and $u \in J$ be an invertible element. Then the k-module J together with the new unit $1_{J^{(u)}} = u^{-1}$ and the new U-operator $U_x{}^{(u)} := U_xU_u$ is a Jordan algebra over k, called the u-isotope of J and is denoted by $J^{(u)}$. We say that two unital Jordan algebras J, J' are isotopic if one is isomorphic to an isotope of the other. Isotopy is an equivalence relation on the class of Jordan algebras (cf. [8]). The set of isotopies from J to itself is a subgroup of GL(J), called the structure group of J and is denoted by Str(J). Note that the structure group contains the automorphism group of J as a subgroup.

3 Tits construction of Albert algebras

Tits has given two rational constructions of Albert algebras, which are exhaustive, that is, all Albert algebras arise from these constructions.

Tits's first construction

Let A be a central simple algebra of degree 3 over a field k and let $\mu \in k^* := k - \{0\}$. For $a, b \in A$, define

$$a.b = \frac{1}{2}(ab + ba), \quad a \times b = a.b - \frac{1}{2}t(a)b - \frac{1}{2}t(b)a + \frac{1}{2}(t(a)t(b) - t(a.b)),$$

here $t = \mathrm{Tr}$ is the reduced trace on A. Further, for $x \in A$, $\overline{x} = \frac{1}{2}(t(x) - x)$. To this data, one attaches an Albert algebra $J(A,\mu)$ as follows: $J(A,\mu) = A_0 \oplus A_1 \oplus A_2$, where $A_i = A$ for $i = 1, 2, 3$, with multiplication,

$$(a_0,a_1,a_2)(b_0,b_1,b_2) = (a_0.b_0+\overline{a_1b_2}+\overline{b_1a_2},\ \overline{a_0}b_1+\overline{b_0}a_1+\mu^{-1}a_2\times b_2,\ a_2\overline{b_0}+b_2\overline{a_0}+\mu a_1\times b_1).$$

With this multiplication, $J(A, \mu)$ is an Albert algebra over k and is referred to as a **first Tits construction** Albert algebra.

Theorem 3.1 ([8], Chapter IX, Theorem 20). *A first construction Albert algebra is either split or division. Moreover, $J(A, \mu)$ is a division algebra if and only if A is a division algebra and μ is not a reduced norm from A.*

Few important characteristics of first Tits construction Albert algebras are given by the below listed theorems.

Theorem 3.2 ([9], Proposition 40.5). *An Albert algebra A over k is a first construction if and only if $\mathrm{Oct}(A)$ is split over k.*

Theorem 3.3. *If A is a first Tits construction, then any isotope of A is isomorphic to A.*

Theorem 3.4 (á la Skolem–Noether, [9], Corollary 40.11). *The first Tits construction Albert algebras $J(A, \mu)$ and $J(A', \mu')$ are isomorphic if and only if $\mu \equiv \mu' N_A(A^*)$.*

Remarks on first Tits construction

1. Let F be a field such that char $F \neq 3$. Let $L = F(\sqrt[3]{\lambda})$, then L is isomorphic to the first Tits construction (F, λ).
2. Any Albert division algebra over an iterated Laurent series field in several variable $\mathbb{C}((t_1, \ldots, t_n))$, $\mathbb{C}$ the complexes is an pure first Tits construction.
3. All Albert algebras become first Tits construction after an appropriate quadratic étale extension.
4. Let A be a central simple algebra of degree 3, then $J(A, 1)$ the corresponding first Tits construction is isomorphic to $J(M_3(F), 1)$.
5. $J(A, 1)$ is an split first construction Albert algebra for every A.
6. Albert algebras are exceptional Jordan algebras.

Tits's second construction

Let K be a quadratic field extension of k and B be a central simple algebra of degree 3 over K and let σ be an involution of the second kind on B. Let $x \mapsto \overline{x}$ be the nontrivial Galois automorphism of K/k. Let $(B, \sigma)_+$ denote the k-subspace of B of σ-symmetric elements in B. Fix a unit u in $(B, \sigma)_+$ such that $N(u) = \mu\overline{\mu}$ for some $\mu \in K^*$, here N denotes the reduced norm on B. Let $J(B, \sigma, u, \mu) = (B, \sigma)_+ \oplus B$. We define a multiplication on $J(B, \sigma, u, \mu)$ as follows:

$$(a_0, a)(b_0, b) = (a_0.b_0 + \overline{au\sigma(b)} + \overline{bu\sigma(a)},\ \overline{a_0}b + \overline{b_0}a + \overline{\mu}(\sigma(a) \times \sigma(b))u^{-1}),$$

where the notation is same as above. With this multiplication, $J(B, \sigma, u, \mu)$ is an Albert algebra over k and is referred to as a **second Tits construction** Albert algebra.

Theorem 3.5 ([9], Theorem 39.18). *$J(B, \sigma, u, \mu)$ is a division algebra if and only if B is a division algebra and μ is not a reduced norm from B.*

Few remarks

1. ([11], Example C.5.2) Example of an Albert division algebra: Let A be a 9-dimensional central simple associative division algebra over a field k (Such do not exist over $\mathbb{R}$ or $\mathbb{C}$, but do exist over $\mathbb{Q}$ and p-adic fields). Then the extension $A(t) = A \otimes_k k(t)$ by the rational function field in one indeterminate remains a division algebra, but does not have the indeterminate t as one of its norms. Then $J(A(t), t)$ is an Albert division algebra over $k(t)$.
2. An Albert algebra over a finite field or a p-adic field is split.
3. ([9], Chapter IX, no. 11) There are exactly three nonisomorphic Albert algebras over the reals $\mathbb{R}$. An Albert algebra over $\mathbb{R}$, is reduced and up to isomorphism are either $H_3(C_a, 1)$, $H_3(C_a, \mathrm{diag}(1, -1, 1))$ or $H_3(C_s, \mathrm{diag}(1, -1, 1))$, where C_a is the nonsplit Cayley algebra over $\mathbb{R}$ and C_s is the split Cayley algebra over $\mathbb{R}$.
4. If the field k is algebraically closed, then there is only one Albert algebra over k, which is the split one.
5. All Albert algebras over iterated Laurent series field $k = \mathbb{R}((X_1, X_2, \ldots, X_n))$ in $n > 1$ variables $X_1, \ldots, X_n$ over the field $\mathbb{R}$ of real numbers are reduced.

4 Invariants of Albert algebras

Every Albert algebra A has a trace form $\mathrm{Tr} : A \to k$ and a quadratic form q_A defined by $q_A(x) = \mathrm{Tr}(x^2)/2$. For a reduced Albert algebra $J = H(C, \Gamma)$, $\Gamma = (\gamma_1, \gamma_2, \gamma_3)$-Hermition 3×3 matrix,

$$q_J \cong \langle 2, 2, 2\rangle \oplus \big(q_3 \otimes \langle -\gamma_1, -\gamma_2, \gamma_1\gamma_2\rangle\big).$$

This quadratic form determines the isomorphism classes of J.

Theorem 4.1 ([3], Theorem 22.4). *Let A be an Albert algebra over k. There exists* 3- *and* 5- *pfister forms q_3 and q_5 over k such that*

$$q_A \oplus q_3 \cong \langle 2, 2, 2\rangle \oplus q_5.$$

Moreover, this property characterizes q_3 and q_5 up to isomorphism and q_5 is divisible by q_3 (i. e., q_5 is tensor product of q_3 with a 2-Pfister form)

To an n-fold Pfister form

$$\langle\langle a_1, a_2, \ldots, a_n\rangle\rangle := \langle 1, -a_1\rangle \otimes \langle 1, -a_2\rangle \otimes \cdots \otimes \langle 1, -a_n\rangle,$$

one attaches an invariant $e_n(\langle\langle a_1, a_2, \ldots, a_n\rangle\rangle) \in H^n(k, \mathbb{Z}/2\mathbb{Z})$, called its Arason invariant (see [14], p. 453), given by

$$e_n(\langle\langle a_1, a_2, \ldots, a_n\rangle\rangle) = (a_1) \cup (a_2) \cup \cdots \cup (a_n),$$

where, for $a \in k^*$, (a) denotes the class of a in $H^1(k, \mathbb{Z}/2\mathbb{Z})$. The mod 2 invariants of A are given by

$$f_3(A) := e_3(q_3) \in H^3(k, \mathbb{Z}/2\mathbb{Z}), \quad f_5(A) := e_5(q_5) \in H^5(k, \mathbb{Z}/2\mathbb{Z}).$$

where q_i, $i = 3, 5$ are the Pfister form attached to the Albert algebra A by the preceding theorem. If $A = \mathcal{H}_3(C, \Gamma)$, is as above, one has

$$f_3(A) = e_3(n_C), \quad f_5(A) = e_5(\langle 1, \gamma_1^{-1}\gamma_2\rangle \otimes \langle 1, \gamma_2^{-1}\gamma_3\rangle \otimes n_C),$$

where n_C is the norm form of the octonion algebra C of A, which is a 3-fold Pfister form. When A is not reduced, there is a cubic extension L of k such that $A \otimes_k L$ is reduced. The descent property of Pfister form due to Rost ([22]), transfers the results from $A \otimes_k L$ down to A and the mod 2 invariants are the same as the mod 2 invariants of its reduced model.

Theorem 4.2 ([22], [[14], 12.7]). *If F has characteristic not 3, there exists a cohomological invariant assigning to each Albert algebra J over F a unique element $g_3(J) \in H^3(F, \mathbb{Z}/3\mathbb{Z})$ which only depends on the isomorphism class of J and satisfies the following two conditions:*

(a) *If $J = J(D, \mu)$ for some central simple associative F-algebra D of degree 3 and some $\mu \in F^*$, then $g_3(J) = [D] \cup [\mu] \in H^3(F, \mathbb{Z}/3\mathbb{Z})$.*

(b) *g_3 commutes with base change, that is, $g_3(J \otimes K) = \mathrm{res}_{K/F}(g_3(J))$ for any field extension K/F.*

Moreover,

(c) *g_3 detects division algebras in the sense that an Albert algebra J over F is a division algebra if and only if $g_3(J) \neq 0$. Hence, $g_3(J) \neq 0$ if and only if J has zero divisors.*

Few remarks

1. ([25], 9.5) Let $k = \mathbb{Q}_p(t)$, for any prime p. For every Albert k-algebra, it is not possible that $f_3(A)$ and $g_3(A)$ are both nonzero, suggesting an association between the mod 2 and mod 3 invariants.
2. The mod 2 invariant $f_3(G)$ "divides" $f_5(G)$. In particular, $f_3(G) = 0$ implies $f_5(G) = 0$.
3. An Albert algebra is split if and only if all its invariants f_3, f_5, g_3 are zero.
4. An Albert algebra A is a first Tits construction if and only if $f_3(A) = 0$ ([9], Proposition 40.5).
5. ([5], Remark 2.1) An Albert algebra A over k is a pure first (resp., second) construction if it cannot be expressed as a second (resp., first) construction. It is known that A is a first construction if and only if $f_3(A) = 0$. Hence if $f_3(A) \neq 0$, then A must be a pure second construction.
6. An Albert algebra $J(B, \tau, u, v)$ is a second Tits construction with τ a distinguished unitary involution of B if and only if $f_5(J) = 0$.

7. Let J_1 and J_2 be two Albert algebras over k which are isotopic. Then $f_3(J_1) = f_3(J_2)$ and $g_3(J_1) = g_3(J_2)$ ([17]). But the invariant f_5 is sensitive to isotopy. For example, if $\sigma' = \mathrm{Int}(v)\sigma$ is distinguished, then $f_5(J(B,\sigma,u,\mu)^{(v)}) = 0$, whereas the f_3 and f_5 invariant for this isotope are the same as for $J(B,\sigma,u,\mu)$ (cf. [28], Remark (4)).

Groups of type F_4

Albert algebras over a field k describe the k-groups of type F_4.

Theorem 4.3 ([26], Section 17.6). *Let G be a group of type F_4 over k. Then there exists an Albert algebra A, unique up to isomorphism, defined over k such that G is k-isomorphic to the group* $\mathbf{Aut}(A)$*, the full group of automorphisms* $\mathbf{Aut}_{\overline{k}}(A \otimes_k \overline{k})$.

Conversely, we have the following.

Theorem 4.4 ([27], Theorem 7.2.1, [17], p. 205). *Let A be an Albert algebra over a field k. Let $G = \mathbf{Aut}(A)$ be the associated algebraic group of automorphisms of A. Then G is a connected simple algebraic group defined over k of type F_4.*

The cohomological invariants of the group $G = \mathbf{Aut}(A)$ coincides with the invariants of the Albert algebra A, that is,

$$f_3(G) = f_3(A), \quad g_3(G) = g_3(A), \quad \text{and} \quad f_5(G) = f_5(A).$$

5 Current research topics and open problems

One of the most important results regarding the cohomological invariants of Albert algebras is that the three invariants f_3, f_5, and g_3 are only possible invariants of an Albert algebra.

Theorem 5.1 ([3], [2]). *Let k be a field of characteristic not 2, 3. The invariants f_3 and f_5 are basically the only invariants mod 2 and g_3 is basically the only invariant mod 3 of Albert algebras over k.*

Since f_3, f_5, and g_3 are the only invariants mod 2 and mod 3, respectively, Jean-Pierre Serre raised the important problem on the classification of Albert algebras asking whether these invariants determine the isomorphism classes of Albert algebras.

1. (cf. [2], open problem 8.7) Is the map

$$g_3 \times f_3 \times f_5 : H^1(k, F_4) \to H^3(k, \mathbb{Z}/3\mathbb{Z}) \times H^3(k, \mathbb{Z}/2\mathbb{Z}) \times H^5(k, \mathbb{Z}/2\mathbb{Z})$$

injective? That is, is an Albert algebra A determined up to isomorphism by its invariants $g_3(A)$, $f_3(A)$, and $f_5(A)$? This will allow to exchange the study of the set $H^1(k, F_4)$ of isomorphism classes of groups of type F_4 over k (equiv. of isomorphism classes of F_4-torsors or of isomorphism classes of Albert algebras) by the abelian group $H^3(k,\mathbb{Z}/3\mathbb{Z}) \oplus H^3(k,\mathbb{Z}/2\mathbb{Z}) \oplus H^5(k,\mathbb{Z}/2\mathbb{Z})$. The question is known to have affirmative answer for the reduced Albert algebras ([17]).

In ([13]), the authors give an affirmative answer to this when the Albert algebras have the form $J(B,\sigma,u,\mu)$ and $J(B,\sigma,u',\mu')$ showing that if the Albert algebras $J(B,\sigma,u,\mu)$ and $J(B,\sigma,u',\mu')$ have the same invariants f_3 and g_3 invariants, then they are isomorphic.

Theorem 5.2 ([13], 2.8). *Let $J = J(B,\sigma,u,\mu)$ and $J' = J(B,\sigma,u',\mu')$ be Albert algebras arising from Tits second construction. Assume $f_3(J) = f_3(J')$ and $g_3(J) = g_3(J')$. Then J is isomorphic to J'.*

Another interesting result proved by Rost in this regard is the following.

Theorem 5.3 ([23], Corollary 4, cf. [20], p. 85). *Let J/F and J_0/F be Albert algebras having the same mod 2 and mod 3 invariants. If F is of characteristic not 2 or 3, then there exists a finite extension E/F whose degree is not divisible by 3 and a finite extension K/F whose degree divides 3 such that $J \otimes E \cong J_0 \otimes E$ and $J \otimes K \cong J_0 \otimes K$.*

2. ([24], [14], Question 13.9) Do the invariants f_3 and g_3 classify Albert algebras up to isotopy?

Thakur in ([28]) proved,

Theorem 5.4 ([28], Theorem 2.1). *Let K be a quadratic extension of k and let B denote a central simple algebra of degree 3 over K, which admits involutions of second kind over k. Let $J = J(B,\sigma,u,\mu)$ and $J' = J(B,\sigma',u',\mu')$ be second Tits construction Albert algebras. Assume that $f_3(J) = f_3(J')$ and $g_3(J) = g_3(J')$. Then J and J' are isotopic.*

Hooda and Thakur, in ([7]), studied this problem via the automorphism groups of the Albert algebras proving the following.

Theorem 5.5 ([7], Theorem 7.2). *Let A_1 and A_2 be Albert division algebras over k. Let $G_i := \mathbf{Aut}(A_i)$, $1 \le i \le 2$. Then the following are equivalent:*
1. *A_1 is isotopic to A_2.*
2. *G_1 and G_2 admit a k-embedding of a common connected reductive algebraic group of type A_2 and the f_3 and g_3 invariants of G_1 and G_2 coincide.*

In ([14]), Petersson showed the following.

Theorem 5.6 ([14], Proposition 13.10). *If f_3, f_5, and g_3 classify Albert algebras up to isomorphism, then f_3 and g_3 classify them up to isotopy.*

Few interesting open problems in this regard are:

3. Is a first Tits construction J determined up to isomorphism by its invariants $f_3(J)$ and $f_5(J)$?

4. ([14], Question 13.6) Does the invariant mod 3 classify first Tits construction Albert algebras up to isomorphism?

Hooda and Thakur, in ([7]), gave an affirmative answer to this for Albert division algebras when they share a common connected, reductive algebraic group of type A_2,

Theorem 5.7 ([7], Corollary 7.1). *Let A_1 and A_2 be a first Tits construction Albert division algebras over k. Let $G_i := \mathbf{Aut}(A_i)$, $1 \le i \le 2$. Assume that G_1 and G_2 admit a k-embedding of a common connected, reductive algebraic group of type A_2 and that $g_3(A_1) = g_3(A_2)$. Then A_1 is isomorphic to A_2 over k.*

Let k be a field with characteristic not 2 and 3, containing the cube roots of unity. Thakur, in ([29]), proves a interesting connection between two Albert division algebras arising from Tits first construction with isomorphic invariants g_3.

Theorem 5.8 ([29], Corollary 2). *Let $J_1 = J(A, \mu)$ and $J_2 = J(B, \nu)$ be Albert division algebras arising from Tits first construction. If $g_3(J_1) = g_3(J_2)$, then $J_2 = J(C, \mu)$ for some central simple algebra C over k of degree 3.*

Factorising of the mod 3 invariant

We have seen that the mod 2 invariant f_5 divides the invariant f_3. The next result gives an understanding of what symbols occur in the decomposition of the mod 3 invariant g_3 of a given Albert algebra A over k. Let A an Albert division algebra over k, a field of characteristic different from 2 and 3 and assume that k contains the cube roots of unity. A nonzero element $x \in A$ is called a Kummer element if $x^3 = \lambda$ for some $\lambda \in k^*$.

Theorem 5.9 ([29], Corollary 1). *Let A be a first Tits construction Albert division algebra and let $x \in A$ be a Kummer element with $x^3 = \lambda$. Then $g_3(A) = [D] \cup [\lambda]$ for some central simple algebra D of degree 3 over k, where $[D] \in H^2(k, \mathbb{Z}/3)$ is the Brauer class of D and $[\lambda] \in H^1(k, \mathbb{Z}/3)$ is the class of μ.*

It will be interesting to obtain a similar result over a general field.

The Skolem–Noether theorem for Albert algebras

In ([13]), the authors prove a Skolem–Noether-type theorem on extensions of isomorphisms on certain types of simple Jordan subalgebras of an Albert algebra. They prove the following.

Theorem 5.10. *Let L/k be a quadratic field extension. Let (B, σ), (B', σ') be degree 3 central simple algebras over L with involutions of the second kind over k. Let $H(B, \sigma)$, $H(B', \sigma')$ denote the 9-dimensional Jordan algebras over k associated to the symmetric elements in (B, σ), (B', σ'), respectively. Suppose that $H(B, \sigma)$ and $H(B', \sigma')$ are subalgebras of an Albert algebra J over k and $\alpha : H(B, \sigma) \simeq H(B', \sigma')$ is an isomorphism of Jordan algebras. Then α extends to an automorphism of J.*

Note that the Skolem–Noether theorem fails to hold for cubic étale subalgebras of Albert algebras, that is, let J be an Albert algebra over a field F and suppose $E, E_0 \subseteq J$

are cubic étale subalgebras. It has been shown that in general an isomorphism $\phi : E \to E_0$ cannot be extended to an automorphism of J ([1], cf. [4]). Another result noteworthy in this context, is a weaker version of this proved by Garibaldi and Petersson in ([4]) by twisting ϕ by the right multiplication of a norm-one element in E. The authors show the following.

Theorem 5.11 ([4], Theorem B). *Let $\phi : E \xrightarrow{\sim} E'$ be an isomorphism of cubic étale subalgebras of an Albert algebra J over a field F. Then there exists an element $w \in E$ satisfying $N_E(w) = 1$ such that $\phi \circ R_w : E \to E'$ can be extended to an element of the structure group of J.*

Note that as pointed out by the author, working with the structure group of J instead of automorphism of J helps formulate a weaker version of the Skolem–Noether theorem but restricts us to homotopies instead of homomorphisms.

Embedding results

Another important research area in the field of Albert algebras are the embedding results. Petersson, in ([15]), motivates this direction of research by recalling that "It is sometimes useful to embed certain elements of a given algebraic structure into a substructure with particularly nice properties. For example, any semi-simple element of a connected linear algebraic group can be embedded into a maximal torus." Albert–Jacobson [[1], Theorem 1] and Jacobson [[8], Theorem IX.11], respectively, showed that over a field of characteristic 2 any element of a reduced Albert algebra can be embedded into a reduced absolutely simple subalgebra of degree 3 and dimension 9. Petersson, in ([4]), generalized this result to a field of an arbitrary characteristic,

Theorem 5.12 ([4], embedding theorem). *Let J be a reduced Albert algebra over an arbitrary field. Then any element of J can be embedded into a reduced absolutely simple subalgebra of degree 3 and dimension 9. Moreover, if J is split, this subalgebra may be chosen to be split as well.*

Another interesting result proved by Petersson and Racine about embedding of a cubic cyclic extension inside a subalgebra for an Albert division algebra arising from first Tits construction is given below.

Theorem 5.13 ([19], 2.7, [16], 4.5). *Let J be an Albert division algebra arising from Tits first construction. Let L/k be a cubic cyclic extension contained in J. Then there exists a subalgebra D^+ of J, for a degree-3 central division algebra D over k, such that $L \hookrightarrow D^+$.*

Few interesting open problems in this regard are:

5. ([9], p. 543) It is unknown if a division Albert algebra J always contains a cyclic cubic field extension. However, this is true if char $F \neq 3$ and F contains a primitive cubic root of unity.

6. ([14], 13.15) Can every element of a first Tits construction Albert division algebra be embedded into a subalgebra isomorphic to D^+, for some central associative division algebra D of degree 3?

Relation between subgroup embeddings and cohomological invariants

An interesting way of looking at the Serre's question (cf. [2], open problem 8.7) is through studying the relationship between the subgroup embedding and cohomological invariants of the groups of type F_4, which arise as the automorphism groups of Albert algebras. Through Theorem 5.5 and Theorem 5.7, we see affirmative answers to the Serre's problem when the automorphism groups of the Albert algebras share a subgroup embedding. This suggests a strong connection between the cohomological invariants of Albert algebras and subgroups embeddings in their automorphism groups. In ([5]) and ([7]), Hooda and Thakur showed that the embeddings of connected simple algebraic groups of type A_2 in groups of type F_4 are controlled by the mod 2 Galois cohomological invariants f_5 and g_3 attached to these groups, suggesting a possible connection between the two invariants. More precisely, we have the following.

Theorem 5.14 ([5], Theorem 3.3). *Let A be an Albert algebra over k. Let $\mathbf{Aut}(A)$ be the algebraic group of type F_4 associated to A. Let H be a connected simple group of type A_2 defined over k. Suppose $H \hookrightarrow \mathbf{Aut}(A)$ over k. Then $f_5(A) = f_3(H) \otimes \tau$ for some two-fold Pfister form τ over k.*

Theorem 5.15 ([7], Theorem 7.1). *Let A be an Albert division algebra and H be a connected simple group of type A_2 defined over k. A necessary condition for H to admit a k-embedding $H \hookrightarrow G = \mathbf{Aut}(A)$ is that $g_2(H)$ divides $g_3(A)$.*

In ([5]), Hooda gives a group theoretic characterization of Albert algebras with zero f_5 invariant.

Theorem 5.16 ([5], Theorem 3.4). *Let A be an Albert algebra over k and $G = \mathbf{Aut}(A)$. Then $f_5(A) = 0$ if and only if there exists a k-embedding $\mathrm{SU}(B, \sigma) \hookrightarrow G$ for some degree 3 central simple algebra B with center a quadratic étale k-algebra K and with a distinguished involution σ.*

Embedding of tori

In ([6]), Hooda showed that in addition to the groups of type A_2, the mod 2 invariants of the groups of type F_4 are also controlled by the unitary k-tori embedded in them. Let L, K be étale algebras of k-dimensions $n, 2$, respectively, and (E, τ) be the K-unitary algebra associated with the pair (L, K). We call the torus $\mathbf{SU}(E, \tau)$ as the K-unitary torus associated to the ordered pair (L, K). With such a K-unitary torus T, we associate the quadratic form $q_T :=< 1, -\alpha\delta >$, where $\mathrm{Disc}(L) = k(\sqrt{\delta})$ and $K = k(\sqrt{\alpha})$. Such tori occur as maximal tori in simple, simply connected groups of type A_{n-1} and G_2.

Definition. A k-torus T is **distinguished** if there exists étale k-algebras L, K be of k-dimensions $3, 2$, respectively, such that $\mathrm{disc}(L) = K$ and $T = \mathbf{SU}(E, \tau)$, where (E, τ) is the K-unitary algebra associated to the pair (L, K).

Theorem 5.17 ([6], Theorem 3.4). *Let A be an Albert algebra over k and $G = \mathbf{Aut}(A)$. Then $f_5(A) = 0$ if and only if G contains a distinguished k-torus.*

Using this result, she further goes on to prove a factorization result of the f_5 invariant of the group of type F_4.

Theorem 5.18 ([6], Theorem 4.7). *Let A be an Albert algebra over k and $G = \mathbf{Aut}(A)$. Let $K = k(\sqrt{\alpha})$ be a quadratic étale k-algebra and L be a cubic étale k-algebra with discriminant δ. Let T be the K-unitary torus associated with the pair (L, K). Suppose $T \hookrightarrow G$ over k. Then $f_5(A) = q_T \otimes \gamma$ for some 4-fold Pfister form γ over k.*

Essential dimension of Albert algebras

Mark L. Macdonald in ([10]) proved an interesting result showing number of independent parameters required to describe an Albert algebra up to isomorphism. Informally speaking, the essential dimension of an algebraic object is the minimal number of algebraically independent parameters one needs to define the object. Let p be a prime integer. The idea of the essential p-dimension is to ignore field extensions of degree prime to p. We say that a field extension K'/K is a prime to p extension if K'/K is finite and the degree $[K' : K]$ is prime to p. For precise definition, see ([12]).

Few remarks (see [10] for details)

1. The essential 2-dimension of F_4 is 5, because any Albert algebra becomes reduced after some degree prime-to-2 extension, and an arbitrary reduced Albert algebra only requires 5 parameters to describe it.
2. The essential 3-dimension of F_4 is 3, because any Albert algebra becomes a first Tits construction after some degree prime-to-3 extension, and they require three parameters to describe them: two parameters for A (the degree 3 central simple algebra, which must be cyclic), and one parameter for λ.
3. Finally, the essential p dimension of F_4 is 0 for all primes $p \geq 5$, because any Albert algebra splits after a degree 6 extension.
4. The essential p-dimensions give lower bounds for the essential dimension. The lower bound for essential dimension of G is the maximum of the lower bounds for the essential p dimension over all $p > 0$. Hence the lower bound for essential dimension of F_4 is 5.

Mark L. Macdonald in ([10]) gave an upper bound for the essential dimension of F_4, proving the following.

Theorem 5.19 ([10], Question 0.1). *Let k be a field of characteristics different from 2 and 3. The algebraically independent parameters (or essential dimension) needed to describe an arbitrary Albert algebra over k up to isomorphism is at most 7.*

Bibliography

[1] Albert AA, Jacobson N. On reduced exceptional simple Jordan algebras. Ann Math. 1957;2(66):400–17.

[2] Garibaldi S. Cohomological Invariants: Exceptional Groups and Spin Groups, With an appendix by Detlev W. Hoffmann. Mem Am Math Soc. 2009;200:937. xii+81.

[3] Garibaldi S, Merkurjev A, Serre J-P. Cohomological invariants in Galois cohomology. University Lecture Series. vol. 28. Providence, RI: American Mathematical Society; 2003.

[4] Garibaldi S, Petersson HP. Outer automorphisms of algebraic groups and a Skolem–Noether theorem for Albert algebras. Doc Math. 2016;21:917–54.

[5] Hooda N. Invariants Mod-2 and subgroups of G_2 and F_4. J Algebra. 2014;411:312–36.

[6] Hooda N. Embeddings of rank-2 tori in algebraic groups. J Pure Appl Algebra. 2018;222(10):3043–57.

[7] Hooda N, Thakur ML. Rational subgroups and invariants of F_4. Isr J Math, TBD, 2019, 1–47.

[8] Jacobson N. Structure and representations of Jordan algebras. AMS Colloquium publications. vol. 39. Providence, Rhode Island: AMS; 1968.

[9] Knus MA, Merkurjev A, Rost M, Tignol JP. The Book of Involutions. Colloquium Publications. vol. 44. Providence: AMS; 1998.

[10] MacDonald ML. Essential dimension of Albert algebras. Bull Lond Math Soc. 2014;46(5):906–14.

[11] McCrimmon K. A Taste of Jordan Algebras. Universitext. New York: Springer; 2004.

[12] Merkurjev AS. Essential dimension. Bull, New Ser, Am Math Soc. 2017;54(4):635–61.

[13] Parimala R, Sridharan R, Thakur ML. A classification theorem for Albert algebras. Trans Am Math Soc. 1998;350(3):1277–84.

[14] Petersson HP. Albert Algebra. Notes Available at the www.fields.utoronto.ca/programs/scientific/11-12/exceptional/Alb.-alg.-Ottawa-2012-Vii-new.pdf.

[15] Petersson HP. An embedding theorem for reduced Albert algebras over arbitrary fields. Commun Algebra. 2015;43(5):2062–88.

[16] Petersson HP, Racine M. A norm theorem for central simple algebras of degree 3. Arch Math. 1984;42:224–8.

[17] Petersson HP, Racine M. Albert algebras. In: Kaup W, McCrimmon, editors. Jordan algebras. vol. 3. Oberwolfach. 1992. Berlin: de Gruyter; 1994. p. 197–207.

[18] Petersson HP, Racine M. Enumeration and classification of Albert algebras: reduced models and the invariant mod 2. In: Non-associative algebra and its applications. Math Appl. vol. 303. Oviedo. 1993. Dordrecht: Kluwer Acad. Publ.; 1994. p. 334–40.

[19] Petersson HP, Racine M. An elementary approach to Serre–Rost invariant of Albert algebras. Indag Math. 1996;7(3):343–65.

[20] Racine M. Exceptional Jordan Algebras. Notes available at the https://www.fields.utoronto.ca/programs/scientific/12-13/torsors/lietheory/Notes/Racine.pdf.

[21] Racine ML. A note on quadratic Jordan algebras of degree 3. Trans Am Math Soc. 1972;164:93–103.

[22] Rost M. A (mod 3) invariant for exceptional Jordan algebras. C R Acad Sci, Sér 1 Math.

1991;313(12):823–7.
[23] Rost. On the classification of Albert algebras. 2002. Preprint. https://www.math.uni-bielefeld.de/~rost/data/albert.pdf.
[24] Serre J-P. Letter to M. L. Racine, 1991.
[25] Serre J-P. Cohomologie galoisienne: progrès et problèmes, Astérisque (1995), no. 227, 229–257, Séminaire Bourbaki. vol. 1993/94, Exp. 783 (= Oe. 166).
[26] Springer TA. Linear Algebraic Groups. 2nd ed. Progress in Mathematics. Boston: Birkhauser; 1998.
[27] Springer TA, Veldkamp FD. Octonions, Jordan Algebras and Exceptional Groups. Springer Monographs in Mathematics. Berlin: Springer; 2000.
[28] Thakur ML. Isotopy and invariants of Albert algebras. Comment Math Helv. 1999;74:297–305.
[29] Thakur ML. Kummer elements and the mod-3 invariant of Albert algebras. J Algebra. 2001;244(2):429–34.

David Joyner and Caroline Melles

Perspectives on p-ary bent functions

Abstract: Over a period of several years, the authors tried to generalize the Dillon and Bernasconi et al. characterizations of Boolean bent functions to the p-ary case, with various degrees of success. A lot of our work was based on a massive number of computational experiments and examples, usually done in SageMath and, for independent confirmation, in Mathematica. Our main theoretical results will be surveyed below.

Keywords: Bent function, strongly regular graph, association scheme

MSC 2010: 05E30

1 Introduction

Boolean bent functions, also called perfectly nonlinear functions, were born in the 1960s, though the first published paper did not appear until the 1970s (see Rothaus [21]). They have a wide variety of applications, from cryptography (e. g., constructing secure stream ciphers) to coding theory (e. g., constructing Kerdock codes) [4]. These bent functions are extremal combinatorial objects in the sense that they have (roughly speaking):

- maximal Walsh–Hadamard transform;
- maximal nonlinearity.

As is often the case in mathematics, extremal objects are especially interesting. Bent functions are no exception to this maxim. Indeed, Boolean bent functions are in natural bijection with:

- certain Hadamard difference sets (the Dillon correspondence);
- certain strongly regular graphs (the Bernasconi–Codenotti–VanderKam correspondence).

The authors discuss these correspondences more precisely below.

The p-ary analogs of these correspondences lack, at the present, even a conjectural formulation. To properly understand the p-ary analog, we introduce a number of new concepts, such as edge-weighted strongly regular graphs and weighted partial difference sets. Given these new concepts, we ask, for example, whether every p-ary

Note: Dedicated to our friend Tony Gaglione on his 70th birthday.

David Joyner, U. S. Naval Academy, Annapolis, MD 21402, USA (retired), e-mail: wdjoyner@gmail.com
Caroline Melles, Mathematics Department, U. S. Naval Academy, Annapolis, MD 21402, USA, e-mail: cgg@usna.edu

https://doi.org/10.1515/9783110638387-012

bent function gives rise to an edge-weighted strongly regular graph. The answer, it turns out, is no (see Examples 53 and 54). Problem 17 and other examples at the end of the paper illustrate some of the twists and turns involved. The authors spent years trying to generalize the Dillon and Bernasconi et al. characterizations of Boolean bent functions, with various degrees of success [7], [17], [19]. A lot of our work was based on a massive number of computational experiments and examples, usually done by David in SageMath and, for independent confirmation, by Caroline in Mathematica [8]. Our main theoretical results will be surveyed below.

2 The Boolean case

We first consider a Boolean function $f\colon GF(2)^n \to GF(2)$. For convenience, we index the components of vectors in $GF(2)^n$ starting at 0, that is, a vector $x \in GF(2)^n$ has components $x_0, x_1, \ldots, x_{n-1}$. We define the *support* of a vector in $GF(2)^n$ as

$$\mathrm{supp}(x) = \{i \mid 0 \le i \le n-1 \text{ and } x_i = 1\}.$$

For any two $x, y \in GF(2)^n$, let $d(x,y)$ denote the *Hamming metric*:

$$d(x,y) = \left|\{i \mid 0 \le i \le n-1 \text{ and } x_i \neq y_i\}\right|. \tag{1}$$

We define the *weight* wt of x to be the number of nonzero coordinates of x, so $d(x,y) = \mathrm{wt}(x-y)$ and $\mathrm{wt}(x) = |\mathrm{supp}(x)|$.

For a given positive integer n, we may identify a Boolean function

$$f\colon GF(2)^n \to GF(2),$$

with its support

$$\Omega_f = \mathrm{supp}(f) = \{x \in GF(2)^n \mid f(x) = 1\}.$$

Let

$$\omega = \omega_f = |\Omega_f|$$

denote the cardinality of the support and let $\mathbf{1}$ denote the all 1s vector. Let $\overline{f} = f + 1$ denote the complementary Boolean function. Note that

$$\Omega_f^c = \Omega_{\overline{f}},$$

where S^c denotes the complement of the set S in $GF(2)^n$.

The *Hamming distance* between two Boolean functions f and g on $GF(2)^n$ is

$$d(f,g) = \left|\{x \in GF(2)^n \mid f(x) \neq g(x)\}\right|.$$

A Boolean function $f\colon \mathrm{GF}(2)^n \to \mathrm{GF}(2)$ of the form

$$f(x_0, x_1, \dots, x_{n-1}) = a_0 x_0 + a_1 x_1 + \dots + a_{n-1} x_{n-1} + c,$$

where $a_0, a_1, \dots, a_{n-1}, c \in \mathrm{GF}(2)$, is called *affine*. The *nonlinearity* of f is its distance from the set $\mathcal{A}_n$ of all n-variable affine functions, that is,

$$\mathrm{nl}(f) = \min_{g \in \mathcal{A}_n} d(f, g).$$

If it is more convenient, a vector in $\mathrm{GF}(2)^n$ may also be identified with an integer in $\{0, 1, \dots, 2^n - 1\}$. Let

$$b : \{0, 1, \dots, 2^n - 1\} \to \mathrm{GF}(2)^n \tag{2}$$

be the binary representation ordered with least significant bit last (so that, e. g., $b(1) = (0, \dots, 0, 1) \in \mathrm{GF}(2)^n$).

For x and y in $\mathrm{GF}(2)^n$, we denote by $x \cdot y$ the scalar product of x and y. For $z \in \mathrm{GF}(2)$, the expression $(-1)^z$ has a well-defined meaning. Let H_n denote the $2^n \times 2^n$ Hadamard matrix defined by $(H_n)_{i,j} = (-1)^{b(i)\cdot b(j)}$, for each i, j such that $0 \le i, j \le n-1$. Inductively, these Hadamard matrices can be defined by

$$H_1 = \begin{pmatrix} 1 & 1 \\ 1 & -1 \end{pmatrix}, \quad \text{and} \quad H_n = \begin{pmatrix} H_{n-1} & H_{n-1} \\ H_{n-1} & -H_{n-1} \end{pmatrix}, \quad \text{for } n > 1.$$

The *Walsh–Hadamard transform* of f is defined to be the vector in $\mathbb{R}^{2^n}$ whose kth component is

$$(Wf)(b(k)) = \sum_{i \in \{0,1,\dots,2^n-1\}} (-1)^{b(i)\cdot b(k) + f(b(i))} = (H_n(-1)^f)_k,$$

where we define $(-1)^f$ as the column vector whose ith component is

$$(-1)^f_i = (-1)^{f(b(i))},$$

for $i = 0, \dots, 2^n - 1$.

Definition 1. For any Boolean function $f\colon \mathrm{GF}(2)^n \to \mathrm{GF}(2)$, we define the *Cayley graph* of f to be the undirected graph $\Gamma = \Gamma_f = (V, E)$ with vertices V and edges E given by

$$V = \mathrm{GF}(2)^n, \quad E = \{(v, w) \in V \times V \mid f(v + w) = 1\}.$$

We shall assume throughout and without further mention that

$$f(0) = 0,$$

so Γ has no loops and we may regard Γ as a simple graph. Indeed, Γ is an ω-regular graph having r connected components, where

$$r = |\mathrm{GF}(2)^n/\mathrm{Span}(\Omega_f)|.$$

For each vertex $v \in V$, the set of neighbors $N(v)$ of v is given by

$$N(v) = v + \Omega_f,$$

where v is regarded as a vector and the addition is induced by the usual vector addition in $\mathrm{GF}(2)^n$. Let $A = (A_{ij})$ be the $2^n \times 2^n$ adjacency matrix of Γ, so

$$A_{ij} = f(b(i) + b(j)), \quad 0 \le i, j \le 2^n - 1.$$

The *spectrum* of the graph Γ is the multiset of eigenvalues λ_k of the (symmetric) adjacency matrix A

$$\mathrm{Spectrum}(\Gamma) = \{\lambda_k \mid 0 \le k \le 2^n - 1\}. \tag{3}$$

For $k = 0, 1, \ldots, 2^n - 1$, let $w_k \in \{\pm 1\}^{2^n}$ be the column vector whose ith component is

$$(w_k)_i = (-1)^{b(k)\cdot b(i)}.$$

Each w_k is an eigenvector of A having eigenvalue $\lambda_k = (H_n \boldsymbol{f})_k$, where H_n is the nth Hadamard matrix, and $H_n \boldsymbol{f}$ means H_n times the vector $\boldsymbol{f}$ of values of f. The equation

$$\lambda_k = \frac{1}{2}(H_n \mathbf{1} - Wf)_k$$

demonstrates the affine relationship between the spectrum of Γ and the Walsh–Hadamard transform.

2.1 Monotone functions

Define an order $0 < 1$ on $\mathrm{GF}(2)$, and define a partial order $\le$ on $\mathrm{GF}(2)^n$ as follows: for each $v, w \in \mathrm{GF}(2)^n$, we say

$$v \le w$$

whenever we have $v_1 \le w_1, v_2 \le w_2, \ldots, v_n \le w_n$, or equivalently $\mathrm{supp}(v) \subseteq \mathrm{supp}(w)$. A Boolean function is called *monotone* (increasing) if whenever we have $v \le w$ then we also have $f(v) \le f(w)$.

For each $v \in \mathrm{GF}(2)^n$, define a monotone function $f = f_v$ to be *atomic based on* v if its support consists of all vectors greater than or equal to v, that is, if

$$\Omega_f = \{x \in GF(2)^n \mid v \leq x\},$$

where $\leq$ is the partial order defined above. We call a monotone function f *atomic* if there is some $v \neq 0$ such that f is atomic based on v.

A graph is said to be *singular* if its adjacency matrix has zero as an eigenvalue.

Theorem 2 ([6, Theorem 3.6]). *Let f be a Boolean atomic monotone function. The associated Cayley graph is singular if and only if the cardinality ω of the support of f is even.*

Definition 3. Let $f\colon GF(2)^n \to GF(2)$ be any monotone function. We say that $\Lambda \subset GF(2)^n$ is the *least support* of f if Λ consists of all vectors in Ω_f which are smallest in the partial ordering $\leq$ on $GF(2)^n$.

A monotone function is atomic if and only if it has only one vector in its least support.

Theorem 4 (Celerier [6]). *Let f be a monotone Boolean function whose least support vectors are given by $\Lambda \subset GF(2)^n$. Then*

$$f(x) = 1 + \prod_{v\in\Lambda}(x^v + 1). \tag{4}$$

The following theorem and conjecture appear in [5].

Theorem 5. *If $n \geq 5$ is odd and f is monotone, then* $\mathrm{nl}(f) \leq 2^{n-1} - 2^{(n-1)/2}$.

Conjecture 6. [1] *If f is monotone and n is a sufficiently large even number, then*

$$\mathrm{nl}(f) \leq 2^{n-1} - 2^{n/2}.$$

2.2 Bent functions

We define a Boolean function $f : GF(2)^n \to GF(2)$ to be *bent* if the absolute value of each component of its Walsh–Hadamard transform is $2^{n/2}$. Equivalently, a function f is bent if and only if its nonlinearity achieves the optimum $2^{n-1} - 2^{n/2-1}$.

Dillon's thesis [11] was one of the first publications to discuss the relationship between bent functions and combinatorial structures such as difference sets. His work concentrated on the Boolean case. Consider functions

$$f : GF(2)^n \to GF(2),$$

where $n > 1$ is an integer. In Dillon's work, it was proven that the "level curve" $f^{-1}(1)$ gives rise to a difference set in $GF(2)^n$ if f is bent.

1 If true, this would provide a 3rd proof that bent $\neq$ monotone. It also suggests, at least to us, that there are some "very nonlinear" monotone functions.

Definition 7. Let G be a finite abelian multiplicative group of order v, and let D be a subset of G with order k. We say D is a (v,k,λ)-*difference set* (DS) if the list of differences $d_1 d_2^{-1}$, for $d_1, d_2 \in D$, represents every nonidentity element in G exactly λ times.

A *Hadamard difference set* is one whose parameters are of the form $(4m^2, 2m^2 - m, m^2 - m)$ or $(4m^2, 2m^2 + m, m^2 + m)$, for some $m > 1$. It is, in addition, *elementary* if G is an elementary abelian 2-group (i. e., isomorphic to $(\mathbb{Z}/2\mathbb{Z})^n$, for some n).

Theorem 8 (Dillon correspondence, [11], Theorem 6.2.10, p. 78). *The function* $f : \mathrm{GF}(2)^n \to \mathrm{GF}(2)$ *with* $f(0) = 0$ *is bent if and only if* $f^{-1}(1)$ *is an elementary Hadamard difference set of* $\mathrm{GF}(2)^n$.

Definition 9. A connected graph $\Gamma = (V,E)$ is a (v,k,λ,μ)-*strongly regular graph* if:
- Γ has v vertices such that each vertex is connected to k other vertices.
- Distinct vertices g_1 and g_2 share edges with either λ or μ common vertices, depending on whether they are neighbors or not.

The following result (slightly sharpened by Stanica) is due to Bernasconi, Codenotti, and VanderKam.

Theorem 10 (BCV correspondence, [1], [2], [3], [22]). *Let* $f : \mathrm{GF}(2)^n \to \mathrm{GF}(2)$ *with* $f(0) = 0$. *The function* f *is bent if and only if the Cayley graph* Γ_f *of* f *is a strongly regular graph having parameters* $(2^n, k, \lambda, \lambda)$ *for some* λ, *where* $k = |\mathrm{supp}(f)|$. *The eigenvalues of* Γ_f *are* $\lambda_1 = \omega_f$, $\lambda_3 = -\lambda_2 = -\sqrt{\omega_f - \lambda}$, *with multiplicities* $m_1 = 1$,

$$m_2 = \frac{\sqrt{\omega_f - \lambda}(2^n - 1) - \omega_f}{2\sqrt{\omega_f - \lambda}}, \quad m_3 = \frac{\sqrt{\omega_f - \lambda}(2^n - 1) + \omega_f}{2\sqrt{\omega_f - \lambda}}.$$

3 The non-Boolean case

Let f be a $\mathrm{GF}(p)$-valued function on $\mathrm{GF}(p)^n$, where $p > 2$ is a prime.

We say f is *even* if $f(x) = f(-x)$ for all $x \in \mathrm{GF}(p)^n$. For simplicity of discussion, we assume in the following that f is even and $f(0) = 0$.

An *atomic p-ary function* is a function $\mathrm{GF}(p)^n \to \mathrm{GF}(p)$ supported at a single point. For $v \in \mathrm{GF}(p)^n$, the *atomic function* supported at v is the function $f_v : \mathrm{GF}(p)^n \to \mathrm{GF}(p)$ such that $f_v(v) = 1$ and, for every $w \in \mathrm{GF}(p)^n$ such that $w \neq v$, $f_v(w) = 0$. A polynomial representation of an atomic function may be obtained from the following theorem.

Theorem 11 (Celerier [7]). *Let* f_v *be an atomic p-ary function. Then*

$$f_v(x) = \prod_{i=0}^{n-1} \left(\frac{1}{(p-1)!} \prod_{j=1}^{p-1} (j + v_i - x_i) \right). \tag{5}$$

Every p-ary function may be expanded in terms of atomic functions (see [7, Corollary 6]). Therefore, every p-ary function has a polynomial representation. The polynomial representation of minimal degree is unique, and is called the *algebraic normal form* (ANF).

3.1 p-ary bent functions

A bent function can be most easily defined in terms of its Walsh transform.

Definition 12. The *Walsh transform* of a function $f : GF(p)^n \to GF(p)$ is a complex-valued function on $GF(p)^n$ that can be defined as

$$W_f(u) = \sum_{x \in GF(p)^n} \zeta^{f(x)-\langle u,x \rangle}, \tag{6}$$

where $\zeta = e^{2\pi i/p}$.

Remark 13. The following are some properties of the Walsh transform:
1. The Walsh coefficients satisfy *Parseval's equation*

$$\sum_{u \in GF(p)^n} |W_f(u)|^2 = p^{2n}.$$

2. If $\sigma = \sigma_k : \mathbb{Q}(\zeta) \to \mathbb{Q}(\zeta)$ is defined by sending $\zeta \mapsto \zeta^k$, then $W_f(u)^\sigma = W_{kf}(ku)$.

We call f *bent* if

$$|W_f(u)| = p^{n/2},$$

for all $u \in GF(p)^n$.

Definition 14. Suppose $f : GF(p)^n \to GF(p)$ is bent. We say f is *regular* if and only if $W_f(u)/p^{n/2}$ is a pth root of unity for all $u \in GF(p)^n$.

Proposition 15 (Hou [15]). *The degree of any bent function $f : GF(p)^n \to GF(p)$, when represented in algebraic normal form, satisfies*

$$\deg(f) \le \frac{n(p-1)}{2} + 1.$$

We call a map $f : GF(p)^n \to GF(p)$ *balanced* if the cardinalities of the "level curves" $|f^{-1}(x)|$ ($x \in GF(p)$) do not depend on x. We call the *signature* of $f : GF(p)^n \to GF(p)$ the list

$$|S_0|,\ |S_1|,\ |S_2|, \dots, |S_{p-1}|,$$

where, for each $a \in GF(p)$,

$$S_a = \{x \mid f(x) = a\} = f^{-1}(a). \tag{7}$$

The following result tells us that f is "balanced away from 0" if its Walsh transform at 0 is rational.

Lemma 16 ([7, Lemma 10]). *If $f : \mathrm{GF}(p)^n \to \mathrm{GF}(p)$ has the property that $W_f(0)$ is a rational number, then*

$$|S_1| = |S_2| = \cdots = |S_{p-1}|,$$

and

$$W_f(0) = |S_0| - |S_1|.$$

In particular,

$$\begin{aligned}|\mathrm{supp}(f)| &= |S_1| + |S_2| + \cdots + |S_{p-1}| \\ &= (p-1)|S_1| = (p-1)(|S_0| - W_f(0)).\end{aligned}$$

Therefore, $W_f(0) \in \mathbb{Q}$ implies $W_f(0) \in \mathbb{Z}$.

The *Cayley graph of f* is defined to be the regular edge-weighted graph

$$\Gamma_f = (\mathrm{GF}(p)^n, E_f), \tag{8}$$

whose vertex set is $V = V(\Gamma_f) = \mathrm{GF}(p)^n$ and whose set of edges is defined by

$$E_f = \{(u, v) \in \mathrm{GF}(p)^n \mid f(u - v) \neq 0\},$$

where the edge $(u, v) \in E_f$ has weight $f(u - v)$. Since f is even, we can (and do) regard Γ_f as a weighted (undirected) graph.

Problem 17. Some natural problems arise. For f even, which graph-theoretic properties of Γ_f can be tied to function-theoretic properties of f?

The following are other examples of natural problems:

1. Which edge-weighted Cayley graphs correspond to p-ary bent functions?
2. Classify the spectrum of Γ_f in terms of the values of the Walsh transform. (See [7, Section 3.3] for a classification of the spectrum in terms of the Fourier transform.)
3. Find necessary and/or sufficient conditions on f for Γ_f to be edge-weighted strongly regular (see Definition 23 below).
4. Can one algebraically (e. g., using spectral data) characterize edge-weighted strongly regular graphs?

Here's an example: It is easy to find necessary and sufficient conditions for Γ_f to be connected (and more generally find a formula for the number of connected components of Γ_f).

Definition 18. Let G be a finite abelian multiplicative group of order v, and let D be a subset of G with order k. The set D (or more precisely, the pair (G, D)) is a (v, k, λ, μ)-*partial difference set* (PDS) if the list of differences $d_1 d_2^{-1}, d_1, d_2 \in D$, represents every nonidentity element in D exactly λ times and every non-identity element in $G \setminus D$ exactly μ times.

Definition 19. Let (G, D) be a PDS. We say it is of *Latin square type* (resp., *negative Latin square type*) if there exist integers $N > 0$ and $R > 0$ (resp., $N < 0$ and $R < 0$) such that

$$(v, k, \lambda, \mu) = (N^2, R(N-1), N + R^2 - 3R, R^2 - R).$$

Let (G, D) be a PDS.

Definition 20. The *Cayley graph* $\Gamma = \Gamma(G, D)$ *associated to the PDS* (G, D) is a graph constructed as follows: let the vertices of the graph be the elements of the group G. Two vertices g_1 and g_2 are connected by a directed edge if $g_2 = dg_1$ for some $d \in D$.

Definition 21. Let W be a weight set of size $|W| = r$, and let $v \in \mathbb{Z}$, $k \in \mathbb{Z}^r$, $\lambda \in \mathbb{Z}^{r^3}$, and $\mu \in \mathbb{Z}^{r^2}$. Let G be a finite abelian multiplicative group of order v, and let D be a subset of G. Suppose that D is a disjoint union $D = D_1 \cup \cdots \cup D_r$, where D_j has order k_j, for $i \leq j \leq r$. We say D is a *weighted* (v, k, λ, μ)*-PDS* if the following properties hold:
- The list of "differences"

$$D_i D_j^{-1} = \{d_1 d_2^{-1} \mid d_1 \in D_i, d_2 \in D_j\},$$

  represents every nonidentity element of D_ℓ exactly $\lambda_{i,j,\ell}$ times and every nonidentity element of $G \setminus D$ exactly $\mu_{i,j}$ times $(1 \leq i, j, \ell \leq r)$.
- For each i, there is a j such that $D_i^{-1} = D_j$ (and if $D_i^{-1} = D_i$ for all i, then we say the weighted PDS is *symmetric*).

How does the above notion of a weighted PDS relate to the usual notion of a PDS?

Lemma 22. *Let* (G, D) *be a symmetric weighted PDS, with parameters* $(v, (k_i), (\lambda_{i,j,\ell}), (\mu_{i,j}))$*, where* $D = D_1 \cup \cdots \cup D_r$ *(disjoint union) is as above. If*

$$\sum_{i,j} \lambda_{i,j,\ell}$$

does not depend on ℓ*, for* $1 \leq \ell \leq r$*, then* D *is also an unweighted PDS with parameters* (v, k, λ, μ)*, where*

$$k = \sum_i k_i, \quad \lambda = \sum_{i,j} \lambda_{i,j,\ell}, \quad \mu = \sum_{i,j} \mu_{i,j}.$$

For an example where $\sum_{i,j} \lambda_{i,j,\ell}$ does depend on ℓ, see [8].

Let Γ be an edge-weighted graph with vertex set V, edge set E, and weight set $W = \mathrm{GF}(p)$.

For each $u \in V$, define:

- $N(u) = N_\Gamma(u)$ to be the set of all neighbors of u in Γ;
- $N(u,a) = N_\Gamma(u,a)$ to be the set of all neighbors v of u in Γ for which the edge $(u,v) \in E$ has weight a (for each $a \in GF(p)^\times = GF(p) - \{0\}$);
- $N(u,0) = N_\Gamma(u,0)$ to be the set of all nonneighbors v of u in Γ (i. e., we have $(u,v) \notin E$).

Suppose that f is a p-ary function, $f: GF(p)^n \to GF(p)$, and $\Gamma = \Gamma_f$ is the Cayley graph of f, with vertex set $V = GF(p)^n$. We define $\mathrm{supp}(f) = \{v \in V \mid f(v) \neq 0\}$ to be the *support* of f. It is clear that $\mathrm{supp}(f) = N(0)$ is the set of neighbors of the zero vector. More generally, for any $u \in V$,

$$N(u) = u + \mathrm{supp}(f), \tag{9}$$

where the last set is the collection of all vectors $u + v$, for some $v \in \mathrm{supp}(f)$. We can extend equation (9) to the more precise statement

$$N(u,a) = u + S_a, \tag{10}$$

for all $a \in GF(p)$. We call $N(u,a)$ the *a-neighborhood* of u.

Definition 23. Let Γ be a connected edge-weighted graph which is regular as a simple (unweighted) graph. Let $W = \{1,2,\dots,p-1\}$ be the set of weights. The graph Γ is called *edge-weighted strongly regular* with parameters v, $k = (k_a)_{a\in W}$, $\lambda = (\lambda_a)_{a\in W^3}$, $\mu = (\mu_a)_{a\in W^2}$, denoted $\mathrm{SRG}_W(v,k,\lambda,\mu)$, if it consists of v vertices such that, for each $a = (a_1,a_2) \in W^2$

$$|N(u_1,a_1) \cap N(u_2,a_2)| = \begin{cases} k_{a_1}, & u_1 = u_2, a_1 = a_2, \\ \lambda_{a_1,a_2,a_3}, & u_1 \in N(u_2,a_3),\ u_1 \neq u_2, \\ \mu_a, & u_1 \notin N(u_2),\ u_1 \neq u_2, \end{cases} \tag{11}$$

where $k = (k_a \mid a \in W) \in \mathbb{Z}^p$, $\lambda = (\lambda_a \mid a \in W^3) \in \mathbb{Z}^{p^3}$, $\mu = (\mu_a \mid a \in W^2) \in \mathbb{Z}^{p^2}$.

The above notion of an edge-weighted strongly regular graph is related to the usual notion of a strongly regular graph by the following result.

Lemma 24. *Let Γ be an edge-weighted strongly regular graph as in Definition 23, with edge-weights W and parameters $(v,(k_a),(\lambda_{a_1,a_2,a_3}),(\mu_{a_1,a_2}))$. If*

$$\sum_{(a_1,a_2)\in W^2} \lambda_{a_1,a_2,a_3}$$

does not depend on a_3, for $a_3 \in W$, then Γ is strongly regular (as an unweighted graph) with parameters (v,k,λ,μ) where

$$k = \sum_{a\in W} k_a, \quad \lambda = \sum_{(a_1,a_2)\in W^2} \lambda_{a_1,a_2,a_3}, \quad \mu = \sum_{(a_1,a_2)\in W^2} \mu_{a_1,a_2}.$$

For examples, see [8].

Let G be an abelian multiplicative group, and let $D \subset G$ be a nonempty subset such that $1 \notin D$ and with disjoint decomposition $D = D_1 \cup D_2 \cup \cdots \cup D_r$.

Definition 25. The *edge-weighted Cayley graph* $\Gamma = \Gamma(G, D)$ *associated to the pair* (G, D) is the edge-weighted graph constructed as follows: let the vertices of the graph be the elements of the group G. Two vertices g_1 and g_2 are connected by a directed edge of weight i if $g_2 = dg_1$ for some $d \in D_i$. If $D_i^{-1} = D_i$ for all i, the graph is undirected.

Theorem 26 ([7, Theorem 29]). *Let G be an abelian multiplicative group, and let $D \subset G$ be a nonempty subset such that $1 \notin D$ and with disjoint decomposition $D = D_1 \cup D_2 \cup \cdots \cup D_r$. The following are equivalent:*

(a) *(G, D) is a symmetric weighted partial difference set having parameters (v, k, λ, μ), where $v = |G|$, $k = \{k_i\}$ with $k_i = |D_i|$, $\lambda = \{\lambda_{i,j,\ell}\}$, and $\mu = \{\mu_{i,j}\}$.*

(b) *$\Gamma(G, D)$ is a strongly regular edge-weighted (undirected) graph with parameters (v, k, λ, μ) as in (a).*

The above result generalizes a well-known result in the unweighted case.

Question. In general, how are the multiplicities of the eigenvalues of the edge-weighted strongly regular graphs $\Gamma(G, D)$ related to the parameters (v, k, λ, μ) of the weighted PDS (G, D)?

We will see in Section 4 that there are functions $f\colon \mathrm{GF}(3)^3 \to \mathrm{GF}(3)$ such that f is bent and regular, yet the Cayley graph Γ_f is not edge-weighted strongly regular. We will also see in Section 4 that there are functions $f\colon \mathrm{GF}(5)^2 \to \mathrm{GF}(5)$ such that f is bent, but the level curves of f do not determine a weighted partial difference set.

Tan, Pott, and Feng [23] describe a relationship between relative difference sets and p-ary bent functions. A (p^n, p, p^n, λ)-relative difference set S in $\mathrm{GF}(p)^n \times \mathrm{GF}(p)$ canonically gives rise to a function $f_S\colon \mathrm{GF}(p)^n \to \mathrm{GF}(p)$ (see [23, p. 672]).

Definition 27. Let G be a finite group of order vm and let N be a subgroup of order m. A subset S of size k is called a (v, m, k, λ)*-relative difference set in G relative to N* if every element of $G \setminus N$ can be represented in the form $s_1 s_2^{-1}$, for s_1, s_2 in S, in exactly λ ways, and no nonidentity element of N may be represented this way.

Proposition 28 (Tan, Pott, and Feng [23, Proposition 2]). *If a subset S of $\mathrm{GF}(p)^n \times \mathrm{GF}(p)$ is a (p^n, p, p^n, λ)-relative difference set, then the corresponding function $f_S\colon \mathrm{GF}(p)^n \to \mathrm{GF}(p)$ is a bent function.*

However, we will concentrate on partial difference sets.

3.2 Association schemes

The following definition is standard, but we give [20] as a reference.

Definition 29. Let S be a finite set, and let $R_0, R_1, \dots, R_s$ denote binary relations on S (subsets of $S \times S$). The *dual* of a relation R is the set

$$R^* = \{(x,y) \in S \times S \mid (y,x) \in R\}.$$

Assume $R_0 = \Delta_S = \{(x,x) \in S \times S \mid x \in S\}$. We say $(S, R_0, R_1, \dots, R_s)$ is an *s-class association scheme on S* if the following properties hold:

- We have a disjoint union

$$S \times S = R_0 \cup R_1 \cup \cdots \cup R_s,$$

  with $R_i \cap R_j = \emptyset$ for all $i \neq j$.
- For each i, there is a j such that $R_i^* = R_j$ (and if $R_i^* = R_i$ for all i, then we say the association scheme is *symmetric*).
- For all i, j and all $(x,y) \in S \times S$, define

$$\rho_{ij}(x,y) = |\{z \in S \mid (x,z) \in R_i, (z,y) \in R_j\}|.$$

  For each k, and for all $(x,y) \in R_k$, the integer $\rho_{ij}(x,y)$ is a constant, denoted ρ_{ij}^k.

These constants ρ_{ij}^k are called the *intersection numbers* of the association scheme.

Next, we recall (see Herman [14]) the matrix-theoretic version of this definition.

Definition 30. Let S be a finite abelian multiplicative group of order m. Let $(S, R_0, \dots, R_s)$ denote a tuple consisting of S with relations R_i for which we have a disjoint union

$$S \times S = R_0 \cup R_1 \cup \cdots \cup R_s,$$

with $R_i \cap R_j = \emptyset$ for all $i \neq j$. Let $A_i \in \mathrm{Mat}_{m\times m}(\mathbb{Z})$ denote the adjacency matrix of R_i, for $i = 0, 1, \dots, s$.

We say that the subring of $\mathbb{Z}[\mathrm{Mat}_{m\times m}(\mathbb{Z})]$ generated by the A_i is an *adjacency ring* (also called the Bose–Mesner algebra) provided the set of adjacency matrices satisfies the following five properties:

- for each integer $i \in \{0, \dots, s\}$, A_i is a $(0,1)$-matrix;
- $\sum_{i=0}^{s} A_i = J$ (the all 1's matrix);
- for each integer $i \in \{0, \dots, s\}$, ${}^tA_i = A_j$, for some integer $j \in \{0, \dots, s\}$;
- there is a subset $K \subset G$ such that $\sum_{j\in K} A_j = I$, and
- there is a set of nonnegative integers $\{\rho_{ij}^k \mid i,j,k \in \{0, \dots, s\}\}$ such that

$$A_i A_j = \sum_{k=0}^{s} \rho_{ij}^k A_k,$$

  for all such i, j.

For his 2014 USNA honors thesis [26], Walsh sifted through lots of data, discovering a pattern regarding intersection numbers for Schur rings of adjacency matrices associated to bent functions. This advance led to conjectural connections between weighted PDSs and bent functions.

Weighted PDSs in the notation of Definition 21 naturally correspond to association schemes of class $r+1$.

Theorem 31. *Let G be a finite abelian multiplicative group, and let D be a subset of G such that $1 \notin D$ and such that D has a disjoint decomposition $D = D_1 \cup D_2 \cup \cdots \cup D_r$. Let $D_0 = \{1\}$, let $D_{r+1} = G \setminus (D \cup D_0)$, and let*

$$R_i = \{(g,h) \in G \times G \mid gh^{-1} \in D_i\}, \quad 0 \le i \le r+1.$$

The following statements are equivalent:
(a) *The set D is a symmetric weighted partial difference set.*
(b) *The graph $\Gamma(G,D)$ is an edge-weighted strongly regular graph with edge weights $\{1,2,\dots,r\}$.*
(c) *The tuple $(G, R_0, R_1, \dots, R_{r+1})$ is a symmetric association scheme of class $r+1$.*

Let $A = (a_{ij})$ denote the $N \times N$ weighted adjacency matrix of $\Gamma = \Gamma(G,D)$, where $i,j \in \{0,1,\dots,N-1\}$ and where

$$a_{ij} = \begin{cases} w, & \text{if } (i,j) \text{ is an edge of weight } w; \\ 0, & \text{if } (i,j) \text{ is not an edge of } \Gamma. \end{cases} \tag{12}$$

From the adjacency matrix A, we can derive weight-specific adjacency matrices as follows. For each weight w of Γ, let $A_w = (a(w)_{ij})$ denote the $N \times N$ $(1,0)$-matrix defined by

$$a(w)_{ij} = \begin{cases} 1, & \text{if } (i,j) \text{ is an edge of weight } w; \\ 0, & \text{if } (i,j) \text{ is not an edge of weight } w. \end{cases} \tag{13}$$

We sometimes extend the weight set of $\Gamma = \Gamma(G,D)$ by imposing the following conventions:
(a) If u and v are distinct vertices of Γ but (u,v) is not an edge of Γ, then we say the *weight* of (u,v) is $w = r+1$.
(b) If $u = v$ is a vertex of Γ (so (u,v) is not an edge, since Γ has no loops), then we say the *weight* of (u,v) is $w = 0$.

This allows us to extend the set of weight-specific adjacency matrices given by Equation (13) to the set $A_0, A_1, \dots, A_r, A_{r+1}$.

Corollary 32. *We use the notation of Theorem 31. The graph $\Gamma(G,D)$ is an edge-weighted strongly regular graph if and only if the (extended) set of weight-specific adjacency matrices form a Bose–Mesner algebra with $K = \{0\}$.*

If the level curves of f form a symmetric weighted PDS, then the edge-weighted Cayley graph corresponding to f agrees with the edge-weighted strongly regular graph associated with the symmetric weighted PDS.

Theorem 33. *We use the notation of Theorem 31. Let $f\colon \mathrm{GF}(p)^n \to \mathrm{GF}(p)$ be an even function such that $f(0) = 0$. Let $G = \mathrm{GF}(p)^n$, and let $D_i = f^{-1}(i)$, for $i = 1, 2, \ldots, p-1$, let $D = D_1 \cup D_2 \cup \cdots \cup D_{p-1}$. If (G, D) is a symmetric weighted partial difference set, then the associated edge-weighted strongly regular graph is the edge-weighted Cayley graph of f, that is, $\Gamma(G, D) = \Gamma_f$.*

Consequently, if $f\colon \mathrm{GF}(p)^n \to \mathrm{GF}(p)$ is an even function with $f(0) = 0$, then saying that the level curves of f give rise to a symmetric weighted PDS is equivalent to saying that the level curves determine a symmetric p-class association scheme. When we refer to the intersection numbers p_{ij}^k of a symmetric weighted PDS, we mean the intersection numbers of the corresponding symmetric association scheme.

Definition 34. Suppose that S is a finite set, and $\{R_0, R_1, \ldots R_s\}$ is a collection of binary relations on S. A collection of binary relations $\{T_0, T_1, \ldots, T_u\}$ is called a *fusion* of $\{R_0, R_1, \ldots R_s\}$ if each T_i is a union of elements of $\{R_0, R_1, \ldots R_s\}$.

An association scheme $(S, R_0, R_1, \ldots, R_s)$ is called *amorphic* if, for every fusion $\{T_0, T_1, \ldots, T_u\}$ of $\{R_0, R_1, \ldots, R_s\}$, $(S, T_0, T_1, \ldots, T_u)$ is also an association scheme.

A symmetric weighted PDS is called *amorphic* if the symmetric association scheme it determines is amorphic.

See Theorem 42 for an example of a relationship between amorphic association schemes and bent functions.

Definition 35. Let $\Gamma = \Gamma(G, D)$ be the edge-weighted Cayley graph associated with a symmetric weighted PDS. We say that Γ is *amorphic* if (G, D) is amorphic, that is, if the association scheme determined by (G, D) is amorphic. If $f\colon \mathrm{GF}(p)^n \to \mathrm{GF}(p)$ is an even function with $f(0) = 0$, then we call f *amorphic* if the level curves D_i of f determine an amorphic association scheme, that is, its associated Cayley graph $\Gamma(G, D)$ is amorphic.[2]

We now describe how amorphic association schemes are essentially equivalent to graph decompositions of a complete graph which are strongly regular of Latin square or negative Latin square type.

Definition 36. A *graph decomposition* of an edge-weighted graph Γ, for this paper, means the graph decomposition determined by the collection $\{\Gamma_a\}$, where for each weight a, we define Γ_a to be the graph with the same vertices as Γ, whose edges are the edges of weight a. There is a corresponding graph decomposition of the complete graph on the vertex set of Γ, consisting of the graphs Γ_a and the complement of Γ. A

2 And thus edge-weighted strongly regular, by Theorem 31.

graph decomposition is said to be a *strongly regular graph decomposition* (see [24]) if the individual graphs of the decomposition are all strongly regular.

The following proposition is a consequence of a theorem from [12] on amorphic association schemes (which we quote from van Dam and Muzychuk [25]).

Proposition 37 (Gol'fand, Ivanov, Klin [12]). *Let $\Gamma = \Gamma(G, D)$ be the edge-weighted Cayley graph associated with a symmetric weighted PDS. Suppose that the corresponding graph decomposition of the complete graph on the vertices of Γ consists of at least 3 (nonempty) graphs. If Γ is amorphic, then the graphs of the decomposition are all strongly regular, and either all of Latin square type, or all of negative Latin square type.*

Remark 38. Suppose that p is a prime greater than 2, and $f\colon \mathrm{GF}(p)^n \to \mathrm{GF}(p)$ is an even function with $f(0) = 0$. Let $\Gamma(G, D)$ be the associated Cayley graph of f. If $\Gamma(G, D)$ is amorphic, and if the complement of $\Gamma(G, D)$ in the complete graph on p^n vertices is not empty, then this complement is strongly regular (and of positive or negative Latin square type). Since the complement of a strongly regular graph is also strongly regular, it follows that $\Gamma(G, D)$, when regarded as an unweighted graph, is strongly regular.

The following result is a consequence of a theorem of van Dam [24, Theorem 3].

Proposition 39 (van Dam). *Let $f\colon \mathrm{GF}(p)^n \to \mathrm{GF}(p)$ be an even function with $f(0) = 0$. If the decomposition by weights of the edge-weighted Cayley graph of f and its complement form a strongly regular graph decomposition of the complete graph, such that the individual graphs are all of Latin square type or all of negative Latin square type, then f is an amorphic function.*

To describe the intersection numbers of an amorphic association scheme corresponding to a p-ary function f in terms of the Latin square parameters of the associated graph decomposition, we will use the following result of Van Dam and Muzychuck [25], based on work of Ito, Munemasa, and Yamada [16]. If $(\mathrm{GF}(p)^n, D_1, D_2, \ldots, D_{p-1})$ is a weighted partial difference set, we will use the notation

$$D_p = \mathrm{GF}(p)^n \setminus \{0\} \cup D_1 \cup \cdots \cup D_{p-1}.$$

Theorem 40. *Let n be even, and suppose that $(\mathrm{GF}(p)^n, D_1, D_2, \ldots, D_{p-1})$ is an amorphic weighted partial difference set, such that*

$$|D_i| = r(N-1) \quad \textit{for } 1 \le i \le p-1, \quad \textit{and} \quad |D_p| = r_p(N-1),$$

where

$$N = \pm p^{\frac{n}{2}}, \quad \textit{and} \quad r = \frac{N}{p}, \quad \textit{and} \quad r_p = r + 1.$$

Then the intersection numbers of the corresponding association scheme are:

(a) $\rho^p_{pp} = N + r_p^2 - 3r_p$;
(b) $\rho^i_{ii} = N + r^2 - 3r$ *if* $1 \le i \le p-1$;
(c) $\rho^i_{pp} = r_p(r_p - 1)$ *if* $1 \le i \le p-1$;
(d) $\rho^i_{jj} = r(r-1)$ *if* i *and* j *are distinct and* $1 \le i \le p$ *and* $1 \le j \le p-1$;
(e) $\rho^p_{jp} = \rho^p_{pj} = r^2$ *if* $1 \le j \le p-1$;
(f) $\rho^i_{ij} = \rho^i_{ji} = r(r-1)$ *if* i *and* j *are distinct and* $1 \le i, j \le p-1$;
(g) $\rho^i_{pi} = \rho^i_{ip} = r^2 - 1$ *if* $1 \le i \le p-1$;
(h) $\rho^i_{jp} = \rho^i_{pj} = r_p(r_p - 1)$ *if* i *and* j *are distinct and* $1 \le i, j \le p-1$.
(i) $\rho^i_{jk} = \rho^i_{kj} = r^2$ *if* $i, j,$ *and* k *are distinct and* $1 \le i \le p$ *and* $1 \le j, k \le p-1$;

3.3 Regular and weakly regular bent functions

Recall from Definition 14 that a bent function $f : GF(p)^n \to GF(p)$ is called regular if, for all $u \in GF(p)^n$, the Walsh transform $W_f(u)$ is of the form $\zeta^b p^{n/2}$ for some integer b, where $\zeta = e^{2\pi i/p}$.

Definition 41. If f is regular, then there is a function $f^* : GF(p)^n \to GF(p)$, called the *dual* (or *regular dual*) of f, such that $W_f(u) = \zeta^{f^*(u)} p^{n/2}$, for all $u \in GF(p)^n$. We call f *weakly regular,*[3] if there is a function $f^* : GF(p)^n \to GF(p)$, called the *dual* (or *μ-regular dual*) of f, such that $W_f(u) = \mu\zeta^{f^*(u)} p^{n/2}$, for some constant $\mu \in \mathbb{C}$ with absolute value 1.

In the case that $p = 3$, more is known about the relationship between amorphic association schemes and weakly regular bent functions.

Theorem 42 (Tan, Pott, Feng [23, Theorem 3]). *Let* $f: GF(3)^{2m} \to GF(3)$ *be an even function such that* $f(0) = 0$. *If* f *is a weakly regular bent function, then the level curves of* f *determine an amorphic association scheme.*

Proposition 43 (Hou [15]). *The degree of any weakly regular bent function* $f : GF(p)^n \to GF(p)$, *when represented in algebraic normal form, satisfies*

$$\deg(f) \le \frac{n(p-1)}{2}.$$

Suppose that $f: GF(p)^n \to GF(p)$, and suppose that $\phi \in GL(n,p)$ is a linear transformation. Let $g(x) = f(\phi(x))$.
- The functions f and g both have the same signature.
- For any $u \in GF(p)^n$, $W_g(u) = W_f({}^t(\phi^{-1})u)$, where t denotes transpose.
- If f is bent, so is g.

3 If μ is fixed and we want to be more precise, we call this *μ-regular*.

- If f is bent and regular, so is g.
- Suppose that f is bent and weakly regular, with μ-regular dual f^*.
  - In this case, f^* is bent and weakly regular, with μ^{-1}-regular dual f^{**} given by $f^{**}(x) = f(-x)$. If f is also even, then f^* is even and $f^{**} = f$.
  - In this case, g is bent and weakly regular, with μ-regular dual g^*, where $g^*(u) = f^*({}^t(\phi^{-1})u)$.

Proposition 44 (Kumar, Scholtz, Welch). *If f is bent then there are functions $f_* : \mathrm{GF}(p)^n \to \mathbb{Z}$ and $f^* : \mathrm{GF}(p)^n \to \mathrm{GF}(p)$ such that*

$$W_f(u)p^{-n/2} = \begin{cases} (-1)^{f_*(u)}\zeta^{f^*(u)}, & \text{if } n \text{ is even, or } n \text{ is odd and } p \equiv 1 \pmod 4, \\ i^{f_*(u)}\zeta^{f^*(u)}, & \text{if } n \text{ is odd and } p \equiv 3 \pmod 4. \end{cases}$$

The above result is known (see [18]), but the form above is due to Helleseth and Kholosha [13] (although we made a minor correction to their statement).

Lemma 45. *Suppose $f : \mathrm{GF}(p)^n \to \mathrm{GF}(p)$ is bent. The following are equivalent:*
- *f is weakly regular.*
- *$W_f(u)/W_f(0)$ is a pth root of unity for all $u \in \mathrm{GF}(p)^n$.*

Suppose $f : \mathrm{GF}(p)^n \to \mathrm{GF}(p)$ is bent and weakly regular. The following are equivalent:
- *f is regular.*
- *$W_f(0)/p^{n/2}$ is a pth root of unity.*

For a proof, see, for example, Lemmas 6.4.5 and 6.4.6 of [17].

Remark 46. Weakly regular bent functions have some interesting relationships to strongly regular graphs and partial difference sets.
- In Chee, Tan, and Zhang [10], it is shown that if n is even, then the unweighted Cayley graphs of homogeneous[4] weakly regular even bent functions $f : \mathrm{GF}(p)^n \to \mathrm{GF}(p)$, with $f(0) = 0$, are strongly regular.
- In Pohill, Tan, Feng, and Ling [20, Corollary 3], it is shown, for all such homogeneous weakly regular bent functions, the level curves give rise to a weighted PDS, the weighted PDS corresponds to an association scheme, and the dual association scheme corresponds to the dual bent function.
- In Tan, Pott, and Feng [23, Theorem 1], it is shown that an even bent function $f : \mathrm{GF}(3)^{2m} \to \mathrm{GF}(3)$ with $f(0) = 0$ is weakly regular if and only if the level curves D_1 and D_2 are partial difference sets of certain sizes.

4 When regarded as a function $f_* : \mathrm{GF}(p^n) \to \mathrm{GF}(p)$ via a "nice" $\mathrm{GF}(p)$-linear isomorphism $\psi : \mathrm{GF}(p^n) \to \mathrm{GF}(p)^n$. By "nice," we mean the trace $\mathrm{tr} : \mathrm{GF}(p^n) \to \mathrm{GF}(p)$ satisfies $\mathrm{tr}(xy) = \langle \psi(x), \psi(y) \rangle$.

3.4 Sizes of level curves

Let $f\colon \mathrm{GF}(p)^{2m} \to \mathrm{GF}(p)$ be an even function such that $f(0) = 0$, where p is a prime greater than 2. Let $D_i = f^{-1}(i)$, for $1 \le i \le p-1$. Let $D_0 = \{0\}$, and let $D_p = f^{-1}(0) \setminus \{0\}$.

Recall from Lemma 16 that if $W_f(0)$ is rational, then the $|D_i|$ all have the same size for $1 \le i \le p-1$.

In particular, this result applies in the case that f is regular and bent.

Lemma 47. *If f is regular bent, then the $|D_i|$ all have the same size for $1 \le i \le p-1$. Moreover, the dual function f^* satisfies $f^*(0) = 0$.*

Proof. Since f is bent, $|W_f(x)| = p^m$. Since f is regular, $W_f(0) = \zeta^b p^m$, for some $b \in \mathbb{Z}$. Therefore,

$$1 + |D_p| + \sum_{i=1}^{p-1} |D_i|\zeta^i = \zeta^b p^m,$$

or

$$1 + |D_p| = \zeta^b p^m - \sum_{i=1}^{p-1} |D_i|\zeta^i.$$

If $\sigma = \sigma_\ell \in \mathrm{Gal}(\mathbb{Q}(\zeta)/\mathbb{Q})$ satisfies $\sigma(\zeta) = \zeta^\ell$, then $\sigma(W_f(x)) = W_{\ell f}(\ell x)$ (see part 2 of Remark 13). Applying σ gives, for each $1 \le \ell \le p-1$,

$$1 + |D_p| = \zeta^{\ell b} p^m - \sum_{i=1}^{p-1} |D_i|\zeta^{\ell i}.$$

Subtracting the equations for $\ell = 1$ and $\ell = j$ gives an equation of the form

$$\sum_{i=1}^{p-1} a_{i,j}\zeta^i = 0.$$

The $a_{i,j}$s are mostly differences of the form $|D_u| - |D_v|$ but some (if $b \neq 0$) also have a $\pm p^m$ term. In any case, applying σ_ℓ, for $0 \le \ell \le p-1$, again to this gives a system of equations which can be written in matrix form $M\boldsymbol{a} = \mathbf{0}$, where M is a Vandermonde matrix. Using the Vandermonde determinant formula, we see $\boldsymbol{a} = \mathbf{0}$, so $a_{i,j} = 0$. If $b = 0$, this means that $W_f(0) = p^m$ and the $|D_i|$ are all the same, as desired, by Lemma 16. If $b \neq 0$, then we must have $\pm p^m \pm |D_u| \pm |D_v| = 0$, for some u, v with $1 \le u, v \le p-1$. This is impossible, since $|D_u|$ and $|D_v|$ are even (since f is an even function) and p^m is odd. Since $f^*(0) = b$ (by definition), we have $f^*(0) = 0$. □

The following proposition on the sizes of level sets is a consequence of results in [19].

Proposition 48. *Suppose that $f : GF(p)^n \to GF(p)$ is an even bent function such that $f(0) = 0$, where $p > 2$ is a prime and $n \geq 2$ is an even integer. Suppose also that the level curves $D_i = \{x \in GF(p)^n \mid f(x) = i\}$, for $1 \leq i \leq p-1$, and $D_p = \{x \in GF(p)^n \setminus \{0\} \mid f(x) = 0\}$ have sizes that are multiples of $p^{\frac{n}{2}} - 1$. Let*

$$s_i = |D_i|,$$

for $1 \leq i \leq p$.

Then

$$W_f(0) = \pm p^{n/2},$$

and consequently

$$s_1 = s_2 = \cdots = s_{p-1}.$$

Similarly, if the level curves D_i for $1 \leq i \leq p-1$ and D_p have sizes that are multiples of $p^{\frac{n}{2}} + 1$, then the same conclusions hold.

In particular, if such an f is weakly regular, then it is regular if and only if $W_f(0) > 0$.

The following result, which refines Proposition 48, will likely appear in Melles–Joyner [19].

Theorem 49. *Let $f : GF(p)^n \to GF(p)$ be an even function such that $f(0) = 0$, where $p > 2$ is a prime and $n \geq 2$ is an even integer. Let $D_i = f^{-1}(i)$, for $1 \leq i \leq p-1$ (where we identify $\{1, 2, \ldots, p-1\}$ with elements of $GF(p)$), let $D_0 = \{0\}$, and let $D_p = GF(p)^n \setminus D_0 \cup D_1 \cup \cdots \cup D_{p-1}$. Let*

$$N = \pm p^{\frac{n}{2}}, \quad \textit{and} \quad r = \frac{N}{p}, \quad \textit{and} \quad r_p = r + 1.$$

Suppose that $(GF(p)^n, D_0, D_1, D_2, \ldots, D_p)$ is an amorphic weighted partial difference set (i. e., a weighted partial difference set such that the corresponding association scheme is amorphic) with

- *$|D_i| = r(N-1)$ for $1 \leq i \leq p-1$, and*
- *$|D_p| = r_p(N-1)$.*

Then f is bent.

Remark 50. Writing out the $N = p^{\frac{n}{2}}$ case (Latin square type) and the $N = -p^{\frac{n}{2}}$ case (negative Latin square type) in the theorem above, we have the following possibilities:

(a) Suppose that the level curves $D_i = \{x \in GF(p)^n \mid f(x) = i\}$ for $1 \leq i \leq p-1$ and $D_p = \{x \in GF(p)^n \setminus \{0\} \mid f(x) = 0\}$ have sizes that are multiples of $p^{\frac{n}{2}} - 1$. Then

$$|D_i| = p^{\frac{n}{2}-1}(p^{\frac{n}{2}} - 1)$$

and

$$|D_p| = (p^{\frac{n}{2}-1} + 1)(p^{\frac{n}{2}} - 1).$$

(b) Suppose that the level curves $D_i = \{x \in GF(p)^n \mid f(x) = i\}$ for $1 \leq i \leq p-1$ and $D_p = \{x \in GF(p)^n \setminus \{0\} \mid f(x) = 0\}$ have sizes that are multiples of $p^{\frac{n}{2}} + 1$. Then

$$|D_i| = p^{\frac{n}{2}-1}(p^{\frac{n}{2}} + 1)$$

and

$$|D_p| = (p^{\frac{n}{2}-1} - 1)(p^{\frac{n}{2}} + 1).$$

Theorem 49 above more or less implies the following.

Remark 51. Let f and $D_1, D_2, \ldots, D_{p-1}$ be defined as in Theorem 49 above. Any two of the following possibilities implies the third.

(a) The graphs in the graph decomposition corresponding to the level curves D_i (as in Proposition 37) are all strongly regular of Latin square type (or all strongly regular of negative Latin square type).
(b) The function f is amorphic.
(c) The function f is bent.

See Propositions 37 and 39 for connections between (a) and (b).

Question. Are the bent functions which arise from (a) and (b) normal (in the sense of [9])?

4 Examples

We will see examples showing the following.

(a) The Cayley graph of a function can be edge-weighted strongly regular, yet the function need not be bent.
(b) The individual graphs of the decompositions corresponding to two bent functions can be isomorphic, yet one function may be weakly regular and the other not.
(c) There are regular bent functions whose Cayley graphs are not edge-weighted strongly regular.

Example 52. Consider the 3-ary functions on $GF(3)^2$ given by

$$f(x_0, x_1) = x_0^2 x_1^2 + 2x_0^2 + x_0 x_1 + 2x_1^2 \quad \text{and} \quad g(x_0, x_1) = 2x_0 x_1.$$

The function g is bent but the function f is not. The Cayley graphs of both f and g are edge-weighted strongly regular. Each function determines a decomposition of the complete graph on 9 vertices into three strongly regular graphs. The decomposition $\Gamma_{f,1}$, $\Gamma_{f,2}$, $\Gamma_{f,3}$ determined by f is shown in Figure 1. The graph $\Gamma_{f,i}$ corresponds to the set $f^{-1}(i)$, for $i = 1, 2$. The graph $\Gamma_{f,3}$ corresponds to the set $f^{-1}(0)\backslash\{0\}$. The graphs Γ_1 and Γ_3 are strongly regular with parameters $(9, 2, 1, 0)$. The graph Γ_2 is strongly regular with parameters $(9, 4, 1, 2)$. The decomposition determined by the bent function g consists of the same three graphs, but with the last two interchanged, that is, $\Gamma_{g,1}$ is isomorphic to $\Gamma_{f,1}$, $\Gamma_{g,2}$ is isomorphic to $\Gamma_{f,3}$, and $\Gamma_{g,3}$ is isomorphic to $\Gamma_{f,2}$. The bent function g is regular (and weakly regular).

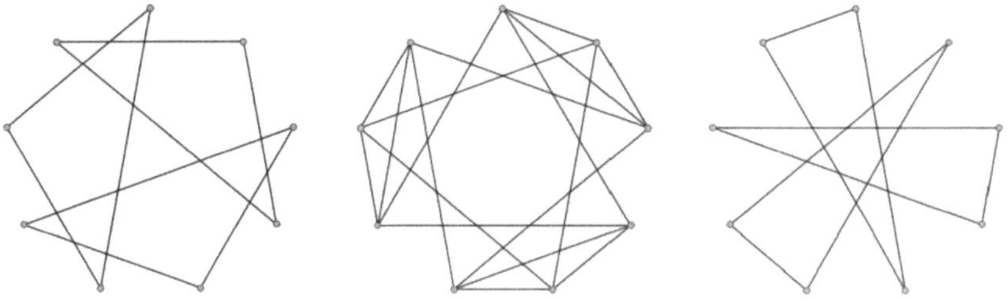

$\Gamma_{f,1} \cong \Gamma_{g,1}$ $\Gamma_{f,2} \cong \Gamma_{g,3}$ $\Gamma_{f,3} \cong \Gamma_{g,2}$

Figure 1: Graph decompositions for the 3-ary functions $f(x_0, x_1) = x_0^2x_1^2 + 2x_0^2 + x_0x_1 + 2x_1^2$ (not bent) and $g(x_0, x_1) = 2x_0x_1$ (bent).

Example 53. Consider the bent 3-ary functions on $\mathrm{GF}(3)^3$ given by

$$f(x_0, x_1, x_2) = 2x_0x_2 + x_1^2 + x_0^2x_1^2 \quad \text{and} \quad g(x_0, x_1, x_2) = x_0x_1 + x_2^2.$$

The functions f and g determine graphs $\Gamma_{f,i}$ and $\Gamma_{g,i}$, for $i = 1, 2, 3$, such that $\Gamma_{f,i}$ is isomorphic to $\Gamma_{g,i}$ for each i. Yet the weighted Cayley graph determined by g is edge-weighted strongly regular, but the weighted Cayley graph determined by f is not. The bent function g is weakly regular, but the bent function f is not:

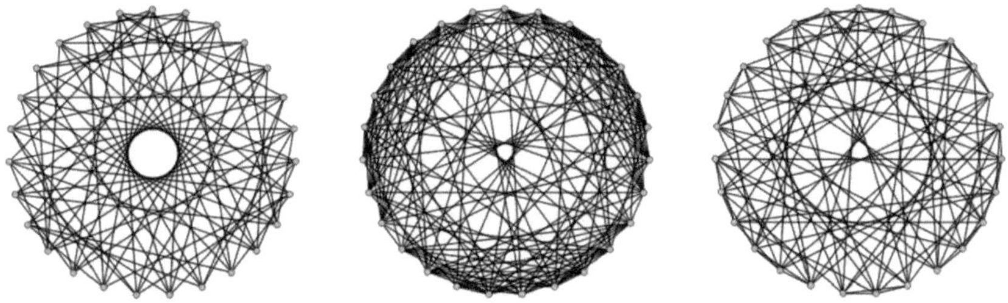

$\Gamma_{f,1} \cong \Gamma_{g,1}$ $\Gamma_{f,2} \cong \Gamma_{g,2}$ $\Gamma_{f,3} \cong \Gamma_{g,3}$

Figure 2: Graph decompositions for the 3-ary bent functions $f(x_0, x_1, x_2) = x_0x_2 + 2x_1^2 + 2x_0^2x_1^2$ (not weakly regular) and $g(x_0, x_1, x_2) = x_0x_1 + x_2^2$ (weakly regular).

Example 54. Consider the bent 5-ary function $f: \mathrm{GF}(5)^2 \to \mathrm{GF}(5)$ given by

$$f(x_0, x_1) = x_0^4 + 2x_0x_1.$$

The function f is regular, but its weighted Cayley graph is not edge-weighted strongly regular.

Example 55. Consider the bent 5-ary function $f: \mathrm{GF}(5)^2 \to \mathrm{GF}(5)$ given by

$$f(x_0, x_1) = -x_0^3x_1 + 2x_1^4.$$

The function f is regular, and its weighted Cayley graph $\Gamma = \Gamma_f$ is edge-weighted strongly regular. The function f determines a decomposition Γ_1, Γ_2, Γ_3, Γ_4, Γ_5 of the complete graph on 25 vertices (see Figure 3), where the graph Γ_i corresponds to the set $f^{-1}(i)$, for $i = 1, 2, 3, 4$, and the graph Γ_5 corresponds to the set $f^{-1}(0) \setminus \{0\}$. The graphs Γ_1, Γ_2, Γ_3, and Γ_4 are strongly regular with parameters $(25, 4, 3, 0)$. The graph Γ_5 is strongly regular with parameters (25, 8, 3, 2).

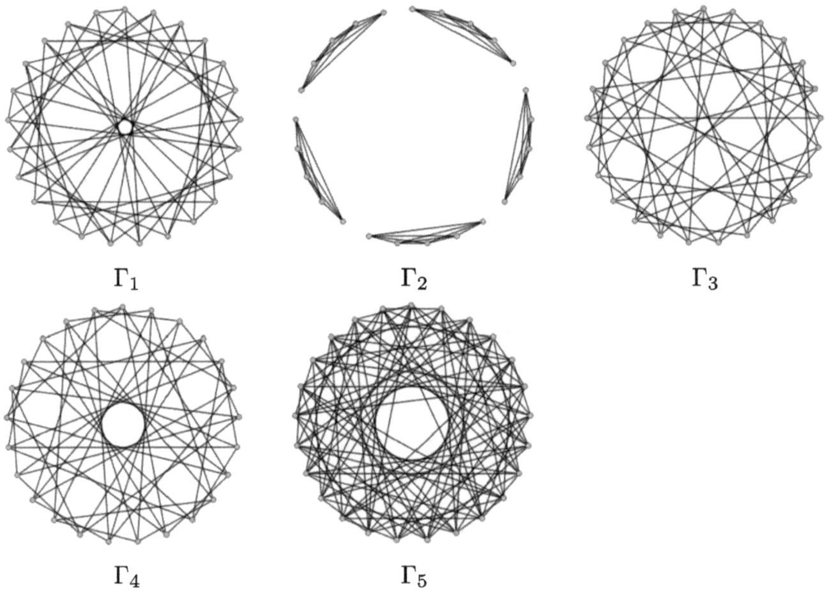

Figure 3: Graph decomposition for the bent 5-ary function $-x_0^3x_1 + 2x_1^4$.

Example 56. Consider the bent 5-ary function $f: \mathrm{GF}(5)^2 \to \mathrm{GF}(5)$ given by

$$f(x_0, x_1) = -x_0^2 + 2x_1^2.$$

The function f is weakly regular, but not regular. Its weighted Cayley graph $\Gamma = \Gamma_f$ is edge-weighted strongly regular. The function f determines the decomposition of the complete graph on 25 vertices shown in Figure 4. The graph Γ_5 corresponding to the set $f^{-1}(0) \setminus \{0\}$ is empty. The graphs of the decomposition are not strongly regular, but the union of each pair of graphs diagonally opposite to each other in the diagram is strongly regular with parameters $(25, 12, 5, 6)$.

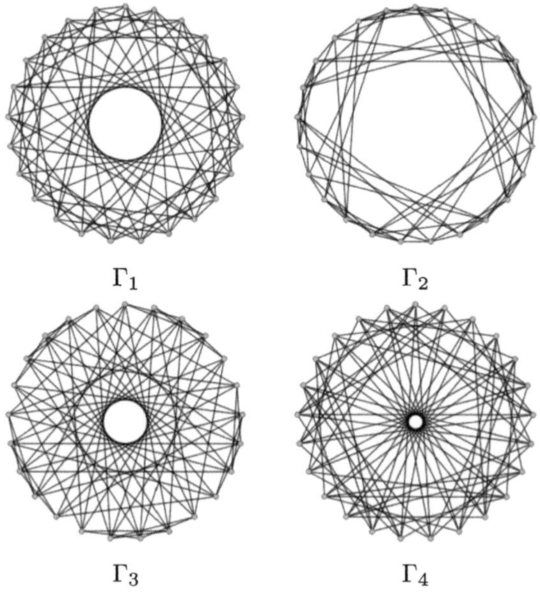

Figure 4: Graph decomposition for the bent 5-ary function $-x_0^2 + 2x_1^2$.

Bibliography

[1] Bernasconi A. Mathematical techniques for the analysis of Boolean functions. Ph. D. dissertation TD-2/98. Università di Pisa-Udine; 1998.

[2] Bernasconi A, Codenotti B. Spectral analysis of Boolean functions as a graph eigenvalue problem. IEEE Trans Comput. 1999;48(3):345–51. http://ilex.iit.cnr.it/codenotti/ps_files/graph_fourier.ps.

[3] Bernasconi A, Codenotti B, VanderKam JM. A characterization of bent functions in terms of strongly regular graphs. IEEE Trans Comput. 2001;50(9):984–5.

[4] Carlet C. Boolean functions for cryptography and error-correcting codes. In: Crama Y, Hammer PL, editors. Boolean Models and Methods in Mathematics, Computer Science, and Engineering. Cambridge: Cambridge University Press; 2010. p. 257–397. http://www1.spms.ntu.edu.sg/~kkhoongm/chap-fcts-Bool.pdf.

[5] Carlet C, Joyner D, Stănică P, Tang D. Cryptographic properties of monotone Boolean functions. J Math Cryptol. 2016;10:1–14.

[6] Celerier C, Joyner D, Melles C, Phillips D. On the Walsh-Hadamard transform of monotone Boolean functions. Tbilisi Math J (Special Issue on Sage and Research). 2012;5:19–35. http://tcms.org.ge/Journals/TMJ/Volume5_2/Volume5_2.html.
[7] Celerier C, Joyner D, Melles C, Phillips D, Walsh S. Edge-weighted Cayley graphs and p-ary bent functions. Integers. 2016;16:A35.
[8] Celerier C, Joyner D, Melles C, Phillips D, Walsh S. Explorations of edge-weighted Cayley graphs and p-ary bent functions. Preprint. https://arxiv.org/abs/1406.1087.
[9] Cesmelioglu A, Meidl W, Pott A. Generalized Maiorana-McFarland class and normality of p-ary bent functions. Finite Fields Appl. 2013;24:105–17.
[10] Chee Y, Tan Y, Zhang X. Strongly regular graphs constructed from p-ary bent functions. 2010. Preprint. http://arxiv.org/abs/1011.4434.
[11] Dillon JF. Elementary Hadamard Difference Sets. Ph. D. thesis. University of Maryland; 1974.
[12] Gol'fand Ja, Ivanov A, Klin M. Amorphic cellular rings. In: Faradžev I, Ivanov A, Klin M, Woldar A, editors. Investigations in Algebraic Theory of Combinatorial Objects. Dordrecht: Kluwer; 1994. p. 167–86.
[13] Helleseth T, Kholosha A. Monomial and Quadratic Bent Functions over the Finite Fields of Odd Characteristic. IEEE Trans Inf Theory. 2006;52:2018–32.
[14] Herman A. Seminar Notes: Algebraic Aspects of Association Schemes and Scheme Rings, seminar notes. 2011. http://uregina.ca/~hermana/ASSR-Lecture9.pdf.
[15] Hou X-D. p-ary and q-ary versions of certain results about bent functions and resilient functions. Finite Fields Appl. 2004;10:566–82.
[16] Ito T, Munemasa A, Yamada M. Amorphous association schemes over Galois rings of characteristic 4. Eur J Comb. 1991;12:513–26.
[17] Joyner D, Melles C. Adventures in graph theory. Basel: Birkhaüser; 2017.
[18] Kumar PV, Scholtz RA, Welch LR. Generalized bent functions and their properties. J Comb Theory, Ser A. 1985;40:90–107.
[19] Melles C, Joyner D. On p-ary bent functions and strongly regular graphs. 2018. Submitted. https://arxiv.org/abs/1904.09359.
[20] Pott A, Tan Y, Feng T, Ling S. Association schemes arising from bent functions. Des Codes Cryptogr. 2011;59:319–31.
[21] Rothaus OS. On bent functions. J Comb Theory, Ser A. 1976;20:300–5.
[22] Stanica P. Graph eigenvalues and Walsh spectrum of Boolean functions. Integers. 2007;7(2):A32.
[23] Tan Y, Pott A, Feng T. Strongly regular graphs associated with ternary bent functions. J Comb Theory, Ser A. 2010;117:668–82.
[24] van Dam E. Strongly regular decompositions of the complete graph. J Algebraic Comb. 2003;17:181–201.
[25] van Dam E, Muzychuk M. Some implications on amorphic association schemes. J Comb Theory, Ser A. 2010;117:111–27.
[26] Walsh S. Combinatorics of p-ary bent functions. United States Naval Academy honors thesis. 2014.

Delaram Kahrobaei, Antonio Tortora, and Maria Tota

Multilinear cryptography using nilpotent groups

Abstract: In this paper, we develop a novel idea of multilinear cryptosystem using nilpotent group identities.

Keywords: Group based cryptotgraphy, multilinear cryptosystem, nilpotent group

MSC 2010: 20F18, 20F45

1 Introduction

In recent years, multilinear maps have attracted attention in cryptography community. The idea has been first proposed by Boneh and Silverberg [1]. For $n > 2$, the existence of n-linear maps is still an open question. One of the main applications of multilinear maps is their use for indistinguishability obfuscation. For example in [5], Lin and Tessaro proved that trilinear maps are sufficient for the purpose of achieving indistinguishability obfuscation. Recently, Huang [3] constructed cryptographic trilinear maps that involve simple, nonordinary abelian varieties over finite fields.

Group-based cryptography has some new direction to offer to answer this question. A bilinear cryptosystem using the discrete logarithm problem in matrices coming from a linear representation of a group of nilpotency class 2 has been proposed in [7].

In this paper, we propose multilinear cryptosystems using identities in nilpotent groups, in which the security is based on the chosen discrete logarithm problem in finite p-groups.

2 Multilinear maps in cryptography

Let n be a positive integer. For cyclic groups G and G_T of prime order p, a map $e : G^n \to G_T$ is said to be a (symmetric) n-linear map (or a multilinear map) if for any

Acknowledgement: The authors were supported by the "National Group for Algebraic and Geometric Structures, and their Applications" (GNSAGA – INdAM). The first author was also partially supported by the ONR (Office of Naval Research) grant N000141512164. This research is also supported by a grant of the University of Campania "Luigi Vanvitelli", in the framework of Programma V:ALERE 2019.

Delaram Kahrobaei, University of York, Deramore Lane, York YO10 5GH, United Kingdom, e-mail: delaram.kahrobaei@york.ac.uk
Antonio Tortora, Dipartimento di Matematica e Fisica, Università della Campania "Luigi Vanvitelli", Caserta, Italy, e-mail: antonio.tortora@unicampania.it
Maria Tota, Dipartimento di Matematica, Università di Salerno, Fisciano (SA), Italy, e-mail: mtota@unisa.it

https://doi.org/10.1515/9783110638387-013

$a_1, \dots, a_n \in \mathbb{Z}$ and $g_1, \dots, g_n \in G$, we have

$$e(g_1^{a_1}, \dots, g_n^{a_n}) = e(g_1, \dots, g_n)^{a_1 \dots a_n}$$

and further e is nondegenerate in the sense that $e(g, \dots, g)$ is a generator of G_T for any generator g of G.

2.1 Fully homomorphic encryption and graded encoding schemes

One of the interesting importance of multilinear maps arises in the notion of one of the revolution which swept the world of cryptography, namely fully homomophic encryption (FHE). The intuition is that FHE ciphertexts behave like the exponents of group elements in a multilinear map, the so called graded encoding scheme [2]. Such a scheme is a family of efficient cyclic groups $G_1, \dots, G_n$ of the same prime order p together with efficient nondegenerate bilinear pairings $e : G_i \times G_j \to G_{i+j}$ whenever $i + j \leq n$. In other words, if we fix a family of generators g_i of the G_i's in such a way that $g_{i+j} = e(g_i, g_j)$, we can add exponents within a given group G_i,

$$g_i^a \cdot g_i^b = g_i^{a+b};$$

and multiply exponents from two groups G_i, G_j as long as $i + j \leq n$:

$$e(g_i^a, g_j^b) = g_{i+j}^{a \cdot b}.$$

This makes g_i^a somewhat similar to an FHE encryption of a.

2.2 Generalization of multilinear maps to any group

Here, we generalize the definition of a multilinear map to arbitrary groups G and G_T. We say that a map $e : G^n \to G_T$ is a (symmetric) n-linear map (or a multilinear map) if for any $a_1, \dots, a_n \in \mathbb{Z}$ and $g_1, \dots, g_n \in G$, we have

$$e(g_1^{a_1}, \dots, g_n^{a_n}) = e(g_1, \dots, g_n)^{a_1 \dots a_n}.$$

Notice that the map e is not necessarily linear in each component. In addition, we say that e is nondegenerate if there exists $g \in G$ such that $e(g, \dots, g) \neq 1$.

3 Preliminaries

3.1 Nilpotent and Engel groups

A group G is said to be nilpotent if it has a finite series

$$\{1\} = G_0 < G_1 < \dots < G_n = G$$

which is central, that is, each G_i is normal in G and G_{i+1}/G_i is contained in the center of G/G_i. The length of a shortest central series is the (nilpotency) class of G. Of course, nilpotent groups of class at most 1 are abelian. A great source of nilpotent groups is the class of finite p-groups, that is, finite groups whose orders are powers of a prime p.

Close related to nilpotent groups is the calculus of commutators. Let $g_1, \ldots, g_n$ be elements of a group G. We will use the following commutator notation: $[g_1, g_2] = g_1^{-1} g_2^{-1} g_1 g_2$. More generally, a simple commutator of weight $n \geq 2$ is defined recursively by the rule

$$[g_1, \ldots, g_n] = [[g_1, \ldots, g_{n-1}], g_n],$$

where by convention $[g_1] = g_1$. A useful shorthand notation is

$$[x,_n g] = [x, \underbrace{g, \ldots, g}_{n}].$$

For the reader convenience, we recall the following property of commutators:

$$[xy, z] = [x, z]^y [y, z] \quad \text{and} \quad [x, yz] = [x, z][x, y]^z \quad \text{for all } x, y, z \in G. \tag{1}$$

For further basic properties of commutators, we refer to [9, 5.1].

It is useful to be able to form commutators of subsets as well as elements. Let $X_1, X_2, \ldots$ be nonempty subsets of a group G. Define the commutator subgroup of X_1 and X_2 to be

$$[X_1, X_2] = \langle [x_1, x_2] \mid x_1 \in X_1, x_2 \in X_2 \rangle.$$

More generally, let

$$[X_1, \ldots, X_n] = [[X_1, \ldots, X_{n-1}], X_n]$$

where $n \geq 2$. Then there is a natural way of generating a descending sequence of commutator subgroups of a group, by repeatedly commuting with G. The result is a series

$$G = \gamma_1(G) \geq \gamma_2(G) \geq \ldots$$

in which $\gamma_{n+1}(G) = [\gamma_n(G), G]$. This is called the lower central series of G and it does not in general reach 1. Notice that $\gamma_n(G)/\gamma_{n+1}(G)$ lies in the center of $G/\gamma_{n+1}(G)$.

A useful characterization of nilpotent groups, in terms of commutators, is the following.

Lemma 1. *A group G is nilpotent of class at most $n \geq 1$ if and only if the identity $[g_1, \ldots, g_{n+1}] = 1$ is satisfied in G, that is, $\gamma_{n+1}(G) = 1$. In particular, in a nilpotent group of class n, the subgroup $\gamma_n(G)$ is central.*

Among the best known generalized nilpotent groups are the so-called Engel groups. A group G is called n-Engel if $[x,_n y] = 1$ for all $x, y \in G$. If G is nilpotent of class n, then G is n-Engel. Also, there are nilpotent groups of class n which are not $(n-1)$-Engel. For example, given a prime p, the wreath product $G = \mathbb{Z}_p \wr \mathbb{Z}_p$ is nilpotent of class p but not $(p-1)$-Engel [4, Theorem 6.2].

Conversely, any finite n-Engel group is nilpotent, by a well-known result of Zorn [9, 12.3.4].

3.2 Nilpotent group identities

The next result is a straightforward application of (1), together with Lemma 1.

Lemma 2. *Let G be a nilpotent group of class $n > 1$ and let a be a nonzero integer. Then, for all $g_1, \ldots, g_n \in G$, we have*

$$[[g_1, \ldots, g_{n-1}]^a, g_n] = [g_1, \ldots, g_n]^a$$

and

$$[g_1, \ldots, g_{n-1}, g_n^a] = [g_1, \ldots, g_n]^a.$$

Then the following proposition holds.

Proposition 3. *Let G be a nilpotent group of class $n > 1$. Then*

$$[g_1, \ldots, g_{i-1}, g_i^{a_i}, g_{i+1}, \ldots, g_n] = [g_1, \ldots, g_{i-1}, g_i, g_{i+1}, \ldots, g_n]^{a_i}, \tag{2}$$

for any $i \in \{1, \ldots, n\}$, $a_i \in \mathbb{Z}\backslash\{0\}$ and $g_i \in G$.

Proof. We argue by induction on n. The case $n = 2$ is true by Lemma 2.

Let $n > 2$. Then $G/\gamma_n(G)$ is nilpotent of class $n-1$. Moreover, $\gamma_n(G)$ is central by Lemma 1. Hence the induction hypothesis gives

$$g := [g_1, \ldots, g_{i-1}, g_i^{a_i}, g_{i+1}, \ldots, g_{n-1}] = [g_1, \ldots, g_{n-1}]^{a_i} \bmod \gamma_n(G).$$

It follows that $g = [g_1, \ldots, g_{n-1}]^{a_i} h$ where $h \in \gamma_n(G)$. Since $\gamma_n(G)$ is central, applying (1), we get

$$[g, g_n] = [[g_1, \ldots, g_{n-1}]^{a_i} h, g_n] = [[g_1, \ldots, g_{n-1}]^{a_i}, g_n]$$

and so

$$[[g_1, \ldots, g_{n-1}]^{a_i}, g_n] = [g_1, \ldots, g_{i-1}, g_i, g_{i+1}, \ldots, g_n]^{a_i}$$

by Lemma 2. □

Let G be a nilpotent group of class $n > 1$ and $g_1, \ldots, g_n \in G$. According to Proposition 3 for any $a_1, \ldots, a_n \in \mathbb{Z}\backslash\{0\}$, we have

$$[g_1^{a_1}, \ldots, g_n^{a_n}] = [g_1, \ldots, g_n]^{\prod_{i=1}^{n} a_i}.$$

Therefore, we can construct the multilinear map $e : G^n \to G$ given by

$$e(g_1, \ldots, g_n) = [g_1, \ldots, g_n].$$

Similarly, given $x \in G$, we can consider the multilinear map $e' : G^{(n-1)} \to G$ given by

$$e'(g_1, \ldots, g_{n-1}) = [x, g_1, \ldots, g_{n-1}].$$

Further, assuming that G is not $(n-1)$-Engel, one can take $x \in G$ in such a way that e' is nondegenerate. In fact, there exists $g \in G$ such that $[x,_{n-1} g] \neq 1$.

4 Multilinear cryptography using nilpotent groups

Here, we propose two multilinear cryptosystems based on the identity (2) in Proposition 3.

4.1 Protocol I

First, we generalize the bilinear map which has been mentioned in [7], to multilinear (n-linear) map for $n + 1$ users. Let $\mathcal{A}_1, \ldots, \mathcal{A}_{n+1}$ be the users with private exponents $a_1, \ldots, a_{n+1}$, respectively. Given an integer $a \neq 0$, the main formula on which our key-exchange protocol is based on, is an identity in a public nilpotent group G of class $n > 1$ (see Proposition 3):

$$[g_1{}^a, g_2, \ldots, g_n] = [g_1, g_2{}^a, \ldots, g_n] = [g_1, g_2, \ldots, g_n{}^a] = [g_1, g_2, \ldots, g_n]^a.$$

The users $\mathcal{A}_j$'s transmit in public channel

$$g_i{}^{a_j}, \quad \text{for } i = 1, \ldots, n;\ j = 1, \ldots, n+1.$$

The key exchange works as follows:
- The user $\mathcal{A}_1$ can compute $[g_1{}^{a_2}, \ldots, g_n{}^{a_{n+1}}]^{a_1}$.
- The user $\mathcal{A}_j$ ($j = 2, \ldots, n$) can compute

$$[g_1{}^{a_1}, \ldots, g_{j-1}{}^{a_{j-1}}, g_j{}^{a_{j+1}}, g_{j+1}{}^{a_{j+2}}, \ldots, g_n{}^{a_{n+1}}]^{a_j}.$$

- The user $\mathcal{A}_{n+1}$ can compute $[g_1{}^{a_1}, \ldots, g_n{}^{a_n}]^{a_{n+1}}$.

The common key is $[g_1, \ldots, g_n]^{\prod_{j=1}^{n+1} a_j}$.

Example: Trilinear cryptography using nilpotent groups of class 3. Let $\mathcal{A}, \mathcal{B}, \mathcal{C}, \mathcal{D}$ be the users with private exponents a, b, c, d, respectively. The users $\mathcal{A}$, $\mathcal{B}$, $\mathcal{C}$, and $\mathcal{D}$ transmit in public channel

$$x^a, y^a, z^a, x^b, y^b, z^b, x^c, y^c, z^c, x^d, y^d, z^d \text{ respectively.}$$

The key exchange works as follows:
- The user $\mathcal{A}$ can compute $[x^b, y^c, z^d]^a$.
- The user $\mathcal{B}$ can compute $[x^a, y^c, z^d]^b$.
- The user $\mathcal{C}$ can compute $[x^a, y^b, z^d]^c$.
- The user $\mathcal{D}$ can compute $[x^a, y^b, z^c]^d$.

The common key is $[x, y, z]^{abcd}$.

4.2 Protocol II

Let G be a public nilpotent group of class $n+1$ which is not n-Engel ($n \geq 1$). Then there exist $x, g \in G$ such that $[x,_n g] \neq 1$. Suppose that $n+1$ users $\mathcal{A}_1, \dots, \mathcal{A}_{n+1}$ want to agree on a shared secret key. Each user $\mathcal{A}_j$ selects a private nonzero integer a_j, computes g^{a_j} and sends it to the other users. Then:
- The user $\mathcal{A}_1$ computes $[x^{a_1}, g^{a_2}, \dots, g^{a_{n+1}}]$.
- The user $\mathcal{A}_j$ ($j = 2\dots, n$), computes $[x^{a_j}, g^{a_1}, \dots, g^{a_{j-1}}, g^{a_{j+1}}, \dots, g^{a_{n+1}}]$.
- The user $\mathcal{A}_{n+1}$ computes $[x^{a_{n+1}}, g^{a_1}, \dots, g^{a_n}]$.

Hence, again by Proposition 3, each user obtains $[x,_n g]^{\prod_{j=1}^{n+1} a_j}$ which is the shared key.

5 Security and platform group

The security of our protocols is based on the discrete logarithm problem (DLP). The ideal platform group for our protocols must be a non-abelian nilpotent group of large order such that the nilpotency class is not too large and the DLP in such a group is hard.

In [10], Sutherland has studied the DLP in finite abelian p-groups, and showed how to apply the algorithms for p-groups to find the structure of any finite abelian group.

In a series of papers by Mahalanobis, the DLP has been studied for finite p-groups but mostly for nilpotent groups of class 2 [6, 8]. In particular, in [7], Mahalanobis and Shinde proposed p-groups of class 2 in which the platform is not practical as showed by the authors.

Bibliography

[1] Boneh D, Silverberg A. Applications of Multilinear Forms to Cryptography. Contemporary Mathematics. vol. 324. Providence: American Mathematical Society; 2003. p. 71–90.

[2] Garg S, Gentry C, Halevi S. Candidate multilinear maps from ideal lattices. In: EUROCRYPT 2013. LNCS. vol. 7881. 2013. p. 1–17.

[3] Huang MA. Trilinear maps for cryptography. 2018. Preprint available at https://arxiv.org/abs/1803.10325.

[4] Liebeck H. Concerning nilpotent wreath products. Proc Camb Philos Soc. 1962;58:443–51.

[5] Lin H, Tessaro S. Indistinguishability Obfuscation from Trilinear Maps and Block-Wise Local PRGs. In: CRYPTO 2017. 2017.

[6] Mahalanobis A. The Diffie–Hellman key exchange protocol and non-abelian nilpotent groups. Isr J Math. 2008;165:161–87.

[7] Mahalanobis A, Shinde P. Bilinear Cryptography Using Groups of Nilpotency Class 2. In: Cryptography and Coding, 16th IMA International Conference, IMACC 2017. Oxford, UK. 2017. p. 127–34.

[8] Mahalanobis A. The MOR cryptosystem and finite p-groups, Algorithmic problems of group theory, their complexity, and applications to cryptography. In: Contemp Math. vol. 633. Providence, RI: Amer. Math. Soc.; 2015. p. 81–95.

[9] Robinson DJS. A course in the Theory of Groups. 2nd ed. New York: Springer; 1996.

[10] Sutherland AV. Structure computation and discrete logarithms in finite abelian p-groups. Math Comput. 2011;80(273):477–500.

Ilya Kapovich

Musings on generic-case complexity

Abstract: We propose a more general definition of generic-case complexity, based on using a random process for generating inputs of an algorithm and using the time needed to generate an input as a way of measuring the size of that input.

Keywords: generic-case complexity, algorithms, group theory

MSC 2010: Primary 03D15, 68Q15, Secondary 20F, 68Q17, 68Q25, 94A

> I have committed the sin of falsely proving Poincare's Conjecture. But that was in another country; and besides, until now no one has known about it.
>
> John R. Stallings [28]

1 Introduction

Paraphrasing one of my mathematical heroes, John R. Stallings, I am committing a mathematical sin. I am writing a paper about a definition, with no theorems, corollaries, lemmas, or propositions. And yet I will try to say something that I think is worth saying.

The notion of *generic-case complexity* was introduced in the 2003 paper, by myself, Alexei Myasnikov, Paul Schupp, and Vladimir Shpilrain [15]. The idea was to define the notion of a complexity class that captures the behavior of an algorithm on "most" inputs of a particular problem. The point of such a notion is to reflect practical behavior of various algorithms. The key difference with an older notion of *average-case complexity* is that generic-case complexity completely ignores the possibly bad behavior of the algorithm on a "negligible" set of inputs, instead of trying to average such behavior against good behavior on typical inputs. Our 2003 paper dealt with generic-case complexity of various group-theoretic algorithmic problems, which subsequently has become an area of active study in group theory; see, for example, [1, 2, 3, 5, 9, 14, 16, 19, 22, 27].

Since then, the notion of generic-case complexity has left the confines of group theory and has been explored in much wider mathematical and computational complexity contexts; see, for example, [8, 14, 18, 20, 23, 24, 25, 26]. Recent work of

Note: To the memory of my advisor, Gilbert Baumslag (1933–2014).

Acknowledgement: The author was supported by the NSF grant DMS-1405146.

Ilya Kapovich, Department of Mathematics and Statistics, Hunter College of CUNY, Room 919/944 East, 695 Park Avenue, New York, NY 10065, USA, e-mail: ik535@hunter.cuny.edu

https://doi.org/10.1515/9783110638387-014

Jockusch-Schupp [13], Downey-Jockusch-Schupp [6], Downey–Jockusch–McNicholl-Schupp [7], Igusa [12], and others, started a systematic abstract development of the theory of generic and coarse computability in the framework of recursion theory. The ideas of generic-case complexity also found applications outside mathematics, particularly in dealing with various big data types; see, for example, [4, 10, 11, 17].

An essential feature of our original definition from [15] included using the concept of "asymptotic density," defined in terms of balls with respect to some sort of a "size" function on the infinite set of inputs Ω, in order to define the notions of "generic" and "negligible" subsets of Ω. Most subsequent more general definitions of generic-case complexity still try to retain the same basic vocabulary of using asymptotic density and balls of increasing radius in the main definitions.

The purpose of this note is to suggest a more organic definition of generic-case complexity, which drops the language of asymptotic density and does not use balls or a size function. Instead, this alternative approach defines genericity in terms of a random process generating inputs for an algorithm, and uses the time needed by a random process to produce an input for measuring the size of that input.

The main new definition, of generic-case time complexity of an algorithm with inputs from a set Ω generated by a discrete time random process

$$\mathcal{W} = W_1, W_2, \ldots, W_n, \ldots$$

is given in Definition 3.2. The definition of generic-case time complexity classes with respect to $\mathcal{W}$ is then given in Definition 3.3. The set-up is sufficiently general to apply to a wide variety of data types as inputs of algorithms. Thus Ω may consists of words in some alphabet, graphs, complexes, labeled diagrams, mechanical configurations, configurations of pixels, etc.

Upon close inspection one observes that almost all the existing results in the literature regarding generic-case complexity are covered by Definition 3.2 and Definition 3.3.

The paper is organized as follows. In Section 2, we recall the original definition of generic-case complexity from [15] and a version of this definition that is currently standard (at least according to such an august source of wisdom and knowledge as Wikipedia). In Section 3, we discuss some drawbacks of these definitions and propose new ones, given in Definition 3.2 and Definition 3.3. We the discuss some features of these new definitions and the various degrees of flexibility for modifying and generalizing these definitions further; see, in particular, Remark 3.4. In Section 4, we discuss some important special cases, particularly the classic computability theory context of $\Omega = \{0,1\}^*$. In Section 5, we discuss some drawbacks and limitations of Definition 3.2 and Definition 3.3 and suggest several possible fixes for addressing these drawbacks.

I am grateful to Paul Schupp for informative and illuminating discussions about complexity theory.

2 The original definition

Notation 2.1 (Running time of an algorithm). Let $\mathfrak{A}$ be a partial deterministic algorithm (in some model of computation) with inputs from a set Ω and outputs in some set U. For an input $w \in \Omega$, denote by $t_{\mathfrak{A}}(w)$ the running time of $\mathfrak{A}$ on the input w to compute an output in U. Namely, $t_{\mathfrak{A}}(w) = n \in \{1, 2, 3, \dots,\}$ if $\mathfrak{A}$, starting with input w, terminates in n steps and outputs a value in U, and $t_{\mathfrak{A}}(w) = \infty$ otherwise. Thus $t_{\mathfrak{A}}(w) = \infty$ precisely when, starting on input w, either $\mathfrak{A}$ runs forever or $\mathfrak{A}$ terminates in finitely many steps but outputs no value in U.

We recall a simplified version of the original definition used in [15].

Convention 2.2 (Basic assumption). For the remainder of this section, unless specified otherwise, we make the following assumption. Let Ω be a countably infinite set (where Ω is understood to be the set of all possible inputs for a particular algorithmic problem). Let $|.| : \Omega \to \mathbb{Z}_{\geq 0}$ be a *size function*. For an integer $n \geq 0$, denote by $B_\Omega(n)$ the set of all $w \in \Omega$ with $|w| = n$. Assume that for all $n \geq 0$ the set $B_\Omega(n)$ is finite.

Definition 2.3. For a subset $S \subseteq \Omega$, define the *lower asymptotic density* $\underline{\rho}_\Omega(S)$ of S in Ω as

$$\underline{\rho}_\Omega(S) = \limsup_{n\to\infty} \frac{\#(B_\Omega(n) \cap S)}{\#(B_\Omega(n))}. \qquad (\dagger)$$

If in the above formula the actual limit of the sequence $\frac{\#(B_\Omega(n)\cap S)}{\#(B_\Omega(n))}$ exists, we call this limit the *asymptotic density* of S in Ω and denote it $\rho_\Omega(S)$. A subset $S \subseteq \Omega$ is called *generic* in Ω if $\rho_\Omega(S) = 1$ (which is equivalent to the condition that $\underline{\rho}_\Omega(S) = 1$). A subset $S \subseteq \Omega$ is called *exponentially generic* in Ω if $\lim_{n\to\infty} \frac{\#(B_\Omega(n)\cap S)}{\#(B_\Omega(n))} = 1$ and the convergence in this limit is exponentially fast.

In [15], the main cases we considered were where Ω is either the set of all freely reduced words or the set of all cyclically reduced words over some group alphabet.

For the purposes of current discussion, the reader should concentrate on the case $\Omega = A^*$, where $A = \{a_1, \dots, a_m\}$ is a finite alphabet with $m \geq 2$ letters, and where for $w \in A^*$ the size $|w|$ is the length of the word w. In this case, $\#(B_\Omega(n)) = 1 + m + m^2 + \dots + m^n = \frac{m^{n+1}-1}{m-1}$, and grows roughly as m^n as $n \to \infty$.

Generic-case complexity is then defined as follows.

Definition 2.4. Let $\mathfrak{A}$ be a partial deterministic algorithm with inputs from the set Ω and outputs in some countable set U.

For a monotone-nondecreasing function $f(n) \geq 0$, we say that $\mathfrak{A}$ has *generic-case time complexity* $\leq f$ (correspondingly, *strong generic-case time complexity* $\leq f$) if there exists a generic subset $S \subseteq \Omega$ (correspondingly, an exponentially generic subset $S \subseteq \Omega$), such that for every $w \in \Omega$ we have $t_{\mathfrak{A}}(w) \leq f(|w|)$.

Remark 2.5. Strictly speaking, Definition 2.4 is only a meta-definition. To make this definition formally precise, we need to fix a mathematical model of computation where there is a well-defined notion of a partial deterministic algorithm with inputs from the set Ω. The first, and most frequently used, way of addressing this issue is to fix an encoding of elements of Ω by finite binary sequences (i. e., natural numbers) or strings in some fixed finite alphabet, and then use the classic definition of a partially computable function (based on Turing machines). Alternatively, one can choose another model of computation with a notion of an algorithm with inputs from the set Ω that is based on a computational device other than a Turing machine.

We will assume that some such choice is made before applying Definition 2.4. In Convention 3.1 in Section 3 below, before proposing a new definition of generic-case complexity, we explicitly assume choosing a particular model of computation and drop the requirement for the set Ω to be countable.

Several limitations of Definition 2.4 quickly became apparent. Definition 2.4 is supposed to capture the practical behavior of $\mathfrak{A}$ on "typical" inputs $\omega \in \Omega$ and implicitly assumes that, for a large $n \geq 1$, a typical input $w \in B_\Omega(n)$ corresponds to the uniform probability distribution on the finite set $B_\Omega(n)$. However, as a practical matter, even in very nice situations, there is often no easy **practical** way of choosing uniformly at random an element $w \in B_\Omega(n)$. Even in cases where such a practical procedure exists, the inputs for $\mathfrak{A}$ may be supplied by a random process which generates a probability distribution on $B_\Omega(n)$ which is far from uniform.

In fact, even in the model case of $A = \{a_1, a_2\}$ and $\Omega = A^* = \{a_1, a_2\}^*$, the "natural" practical random process is given by a sequence of i.i.d random variables $X_1, X_2, \dots, X_n, \dots$ with $X_i \in A$ having the uniform probability distribution on A. After n steps, this process generates a word $W_n = X_1 X_2 \dots X_n \in A^n$ of length n and gives a uniform probability distribution ν_n on the n-sphere $S(n) = A^n$ of cardinality 2^n in Ω. To get a uniform probability distribution on the ball $B_\Omega(n)$ of cardinality $2^{n+1} - 1$, we then have to take the weighted sum $\sum_{i=0}^{n} \frac{2^i}{2^{n+1}-1} \nu_i$. That is, in order to pick uniformly at random an element from the ball $B_\Omega(n)$, we first need to pick a random integer $j \in \{0, 1, \dots, n\}$, where the probability of picking i is $\frac{2^i}{2^{n+1}-1}$, and then pick uniformly at random an element w from the j-sphere A^j. While it is possible to program an actual computer simulation of such a process, that is not what one usually does in practice when choosing elements of A^*. Instead, one usually just runs the above mentioned sequence of i. i. d. s $X_1, X_2, \dots, X_n$ for n steps and by concatenating generates a uniformly random word W_n of length n. That is, one works with a random process that after n-steps generates a probability distribution on the ball $B_\Omega(n)$ whose support is actually just the n-sphere $S(n) = A^n$. Thus in the nice setup of $\Omega = A^* = \{a_1, a_2\}^*$ working with uniform probability distributions on n-spheres is more natural than with uniform probability distributions on balls, although in this case both of these approaches produce the same notion of a generic subset of A^* (see [15] for details).

More importantly, in most other situations, various natural random processes generating inputs in some set Ω as in Convention 2.2, result in distributions on $B_\Omega(n)$ that are far from uniform. This happens, for example, for random walks on groups and graphs, various random processes generating planar diagrams, higher dimensional simplicial or cell complexes, etc. For example, if Γ is a finite connected graph, then a simple random walk of length n on Γ starting at a particular vertex v_0 can be used to generate paths of length n starting at x_0. However, unless the graph happens to be highly symmetric, the distribution on the set of all paths of length n in Γ starting at x_0 determined by this random walk will be far from uniform. Moreover, there will be no alternative practical way of picking such a path uniformly at random.

The standard way of addressing this issue is captured by the following generalization of Definition 2.3.

Definition 2.6. Let Ω and $|.|$ be as in Convention 2.2. Let μ_n, where $n = 1, 2, \dots$, be a sequence of probability distributions on $B_\Omega(n)$. A subset $S \subseteq \Omega$ is *generic* for $(\mu_n)_n$ if $\lim_{n\to\infty} \mu_n(S \cap B_\Omega(n)) = 1$. If, moreover, the convergence in this limit is exponentially fast, then S is said to be *exponentially generic* for $(\mu_n)_n$.

One can then modify Definition 2.4 by using Definition 2.6 instead of Definition 2.3 to get a more general notion of generic-case time complexity, with respect to $(\mu_n)_n$, for an algorithm $\mathfrak{A}$.

Definition 2.7. Let Ω and $|.|$ be as in Convention 2.2. Let μ_n, where $n = 1, 2, \dots,$, be a sequence of probability distributions on $B_\Omega(n)$.

Let $\mathfrak{A}$ be a partial deterministic algorithm with inputs from the set Ω and outputs in some countable set U.

For a monotone-nondecreasing function $f(n) \geq 0$, we say that $\mathfrak{A}$ has *generic-case time complexity* $\leq f$ (correspondingly, *strong generic-case time complexity* $\leq f$) with respect to $(\mu_n)_n$ if there exists a subset $S \subseteq \Omega$, which is generic for $(\mu_n)_n$ (correspondingly, exponentially generic for $(\mu_n)_n$) and such that for every $w \in \Omega$ we have $t_{\mathfrak{A}}(w) \leq f(|w|)$.

3 A more organic approach

Definition 2.6 is technically sufficient to account for a wide variety of situations where one wants to talk about generic-case complexity. Nevertheless, Definition 2.6 still has several philosophical and practical drawbacks.

First, talking about generic subsets is not the most natural thing to do in probability theory. Rather, from the probabilistic point of view, it is more natural to talk about **events.** If X is a measure space with a probability measure μ, saying that some event happens μ-almost surely does mean that this event corresponds to a subset of $S \subseteq X$ with $\mu(S) = 1$. However, most mathematical arguments in probability theory are phrased in terms of estimating or computing probabilities that certain events occur (rather than in terms of talking about such events as specific subsets of the sample space). Thinking in terms of events rather than explicitly defined subsets is crucial for making probability theory work.

Moreover, even more crucially, if we think of $(\mu_n)_{n\geq 1}$ as measures on Ω corresponding to a sequence of "random" choices of elements of Ω, the sample space Ω_* of such a sequence is (a subset of) $\Omega^{\mathbb{N}}$ rather than Ω. Events corresponding to such sequences of random choices of elements of Ω are subsets of Ω_*. Thus genericity should really be understood in terms of subsets of Ω_* rather than of Ω. Consider, for example, the situation where $\Omega = \{a, b\}$ and where $\mathfrak{A}$ is a partial algorithm with inputs from Ω such that $t_{\mathfrak{A}}(a) = 1$ and $\mathfrak{A}(b) = \infty$. Let $\mathcal{W} = W_1, W_2, \ldots$ be a sequence of i. i. d. Ω-valued random variables where each W_i having the uniform probability distribution μ on Ω (i. e., $\mu(a) = \mu(b) = 1/2$.) Then, by the law of large numbers, with probability tending to 1 as $n \to \infty$ (i. e., "generically"), for a sequence $W_1, \ldots, W_n$ generated by $\mathcal{W}$ the algorithm $\mathfrak{A}$ terminates in a single step on at least $n/3$ of the inputs $W_1, \ldots, W_n$. Yet, it is impossible to express this statement in terms of genericity of subsets of Ω itself.

Second, the insistence on the fact that μ_n be supported on the n-ball $B_\Omega(n)$ with respect to some specific "size function" $|.|$ is not always natural and ultimately unnecessary. For various kinds of "growth" random processes, generating planar graphs, van Kampen diagrams, complexes, etc., there is not necessarily a clear choice of a specific size function $|.|$ on Ω (and often several possible choices make sense).

To rectify these issues, we propose Definition 3.2 below. Before stating this definition, we adopt the following convention.

Convention 3.1. Fix a model of computation $\mathcal{M}$, which involves specifying the set Ω of all possible inputs for an algorithm and a description of allowable computational devices, such as Turing machines, Blum–Shub–Smale machine (for computations with real numbers), etc. We do not require Ω to be countable but, to simplify exposition, we restrict ourselves to deterministic models of computation. We also assume that the computational devices in $\mathcal{M}$ operate in discrete time, which a single computational step taking exactly one unit of time. As in notation 2.1, if $\mathfrak{A}$ is a partial deterministic algorithm with inputs from the set Ω, then for $w \in \Omega$ the running time of $\mathfrak{A}$ on the input w to produce an output in U is denoted by $t_{\mathfrak{A}}(w)$. Thus $t_{\mathfrak{A}}(w)$ is either a positive integer or ∞, where the latter happens exactly when the algorithm $\mathfrak{A}$ starting on the input w either runs forever or terminates in finite time without producing an output in U.

Our main definition is the following.

Definition 3.2 (Generic-case complexity of an algorithm with respect to a random process). Let $\mathfrak{A}$ be a partial deterministic algorithm with inputs from the set Ω and values in some set U.

Let $\mathcal{W} = W_1, W_2, \ldots, W_n, \ldots$ be a discrete time random process that, after n steps, generates an input $W_n \in \Omega$.

For a monotone-nondecreasing function $f(n) \geq 0$, we say that $\mathfrak{A}$ has *generic-case time complexity* $\leq f$ *with respect to* $\mathcal{W}$ if

$$\lim_{n\to\infty} \Pr(t_{\mathfrak{A}}(W_n) \leq f(n)) = 1. \qquad (\ddagger)$$

If, moreover, the convergence in this limit is exponentially fast, we say that $\mathfrak{A}$ has *strong generic-case time complexity* $\leq f$ *with respect to* $\mathcal{W}$.

In the above definition, we take the view that the time n, needed to generate an input W_n after n steps of $\mathcal{W}$, serves as a reasonable way of measuring the size of W_n. As a practical matter, for many computationally natural random processes $\mathcal{W}$, there is (often more than one) choice of a size function $|.|$ such that we always have $|W_n| \leq O(n)$ or perhaps $|W_n| \leq O(n \log n)$, or something similar. However, explicitly specifying such a size function and making it a part of the definition of generic-case complexity is not really necessary.

Definition 3.2 makes it unnecessary to first define the notion of a generic subset of Ω and phrases condition ($\ddagger$) in terms of probabilities of events rather than in terms of existence of generic sets. Formally, $(W_n)_{n\geq 1}$ is a sequence of Ω-valued random variables, so that for each $n \geq 1$ W_n gives a probability distribution μ_n on Ω.

In practice, one would want to concentrate on the situation where each μ_n is finitely supported, and where the random process $\mathcal{W} = W_1, W_2, \ldots, W_n, \ldots$ can be relatively easily programmed on a computer. Moreover, Definition 3.2 best reflects the idea of a "practical" behavior of an algorithm in the case where $\mathcal{W}$ is a Markov process, and where we use the input W_n to "construct" the next input W_{n+1}. However, formally, we do not need to impose these requirements as a part of the above definition.

Definition 3.2 also makes it more conceptually clear that the notion of generic-case complexity does depend (and crucially so) on the choice of a random process $\mathcal{W}$ generating the inputs of a problem. This key point often easily gets lost in the contexts of discussing versions of generic-case complexity based on asymptotic density considerations.

One can then define the notion of a generic complexity class.

Definition 3.3 (Generic-case complexity class). Let $\mathcal{C}$ be a deterministic time complexity class (such as linear time, quadratic time, polynomial time, exponential time, etc.) for our computational model $\mathcal{M}$. Let $h : \Omega \to U$ be a function, where U is another countable set. Let $\mathcal{W} = W_1, W_2, \ldots, W_n, \ldots$ be a discrete time random process that, after n steps, generates an input $W_n \in \Omega$.

(1) Let $\mathfrak{A}$ be a deterministic partial algorithm with inputs from the set Ω and values in U such that $\mathfrak{A}$ is *correct* for h, that is, whenever $\mathfrak{A}$ actually outputs some value on $w \in \Omega$, that value is equal to $h(w)$.
We say that $\mathfrak{A}$ *computes* h *with generic-case time complexity* $\mathcal{C}$ *with respect to* $\mathcal{W}$ if there exists a monotone nondecreasing function $f(n) \geq 0$ satisfying the time constraints of $\mathcal{C}$ such that $\mathfrak{A}$ has generic-case time complexity $\leq f$ with respect to $\mathcal{W}$.

(2) We say that h is *computable with generic-case time complexity $\mathcal{C}$ with respect to $\mathcal{W}$* if there exists a correct partial algorithm $\mathfrak{A}$ for h such that $\mathfrak{A}$ computes h with generic-case time complexity $\mathcal{C}$ with respect to $\mathcal{W}$.
(3) For given Ω, $\mathcal{W}$, and $\mathcal{C}$, we denote by $\mathrm{Gen}_{\mathcal{W}}(\mathcal{C})$ the set of all $h : \Omega \to U$ that are computable with generic-case time complexity $\mathcal{C}$ with respect to $\mathcal{W}$. The notion of a function *h computable with strong generic-case time complexity $\mathcal{C}$ with respect to $\mathcal{W}$* and the corresponding complexity class $\mathrm{SGen}_{\mathcal{W}}(\mathcal{C})$ are defined similarly.

Remark 3.4. We again stress that Definition 3.2 and Definition 3.3 do require first choosing a model of computation $\mathcal{M}$, as specified in Convention 3.1.

Definition 3.2 and Definition 3.3 are flexible enough to allow for a straightforward modification of generic-case complexity $\mathcal{C}$ with respect to $\mathcal{W}$, where $\mathcal{C}$ is a deterministic complexity class with a resource bound on space (such as log-space, linear space, etc) or on combination of time and space. One can also easily modify these definitions to allow for dealing with nondeterministic algorithms and complexity classes. If $\mathfrak{A}$ is a nondeterministic algorithm with inputs from Ω, one just needs to define $t_{\mathfrak{A}}(w)$ as the shortest length of a computational path for $\mathfrak{A}$ that starts with an input w.

Other variations of these definitions are possible. In particular, one can relax the assumption that the limit in (‡) be equal to 1 and instead require this limit (or the corresponding liminf) to be positive. One can also relax the requirement in part (2) of Definition 3.3 that the algorithm $\mathfrak{A}$ be correct for h, and instead require that $\mathfrak{A}$ produce correct values of h with asymptotically positive probability or probability tending to 1 as $n \to \infty$. These two types of generalizations were introduced by Jockusch–Schupp [13] in the model context of $\Omega = \{0,1\}^*$ and W_n picking uniformly at random a binary string of length n, leading to the notions of "generic computability at density d" (where $0 < d \le 1$) and of "coarse computability."

Also, it is fairly straightforward to adapt Definition 3.2 and Definition 3.3 to deal with continuous-time random-processes $\mathcal{W} = (W_t)_{t\ge 0}$ and to drop the requirement for Ω to be countable. One just needs to work in a computational model with a well-defined notion of an algorithm (deterministic or nondeterministic) with inputs from Ω.

For Definition 3.2 and Definition 3.3, the case where $U = \{0,1\}$ and the function $h : \Omega \to \{0,1\}$ is the characteristic function of some subset $D \subseteq \Omega$, corresponds to the *decision problem* of determining whether an element w of Ω belongs to D. The definition of generic-case complexity considered in [15] was limited to considering decision problems.

We stress that the dependence of Definition 3.2 and Definition 3.3 on the random process $\mathcal{W}$ is an essential feature of these notions which reflects the fact that the practical behavior of various algorithms **does depend** on the choice of random processes used to generate inputs for an algorithm.

We demonstrate this point on the following simple example.

Example 3.5. Let G be a finitely generated group that splits as a HNN-extension of another finitely generated group H with stable letter t and associated isomorphic subgroups $L_1, L_2 \leq H$, so that $G = \langle H, t | t^{-1}L_1 t = L_2 \rangle$. Let $\{b_1, \ldots, b_m\}$ be a finite generating set for H (where $m \geq 2$), let $X = \{t, b_1, \ldots, b_m\}^{\pm 1}$, and let $B = \{b_1, \ldots, b_m\}^{\pm 1}$. Let $\Omega = X^*$ be the set of all words over X and let $D \subseteq X^*$ be the set of all words w over X such that $w =_G 1$.

Let $\mathcal{W}$ be the random process such that at time n $W_n \in A^n$ is a word over X of length n chosen uniformly at random. Let $\mathcal{W}'$ be the random process such that at time n $W_n' \in B^n$ is a word over B of length n chosen uniformly at random.

Let $\mathfrak{A}$ be the partial algorithm with inputs from $\Omega = X^*$ which proceeds as follows. Given a word $w \in X^*$, the algorithm first computes the exponent sum $\sigma_t(w)$ on t in w. If $\sigma_t(w) \neq 0$, the algorithm declares that $w \notin D$ and terminates. If $\sigma_t(w) = 0$, the algorithm $\mathfrak{A}$ runs forever. Note that for every $w \in B^*$ we have $t_{\mathfrak{A}}(w) = \infty$.

Then, as explained in [15], $\mathfrak{A}$ solves the decision problem of belonging to D with linear time generic-case complexity with respect to $\mathcal{W}$. On there other hand, there is no time complexity class $\mathcal{C}$ such that $\mathfrak{A}$ solves the decision problem of belonging to D with generic-case complexity in $\mathcal{C}$ with respect to $\mathcal{W}'$.

Note that, as a practical matter, in the above example both $\mathcal{W}$ and $\mathcal{W}'$ are equally valid ways of generating inputs in $\Omega = A^*$ because both these processes can be easily programmed with a computer and implemented in practice.

Definition 3.2 fits well with the notion of a "random" element $w \in \Omega$ having some particular property. Thus let $D \subseteq \Omega$ be the set of all elements satisfying some property (in which case we also refer to $D \subseteq \Omega$ as a *property* of elements of Ω). Let $\mathcal{W} = W_1, W_2, \ldots, W_n, \ldots$ be a discrete time random process that, after n steps, generates an input $W_n \in \Omega$. We say that an element of Ω *has property D generically with respect to $\mathcal{W}$* if

$$\lim_{n\to\infty} \Pr(W_n \in D) = 1.$$

4 Important special cases

There are several situations where there is one especially natural choice of a random process $\mathcal{W}$ generating elements of Ω, and it makes sense to fix that choice. In particular, if $A = \{a_1, \ldots, a_m\}$ is a finite alphabet with $m \geq 2$ letters and $\Omega = A^*$, then it is particularly natural to use $\mathcal{W} = W_1, W_2, \ldots, W_n, \ldots$ where W_n picks uniformly at random a word $w \in A^n$ of length n over A. Thus W_n induces the uniform probability distribution μ_n on A^n, and, as noted earlier, it is easy to practically implement the process $\mathcal{W}$ by using a concatenation of n i. i. d. random variables with uniform probability distribution on A. Similarly, in the context of group-theoretic decision problems originally considered in [15], we are dealing with a finitely generating group G with a finite generating set $A = \{a_1, \dots, a_m\}$. The inputs for such decision problems are freely reduced words in the group alphabet $X = \{a_1, \dots, a_m\}^{\pm 1}$, so that $\Omega = F(a_1, \dots, a_m)$. It is then particularly natural to use the random process $\mathcal{W} = W_1, W_2, \dots, W_n, \dots$ where W_n picks uniformly at random a freely reduced word of length n over X. Again, $\mathcal{W}$ is easy to implement in practice by considering the simple nonbacktracking random walk on $F(a_1, \dots, a_m)$. In both of the above situations, the original asymptotic density approach from Definition 2.3 produces the notions of genericity and generic-case complexity equivalent to those provided by Definition 3.2 and Definition 3.3. The reason comes from Stolz' theorem which explains why in these cases computing asymptotic density via balls of radius n yields the same notion of a generic set as when using uniform probability distributions on spheres. See Lemma 5.5 and Lemma 5.6 in [15] for details.

A particularly important instance of the above situation is the case where $A = \{0, 1\}$ and where $\Omega = A^* \setminus \{\varepsilon\}$ is the set of all finite nonempty binary sequences, which is naturally identified, via binary expansion, with the set $\mathbb{Z}_{\geq 0}$. Then again it is especially natural to consider the random process $\mathcal{W} = W_1, W_2, \dots, W_n, \dots$ where W_n picks uniformly at random a binary sequence $w \in \{0, 1\}^n$ of length n. Using this Ω and this choice of $\mathcal{W}$ and studying in-depth and detail the corresponding generic complexity classes, $\mathrm{Gen}_{\mathcal{W}}(\mathcal{C})$ provides an ideal model setting for studying the concepts provided by Definition 3.2 and Definition 3.3 (and, as noted above, in this situation these definitions provide the same notions of generic-case complexity as those given by Definition 2.4). This point of view is taken in the recent work of Jockusch–Schupp [13], Downey–Jockusch–Schupp [6] and Downey–Jockusch–McNicholl–Schupp [7] where a systematic development of the theory of generic and coarse computability is pursued in this setting. "Coarse computability" refers to relaxing the requirement in Definition 3.3 that the algorithm $\mathfrak{A}$ be correct for the function h, and allowing $\mathfrak{A}$ to produce incorrect answers with small probability.

5 Limitations of the new definition and possible fixes

Definition 3.2 is based on the premise that for each $n = 1, 2, \dots$ the random process $\mathcal{W}$ generates an element $W_n \in \Omega$ that can be used as an input for the algorithm $\mathfrak{A}$, and that n serves as a reasonable measure of the "size" of W_n. This assumption works well in many natural situations, where $\mathcal{W}$ proceeds by performing "bounded local perturbations." By that we mean that during the nth step of the process, the configuration $W_{n-1} \in \Omega$ is modified by some sort of a bounded local change to produce a new configuration $W_n \in \Omega$, which again constitutes a valid input for $\mathfrak{A}$.

However, there are many situations where such an assumption about the nature of the random process $\mathcal{W}$ is not reasonable. For example, it may be that after some number n of steps the random process does generate a valid input $W_n \in \Omega$. Then it performs k auxiliary processing steps (each consuming a single unit of time and each constituting a "bounded local perturbation"), so that $W_{n+1}, \ldots, W_{n+k}$ are "auxiliary" objects/configurations that do not belong to Ω and that are used to produce a valid input $W_{n+k+1} \in \Omega$ which can now be fed into $\mathfrak{A}$. In general, k itself may depend in either random or deterministic way in the previous "valid" input $W_n \in \Omega$, or even of the entire trajectory $W_1, .., W_n$ up to time n. Moreover, k need not be uniformly bounded above by a constant independent of n. In this situation, the set-up used in Definition 3.2 is not suitable.

In this case, the random process $\mathcal{W} = W_1, W_2, \ldots$ takes values $W_n \in \Omega'$, where $\Omega' = \Omega \sqcup \Omega^{\mathrm{aux}}$ is a bigger set, with Ω^{aux} being the set of "auxiliary" configurations.

There are several different ways in which Definition 3.2 can be adapted to deal with this more general set-up.

Case 1. Suppose there is $n_0 \geq 1$ such that for every $n \geq n_0$ we have $\Pr(W_n \in \Omega) > 0$. We can then modify Definition 3.2 by rephrasing it terms of conditional probabilities and replacing (‡) by requiring that

$$\lim_{n\to\infty} \Pr(t_{\mathfrak{A}}(W_n) \leq f(n) \mid W_n \in \Omega) = 1. \qquad (\spadesuit)$$

Case 2. For many natural examples of $\mathcal{W}$ (such as the example with generating random graphs discussed below), there is some deterministic sequence of times $\mathbf{n} = (n_i)_i$, $1 \leq n_1 < n_2 < \cdots$ (independent of the trajectory of $\mathcal{W}$) such that for each $i \geq 1$ we always have $W_{n_i} \in \Omega$ and for each $n \notin \{n_1, n_2, \ldots,\}$ we always have $W_n \in \Omega^{\mathrm{aux}}$.

In such a situation, we can modify Definition 3.2 and replace (‡) by requiring that

$$\lim_{i\to\infty} \Pr(t_{\mathfrak{A}}(W_{n_i}) \leq f(n_i)) = 1. \qquad (\clubsuit)$$

Example 5.1. Consider the situation where we are trying to generate random graphs on n vertices with $n \to \infty$. The set Ω is defined as $\Omega = \bigsqcup_{n=1}^{\infty} \Omega[n]$ where $\Omega[n]$ consists of all simple graphs with vertex set $\{1, 2, \ldots, n\}$.

If at some stage of the process, we have constructed a graph $\Gamma \in \Omega[n]$; we then construct a graph $\Delta \in \Omega[n+1]$ as follows. We first add a new vertex $n+1$ to Γ and then perform n independent flips of a fair coin n to decide whether to put an edge between vertex $n+1$ and vertices $1, \ldots, n$. Thus we used $n+1$ extra steps to produce a sequence $\Gamma_1, \ldots, \Gamma_{n+1} = \Delta$, where we view $\Gamma_{n+1} = \Delta$ as an element of $\Omega[n+1]$, and view the "auxiliary" graphs $\Gamma_1, \ldots, \Gamma_n$ as elements of Ω^{aux}. (Although it is possible to think of $\Gamma_1, \ldots, \Gamma_n$ as elements of $\Omega[n+1]$, we can formally enforce the condition $\Gamma_1, \ldots, \Gamma_n \notin \Omega[n+1]$ by adding a "flag" register to each of these graphs.)

A reasonable choice of $\mathcal{W}$ here would proceed as follows. We always put W_1 to be the graph consisting of a single vertex. We then start applying the above procedure iteratively. This defines a random process $\mathcal{W} = W_1, W_2, \ldots$ such that for $i = 1, 2, \ldots$ and $n_i = 1 + 2 + \cdots + i = i(i+1)/2$ we have $W_{n_i} \in \Omega[i]$ and for each $n \notin \{n_1, n_2, \ldots,\}$ we always have $W_n \in \Omega^{\mathrm{aux}}$.

Example 5.2. There are situations where neither (♠) nor (♣) gives a suitable generalization of Definition 3.2. For example, let $\Omega = \bigsqcup_{n=1}^{\infty} S_n$ be the disjoint union of symmetric groups S_n. Suppose we are trying to devise a random process that, as $n \to \infty$, produces uniform probability distributions μ_n on S_n. For a given n, we can generate a uniformly random permutation $\sigma \in S_n$ in a reasonable way: we create a random rearrangement $j_1, \dots, j_n$ of the numbers $1, \dots, n$ by first picking uniformly at random $j_1 \in \{1, \dots, n\}$, then picking uniformly at random $j_2 \in \{1, \dots, n\} \setminus \{j_1\}$, then picking uniformly at random $j_3 \in \{1, \dots, n\} \setminus \{j_1, j_2\}$, and so on. Then we define $\sigma \in S_n$ as $\sigma(i) = j_i$ for $i = 1, \dots, n$. However, unlike in the example with generating random graphs given above, having chosen uniformly at random $\sigma \in S_n$ does not allow us to use σ as a starting block for building a uniformly random element of S_{n+1}. Formally, we can still use the approach of Case 2 given by (♣) if, after having chosen a random $\sigma \in S_n$, we forget this choice completely and start building a random permutation in S_{n+1} from scratch, using the above procedure. The method of Case 2 is applicable since the number of steps needed to generate a random element of S_n from scratch depends only on n. However, applying the approach of Case 2 in this situation does seem fairly artificial, and it may be better to use (a version of) Definition 2.7 directly in this case.

Bibliography

[1] Bassino F, Nicaud C, Weil P. Random generation of finitely generated subgroups of a free group. Int J Algebra Comput. 2008;18(2):375–405.

[2] Borovik A, Myasnikov A, Remeslennikov V. Generic complexity of the conjugacy problem in HNN-extensions and algorithmic stratification of Miller's groups. Int J Algebra Comput. 2007;17(5–6):963–97.

[3] Borovik A, Myasnikov A, Remeslennikov V. The conjugacy problem in amalgamated products. I. Regular elements and black holes. Int J Algebra Comput. 2007;17(7):1299–333.

[4] Boselli R, Cesarini M, Mercorio F, Mezzanzanica M. Data quality on KDD: A real-life scenario. In: 22nd Italian Symposium on Advanced Database Systems. 2014. p. 378–85.

[5] Diekert V, Myasnikov AG, Weiss A. Conjugacy in Baumslag's group, generic case complexity, and division in power circuits. Lecture Notes in Computer Science (including subseries Lecture Notes in Artificial Intelligence and Lecture Notes in Bioinformatics). vol. 8392. 2014. p. 1–12.

[6] Downey RG, Jockusch CG, Schupp PE. Asymptotic density and computably enumerable sets. J Math Log. 2013;13:2.

[7] Downey RG, Jockusch CG, McNicholl TH, Schupp PE. Asymptotic density and the Ershov hierarchy. Math Log Q. 2015;61(3):189–95.

[8] Gilman R, Miasnikov AG, Myasnikov AD, Ushakov A. Report on Generic Case Complexity, Herald of Omsk University, 2007. Special Issue, p. 103–110; arXiv:0707.1364.

[9] Gilman R, Miasnikov A, Osin D. Exponentially generic subsets of groups. Ill J Math. 2010;54(1):371–88.

[10] Holzinger A. On knowledge discovery and interactive intelligent visualization of biomedical data: Challenges in Human-Computer Interaction & Biomedical Informatics. In: SECRYPT 2012 – Proceedings of the International Conference on Security and Cryptography. 2012. p. IS9–IS20.

[11] Holzinger A, Stocker C, Dehmer M. Big complex biomedical data: Towards a taxonomy of data. Commun Comput Inf Sci. 2014;455:3–18.
[12] Igusa G. Nonexistence of minimal pairs for generic computability. J Symb Log. 2013;78(2):511–22.
[13] Jockusch CG, Schupp PE. Generic computability, Turing degrees, and asymptotic density. J Lond Math Soc (2). 2012;85(2):472–90.
[14] Kambites M. Generic complexity of finitely presented monoids and semigroups. Comput Complex. 2011;20(1):21–50.
[15] Kapovich I, Myasnikov A, Schupp P, Shpilrain V. Generic-case complexity, decision problems in group theory and random walks. J Algebra. 2003;264(2):665–94.
[16] Kapovich I, Schupp P, Shpilrain V. Generic properties of Whitehead's algorithm and isomorphism rigidity of random one-relator groups. Pac J Math. 2006;223(1):113–40.
[17] Mezzanzanica M, Boselli R, Cesarini M, Mercorio F. Improving data cleansing accuracy a model-based approach. In: DATA 2014 – Proceedings of 3rd International Conference on Data Management Technologies and Applications. 2014. p. 189–201.
[18] Myasnikov AD. Generic case complexity and one-way functions. Groups Complex Cryptol. 2009;1(1):13–31.
[19] Myasnikov A, Shpilrain V, Ushakov A. Group-based cryptography. Advanced Courses in Mathematics. CRM Barcelona. Basel: Birkhäuser Verlag; 2008.
[20] Myasnikov AG, Rybalov A. Generic complexity of undecidable problems. J Symb Log. 2008;73(2):656–73.
[21] Myasnikov AG, Ushakov A. Random subgroups and analysis of the length-based and quotient attacks. J Math Cryptol. 2008;2(1):29–61.
[22] Myasnikov AG, Ushakov A. Random van Kampen diagrams and algorithmic problems in groups. Groups Complex Cryptol. 2011;3(1):121–85.
[23] Rybalov A. On the strongly generic undecidability of the halting problem. Theor Comput Sci. 2007;377(1–3):268–70.
[24] Rybalov AN. Generic complexity of Presburger arithmetic. Theory Comput Syst. 2010;46(1):2–8.
[25] Rybalov AN. Generic complexity of first-order theories. Sib Èlektron Mat Izv. 2011;8:168–78 (Russian).
[26] Rybalov A. Generic complexity of the Diophantine problem. Groups Complex Cryptol. 2013;5(1):25–30.
[27] Shpilrain V. Sublinear time algorithms in the theory of groups and semigroups. Ill J Math. 2010;54(1):187–97.
[28] Stallings J. How not to prove the Poincaré conjecture. In: Topology Seminar. Wisconsin. 1965. Ann of Math Stud. vol. 60. Princeton, NJ: Princeton Univ. Press; 1966. p. 83–8.

Martin Kreuzer and Xingqiang Xiu

Noncommutative Gebauer–Möller criteria

Abstract: For an efficient implementation of Buchberger's algorithm, it is essential to avoid the treatment of as many unnecessary critical pairs or obstructions as possible. In the case of the commutative polynomial ring, this is achieved by the Gebauer–Möller criteria. Here, we present an adaptation and extension of the Gebauer–Möller criteria for noncommutative polynomial rings, that is, for free associative algebras over fields. The essential idea is to detect unnecessary obstructions using other obstructions with or without overlap. Experiments show that the new criteria are able to detect almost all unnecessary obstructions during the execution of Buchberger's procedure.

Keywords: Gröbner basis, free associative algebra, obstruction, Buchberger procedure

MSC 2010: 16-08, 20-04, 13P10

1 Introduction

Gröbner basis computations for noncommutative algebras, for instance, for group rings, have been used to find new theorems, to study conjectures, to complete classification results, and to construct counterexamples. This technique has been successful for a variety of problems and resulted in a plethora of results. (For contributions of the authors in this direction see, for instance, [9], [12], [15], and [25].) In particular, in the cases of free groups and free monoids, efficient algorithms for explicit computations of examples are extremely valuable tools. Numerous approaches have been developed to compute noncommutative Gröbner bases. Among them, let us mention algorithms for computing Gröbner bases for path algebras in [10], various methods based on the letterplace technique in the package `Plural` of `Singular` (see [16], [17]), and an algorithm based on a noncommutative version of the F4 Algorithm in `Magma` (see [3]). As part of his dissertation [25], the second author developed and optimized a Gröbner basis procedure for the monoid ring of a finitely generated free monoid, that is, for the noncommutative polynomial ring, in a package for the computer algebra system

Acknowledgement: The second author is grateful to the Chinese Scholarship Council (CSC) for providing partial financial support. Both authors thank G. Studzinski for valuable discussions about noncommutative Gröbner bases. Special thanks go to the anonymous referees for their very careful reading of the paper and many constructive suggestions.

Martin Kreuzer, Fakultät für Informatik und Mathematik, Universität Passau, D-94030 Passau, Germany, e-mail: martin.kreuzer@uni-passau.de
Xingqiang Xiu, College of Mathematics and Statistics, Hainan University, 99 Longkun Road, Haikou 571158, China, e-mail: xiuxingqiang@gmail.com

https://doi.org/10.1515/9783110638387-015

ApCoCoA (see [1]). This procedure is based on the noncommutative version of the classical algorithm by B. Buchberger, and in this paper we describe some techniques to speed up the computations.

In commutative algebra and algebraic geometry, Gröbner bases have become a fundamental tool for computations ever since B. Buchberger's thesis [4]. The most time-consuming part in Buchberger's algorithm is the computation of the normal remainder of an S-polynomial corresponding to a critical pair. Therefore, a significant amount of energy has been spent on reducing the number of critical pairs which have to be treated. After the discovery of various criteria for discarding critical pairs ahead of time by B. Buchberger and H. M. Möller (see [5], [6], and [18]), this subject found an initial resolution via the *Gebauer–Möller installation* presented in [11], which offers a good compromise between efficiency and the success rate for detecting unnecessary critical pairs. Later, in [7], it was shown that the criteria used by Gebauer and Möller were nearly optimal in comparison to treating a set of critical pairs which corresponds to a minimal system of generators of the syzygy module of the leading term ideal.

A somewhat different picture presents itself for the computation of Gröbner bases for two-sided ideals in noncommutative polynomial rings. The basic Gröbner basis theory in this case was described by G. H. Bergman (see [2]), T. Mora (see [19] and [20]) and others, and obstructions, the noncommutative analogue of critical pairs, were studied in [20]. Some noncommutative analogues of the criteria used by Gebauer and Möller in the commutative case were developed in previous papers, but a systematic study seems to be missing. In this paper, we try to provide such a study.

Furthermore, a number of authors endeavoured to implement efficient versions of Buchberger's procedure for noncommutative polynomial rings. In many cases, they seem to have developed specialized versions of the Gebauer–Möller criteria. For instance, the package `Plural` of the computer algebra system `Singular` is based on the letterplace technique and implements versions of the product and the chain criterion (see [24]). On the other hand, the system `Magma` is based on a variant of the F4 Algorithm which does not use criteria for unnecessary obstructions. For an overview on some rules which have been implemented, see also [8]. However, the precise rules and criteria which have been used for particular systems are frequently not available.

In this paper, we present generalizations of the Gebauer–Möller criteria for noncommutative polynomials. They cover not only the known cases of useless obstructions discussed in [20], Lemma 5.11 and [8], but form a complete analogue of the results in the commutative case. One of the key ingredients we use for this purpose is the consideration of obstructions without overlaps. We detect useless obstructions, that is, obstructions that can be represented by other obstructions, using not only obstructions with overlaps but using also those without overlaps. We show that this approach does not increase unnecessary computations, since a Gröbner representation is inherent in the S-polynomial of every obstruction without overlaps. Consequently, we reduce the number of obstructions more efficiently and obtain improved noncommutative versions of the Gebauer–Möller criteria.

This paper is organized as follows. In Section 2, we recall the basic theory of Gröbner bases for two-sided ideals in noncommutative polynomial rings. In particular, we introduce and study obstructions (see Definitions 2.5 and 2.13, and Lemmas 2.8 and 2.12), present the Buchberger criterion (see Proposition 2.14), and formulate the Buchberger procedure (see Theorem 2.15). The noncommutative analogues of the Gebauer–Möller criteria are developed in Section 3. They are based on a careful study of the set of newly constructed obstructions which are produced during the execution of Buchberger's procedure. As a result, we are able to formulate the noncommutative multiply criterion (see Proposition 3.5), the noncommutative leading word criterion (see Proposition 3.6), noncommutative tail reduction (see Proposition 3.8), and the noncommutative backward criterion (see Proposition 3.13). When we combine these criteria, the result is a new Improved Buchberger procedure 4.1 in Section 4. Then, in Section 5, we compare the versions of the noncommutative Gebauer–Möller criteria studied previously by T. Mora in [20] and [22] to our new rules. In particular, we show that the cases of Mora's lemma (cf. [20], Lemma 5.11) are a strict subset of the cases covered by our noncommutative backward criterion.

The second author has implemented a version of the Buchberger procedure for noncommutative polynomial rings in a package for the computer algebra system ApCoCoA which includes the noncommutative Gebauer–Möller criteria developed here (see [1]). All examples were computed using this package. In the last section, we present some experimental results illustrating the efficiency of the criteria for some cases of moderately difficult Gröbner basis computations arising in group theory. Unless mentioned otherwise, we adhere to the definitions and terminology given in [13] and [14].

2 Gröbner bases in $K\langle X\rangle$

In the following, we let $X = \{x_1, \ldots, x_n\}$ be a finite set of indeterminates (or a finite alphabet), and $\langle X\rangle$ the monoid of all **words** (or **terms**) $x_{i_1} \cdots x_{i_l}$ where the multiplication is concatenation of words. The empty word will be denoted by 1. Furthermore, let K be a field, and let

$$K\langle X\rangle = \{c_1 w_1 + \cdots + c_s w_s \mid s \in \mathbb{N}, c_i \in K \setminus \{0\}, w_i \in \langle X\rangle\}$$

be the noncommutative polynomial ring generated by X over K (or the free associative K-algebra generated by X). Let us introduce some fundamental notions of Gröbner basis theory in this setting.

Definition 2.1. A **word ordering** on $\langle X\rangle$ is a well-ordering σ which is compatible with multiplication, that is, for which $w_1 \geq_\sigma w_2$ implies $w_3 w_1 w_4 \geq_\sigma w_3 w_2 w_4$ for all words $w_1, w_2, w_3, w_4 \in \langle X\rangle$.

In the commutative case, a word ordering is usually called a **term ordering** or a **monomial ordering**. For instance, the *length-lexicographic ordering* LLex is a word ordering. It first compares the length of two words and then breaks ties using the noncommutative lexicographic ordering with respect to $x_1 >_{\mathtt{LLex}} \cdots >_{\mathtt{LLex}} x_n$. Note that the noncommutative lexicographic ordering by itself is not a word ordering, since it is neither a well-ordering nor compatible with multiplication.

Definition 2.2. Let σ be a word ordering on $\langle X\rangle$.

(a) Given a polynomial $f \in K\langle X\rangle \setminus \{0\}$, there exists a unique representation $f = c_1 w_1 + \cdots + c_s w_s$ with $c_1, \ldots, c_s \in K \setminus \{0\}$ and $w_1, \ldots, w_s \in \langle X\rangle$ such that $w_1 >_\sigma \cdots >_\sigma w_s$. The word $\mathrm{Lw}_\sigma(f) = w_1$ is called the **leading word** of f with respect to σ. The element $\mathrm{Lc}_\sigma(f) = c_1$ is called the **leading coefficient**. We let $\mathrm{Lm}_\sigma(f) = c_1 w_1$ and call it the **leading monomial** of f.

(b) Let $I \subseteq K\langle X\rangle$ be a two-sided ideal. The set $\mathrm{Lw}_\sigma\{I\} = \{\mathrm{Lw}_\sigma(f) \mid f \in I \setminus \{0\}\} \subseteq \langle X\rangle$ is called the **leading word set** of I. The two-sided ideal $\mathrm{Lw}_\sigma(I) = \langle \mathrm{Lw}_\sigma(f) \mid f \in I \setminus \{0\}\rangle \subseteq K\langle X\rangle$ is called the **leading word ideal** of I.

(c) A subset G of a two-sided ideal $I \subseteq K\langle X\rangle$ is called a **σ-Gröbner basis** of I if the set of the leading words $\mathrm{Lw}_\sigma\{G\} = \{\mathrm{Lw}_\sigma(f) \mid f \in G \setminus \{0\}\}$ generates the leading word ideal $\mathrm{Lw}_\sigma(I)$.

In the following, we focus on computations of Gröbner bases for two-sided ideals in $K\langle X\rangle$. For readers who want to know further properties and applications of noncommutative Gröbner bases, we refer to [20] and [25]. Throughout this paper, we assume that σ is a word ordering on $\langle X\rangle$. The next algorithm is a central part of all Gröbner basis computations.

Theorem 2.3 (The division algorithm). *Let $f \in K\langle X\rangle$, $s \geq 1$, and $G = \{g_1, \ldots, g_s\} \subseteq K\langle X\rangle \setminus \{0\}$. Consider the following sequence of instructions:*

(D1) *Let $k_1 = \cdots = k_s = 0$, $p = 0$, and $v = f$.*

(D2) *Find the smallest index $i \in \{1, \ldots, s\}$ such that $\mathrm{Lw}_\sigma(v) = w\,\mathrm{Lw}_\sigma(g_i)w'$ for some words $w, w' \in \langle X\rangle$. If such an i exists, increase k_i by 1, set $c_{ik_i} = \frac{\mathrm{Lc}_\sigma(v)}{\mathrm{Lc}_\sigma(g_i)}$, $w_{ik_i} = w$, $w'_{ik_i} = w'$, and replace v by $v - c_{ik_i} w_{ik_i} g_i w'_{ik_i}$.*

(D3) *Repeat step (D2) until there is no more $i \in \{1, \ldots, s\}$ such that $\mathrm{Lw}_\sigma(v)$ is a multiple of $\mathrm{Lw}_\sigma(g_i)$. If now $v \neq 0$, then replace p by $p + \mathrm{Lm}_\sigma(v)$ and v by $v - \mathrm{Lm}_\sigma(v)$, and continue with step (D2).*

(D4) *Return the tuples $(c_{11}, w_{11}, w'_{11}), \ldots, (c_{sk_s}, w_{sk_s}, w'_{sk_s})$ and p.*

This is an algorithm which returns tuples $(c_{11}, w_{11}, w'_{11}), \ldots, (c_{sk_s}, w_{sk_s}, w'_{sk_s})$ and a polynomial $p \in K\langle X\rangle$ such that the following conditions are satisfied:

(a) *We have $f = \sum_{i=1}^{s} \sum_{j=1}^{k_i} c_{ij} w_{ij} g_i w'_{ij} + p$.*

(b) *No element of $\mathrm{Supp}(p)$ is contained in $\langle \mathrm{Lw}_\sigma(g_1), \ldots, \mathrm{Lw}_\sigma(g_s)\rangle$.*

(c) *For all* $i \in \{1,\dots,s\}$ *and all* $j \in \{1,\dots,k_i\}$, *we have* $\mathrm{Lw}_\sigma(w_{ij}g_iw'_{ij}) \leq_\sigma \mathrm{Lw}_\sigma(f)$. *If* $p \neq 0$, *we have* $\mathrm{Lw}_\sigma(p) \leq_\sigma \mathrm{Lw}_\sigma(f)$.

(d) *For all* $i \in \{1,\dots,s\}$ *and all* $j \in \{1,\dots,k_i\}$, *we have* $\mathrm{Lw}_\sigma(w_{ij}g_iw'_{ij}) \notin \langle \mathrm{Lw}_\sigma(g_1),\dots,\mathrm{Lw}_\sigma(g_{i-1})\rangle$.

A polynomial $p \in K\langle X\rangle$ obtained in Theorem 2.3 is called a **normal remainder** of f with respect to G and is denoted by $\mathrm{NR}_{\sigma,G}(f)$. Note that the resulting tuples $(c_{11}, w_{11}, w'_{11}), \dots, (c_{sk_s}, w_{sk_s}, w'_{sk_s})$ and polynomial p satisfying conditions (a)–(d) are *not* unique. This is due to the fact that in step (D2) of the division algorithm there might exist more that one pair (w, w') satisfying $\mathrm{Lw}_\sigma(v) = w\,\mathrm{Lw}_\sigma(g_i)w'$. The following example is a case in point.

Example 2.4. Consider the free monoid ring $\mathbb{Q}\langle x, y, z\rangle$ equipped with the word ordering $\sigma = \mathtt{LLex}$. We use the division algorithm to divide $f = zx^2yx$ by the tuple (g_1, g_2) where $g_1 = xy + x$ and $g_2 = x^2 + xz$. Note that $\mathrm{Lw}_\sigma(g_1) = xy$ and $\mathrm{Lw}_\sigma(g_2) = x^2$. Let us follow the steps of the algorithm:

(1) Let $k_1 = k_2 = 0$, $p = 0$, and $v = f = zx^2yx$.

(2) Since $\mathrm{Lw}_\sigma(v) = zx\,\mathrm{Lw}_\sigma(g_1)x$, we set $k_1 = 1$, $c_{11} = 1$, $w_{11} = zx$, $w'_{11} = x$, and replace v by $v - c_{11}w_{11}g_1w'_{11} = -zx^3$.

(2) Since $\mathrm{Lw}_\sigma(v) = zx^3 = z\,\mathrm{Lw}_\sigma(g_2)x$, we set $k_2 = 1$, $c_{21} = -1$, $w_{21} = z$, $w'_{21} = x$, and replace v by $v - c_{21}w_{21}g_2w'_{21} = zxzx$.

(3) Since $\mathrm{Lw}_\sigma(v)$ is not a multiple of $\mathrm{Lw}_\sigma(g_1)$ or $\mathrm{Lw}_\sigma(g_2)$, we replace p by $p + zxzx$ and v by $v - zxzx = 0$.

(4) The algorithm returns the tuples $(c_{11}, w_{11}, w'_{11})$, $(c_{21}, w_{21}, w'_{21})$, and the polynomial $p = zxzx$.

Thus we get the representation $f = zxg_1x - zg_2x + zxzx$. Observe that there was another choice for (w_{21}, w'_{21}) in the second iteration of Step (2), namely $(zx, 1)$. It would have given the representation $f = zxg_1x - zxg_2 + zg_2z - zxz^2$.

For $s \geq 1$, we let $F_s = (K\langle X\rangle \otimes_K K\langle X\rangle)^s$ be the free two-sided $K\langle X\rangle$-module of rank s with the canonical basis $\{e_1,\dots,e_s\}$, where $e_i = (0,\dots,0,1\otimes 1,0,\dots,0)$ with $1\otimes 1$ occurring in the ith position for $i = 1,\dots,s$, and we let $\mathbb{T}(F_s)$ be the set of terms in F_s, that is, $\mathbb{T}(F_s) = \{we_iw' \mid i \in \{1,\dots,s\}, w, w' \in \langle X\rangle\}$.

Definition 2.5. Let $G = \{g_1,\dots,g_s\} \subseteq K\langle X\rangle \setminus \{0\}$ with $s \geq 1$, and let $i, j \in \{1,\dots,s\}$ such that $i \leq j$.

(a) If there exist some words $w_i, w'_i, w_j, w'_j \in \langle X\rangle$ such that $w_i\,\mathrm{Lw}_\sigma(g_i)w'_i = w_j\,\mathrm{Lw}_\sigma(g_j)w'_j$, then we call the element

$$\mathbf{o}_{i,j}(w_i, w'_i; w_j, w'_j) = \frac{1}{\mathrm{Lc}_\sigma(g_i)}w_ie_iw'_i - \frac{1}{\mathrm{Lc}_\sigma(g_j)}w_je_jw'_j \in F_s$$

an **obstruction** of g_i and g_j, provided it is nonzero. If $i = j$, it is called a **self obstruction** of g_i. We denote the set of all obstructions of g_i and g_j by $\mathrm{Obs}(i,j)$.

(b) Let $\mathbf{o}_{i,j}(w_i, w_i'; w_j, w_j') \in \mathrm{Obs}(i,j)$ be an obstruction of g_i and g_j. The polynomial

$$S_{i,j}(w_i, w_i'; w_j, w_j') = \frac{1}{\mathrm{LC}_\sigma(g_i)} w_i g_i w_i' - \frac{1}{\mathrm{LC}_\sigma(g_j)} w_j g_j w_j' \in K\langle X\rangle$$

is called the **S-polynomial** of $\mathbf{o}_{i,j}(w_i, w_i'; w_j, w_j')$.

Let us look at some sets of obstructions in a concrete case.

Example 2.6. Consider the free monoid ring $\mathbb{Q}\langle x,y,z\rangle$ equipped with the word ordering σ = LLex. Let $g_1 = 2x^2 + yx$ and $g_2 = xy + zy$. Then we have $\mathrm{Lw}_\sigma(g_1) = x^2$ and $\mathrm{Lw}_\sigma(g_2) = xy$. This leads to the following sets of obstructions:

(a) We have $\mathrm{Obs}(1,1) = \{\mathbf{o}_{1,1}(x,1;1,x)\} \cup \{\mathbf{o}_{1,1}(x^2w,1;1,x^2w) \mid w \in \langle X\rangle\}$.

(b) We have $\mathrm{Obs}(1,2) = \{\mathbf{o}_{1,2}(1,y;x,1)\} \cup \{\mathbf{o}_{1,2}(xyw,1;1,wx^2) \mid w \in \langle X\rangle\} \cup \{\mathbf{o}_{1,2}(1,wxy; x^2w,1) \mid w \in \langle X\rangle\}$.

(c) We have $\mathrm{Obs}(2,2) = \{\mathbf{o}_{2,2}(xyw,1;1,wxy) \mid w \in \langle X\rangle\}$.

Using these definitions, we can characterize Gröbner bases in the following way which is a noncommutative analog of the famous *Buchberger criterion*.

Proposition 2.7. *Let $G = \{g_1,\dots,g_s\} \subseteq K\langle X\rangle \setminus \{0\}$ be a set of polynomials which generate a two-sided ideal $I = \langle G\rangle \subseteq K\langle X\rangle$. Then the following conditions are equivalent:*

(a) *The set G is a σ-Gröbner basis of I.*

(b) *For every obstruction $\mathbf{o}_{i,j}(w_i, w_i'; w_j, w_j')$ in the set $\bigcup_{1\le i\le j\le s} \mathrm{Obs}(i,j)$, its S-polynomial $S_{i,j}(w_i, w_i'; w_j, w_j')$ has a representation*

$$S_{i,j}(w_i, w_i'; w_j, w_j') = \sum_{k=1}^{\mu} c_k w_k g_{i_k} w_k'$$

with $c_k \in K$, $w_k, w_k' \in \langle X\rangle$, and $g_{i_k} \in G$ for all $k \in \{1,\dots,\mu\}$ such that $\mathrm{Lw}_\sigma(w_j g_j w_j') >_\sigma \mathrm{Lw}_\sigma(w_k g_{i_k} w_k')$ if $c_k \neq 0$ for some $k \in \{1,\dots,\mu\}$.

Proof. Standard proofs are available, for instance, in [8], Theorem 1.5 and [20], Theorem 3.7. For a proof using the lifting of syzygies (in the spirit of [13], Chapter 2), see [25], Proposition 4.1.2. □

A presentation of $S_{i,j}(w_i, w_i'; w_j, w_j')$ as in Proposition 2.7.b is called a (weak) **Gröbner representation** of the S-polynomial $S_{i,j}(w_i, w_i'; w_j, w_j')$ in terms of G.

Observe that there are infinitely many obstructions in each set $\mathrm{Obs}(i,j)$, due to the following two types of **trivial obstructions**:

(T1) If $\mathbf{o}_{i,j}(w_i, w_i'; w_j, w_j') \in \mathrm{Obs}(i,j)$, then, for all $w, w' \in \langle X\rangle$, we have $\mathbf{o}_{i,j}(ww_i, w_i'w'; ww_j, w_j'w') \in \mathrm{Obs}(i,j)$.

(T2) For all $w \in \langle X\rangle$, we have $\mathbf{o}_{i,j}(\mathrm{Lw}_\sigma(g_j)w, 1; 1, w\,\mathrm{Lw}_\sigma(g_i))$, $\mathbf{o}_{i,j}(1, w\,\mathrm{Lw}_\sigma(g_j); \mathrm{Lw}_\sigma(g_i)w, 1) \in \mathrm{Obs}(i,j)$.

Before going on, let us get rid of these two types of trivial obstructions. The following lemma handles trivial obstructions of type (T1).

Lemma 2.8. *If the S-polynomial of $\mathbf{o}_{i,j}(w_i, w_i'; w_j, w_j') \in \mathrm{Obs}(i,j)$ has a Gröbner representation in terms of G, then for all $w, w' \in \langle X\rangle$, the S-polynomial of $\mathbf{o}_{i,j}(ww_i, w_i'w'; ww_j, w_j'w')$ also has a Gröbner representation in terms of G.*

Proof. Without loss of generality, we assume that $S_{i,j}(w_i, w_i'; w_j, w_j')$ is non-zero. We write $S_{i,j}(w_i, w_i'; w_j, w_j') = \sum_{k=1}^{\mu} c_k w_k g_{i_k} w_k'$, where $c_k \in K \setminus \{0\}$, $w_k, w_k' \in \langle X\rangle$, and $g_{i_k} \in G$ such that $\mathrm{Lw}_\sigma(w_j g_j w_j') >_\sigma \mathrm{Lw}_\sigma(w_k g_{i_k} w_k')$ for all $k \in \{1, \dots, \mu\}$. For all $w, w' \in \langle X\rangle$, it is clear that $S_{i,j}(ww_i, w_i'w'; ww_j, w_j'w') = \sum_{k=1}^{\mu} c_k ww_k g_{i_k} w_k'w'$. Since the word ordering σ is compatible with multiplication, we have $w\,\mathrm{Lw}_\sigma(w_j g_j w_j')w' >_\sigma w\,\mathrm{Lw}_\sigma(w_k g_{i_k} w_k')w'$ for all $k \in \{1, \dots, \mu\}$. Hence we have $\mathrm{Lw}_\sigma(ww_j g_j w_j'w') >_\sigma \mathrm{Lw}_\sigma(ww_k g_{i_k} w_k'w')$ for all $k \in \{1, \dots, \mu\}$ and $S_{i,j}(ww_i, w_i'w'; ww_j, w_j'w') = \sum_{k=1}^{\mu} c_k ww_k g_{i_k} w_k'w'$ is a Gröbner representation in terms of G. □

To deal with trivial obstructions of type (T2), we introduce some terminology as follows.

Definition 2.9. Let $G = \{g_1, \dots, g_s\} \subseteq K\langle X\rangle \setminus \{0\}$ with $s \geq 1$.

(a) Let $w_1, w_2 \in \langle X\rangle$ be two words. If there exist some words $w, w', w'' \in \langle X\rangle$ and $w \neq 1$ such that $w_1 = w'w$ and $w_2 = ww''$, or $w_1 = ww'$ and $w_2 = w''w$, or $w_1 = w$ and $w_2 = w'ww''$, or $w_1 = w'ww''$ and $w_2 = w$, then we say w_1 and w_2 have an **overlap** at w. Otherwise, we say that w_1 and w_2 have **no overlap**.

(b) Let $\mathbf{o}_{i,j}(w_i, w_i'; w_j, w_j') \in \mathrm{Obs}(i,j)$ be an obstruction. We say that this obstruction has **no overlap** if $\mathrm{Lw}_\sigma(g_i)$ and $\mathrm{Lw}_\sigma(g_j)$ have no overlap inside $w_i \,\mathrm{Lw}_\sigma(g_i) w_i' = w_j \,\mathrm{Lw}_\sigma(g_j) w_j'$, in the sense that there is a word w'' such that $w_i \,\mathrm{Lw}_\sigma(g_i) w_i' = w_i \,\mathrm{Lw}_\sigma(g_i) w'' \,\mathrm{Lw}_\sigma(g_j) w_j'$.

Clearly, if $\mathrm{Lw}_\sigma(g_i)$ and $\mathrm{Lw}_\sigma(g_j)$ have no overlap, then every obstruction of g_i and g_j has no overlap. The following example, given in [8], shows that the converse is not true in general.

Example 2.10. Let $g_i, g_j \in K\langle x_1, x_2, x_3\rangle$ be polynomials such that $\mathrm{Lw}_\sigma(g_1) = x_1x_2$ and $\mathrm{Lw}_\sigma(g_2) = x_2x_3$. Clearly, these two leading words have an overlap, namely x_2. However, an obstruction of the form $\mathbf{o}_{12}(1, x_2^i x_3; x_1x_2^i, 1)$ with $i \geq 1$ yields $w_1 \,\mathrm{Lw}_\sigma(g_1) w_1' = x_1 x_2^{i+1} x_3$ and the two leading words do not overlap inside this word.

Let us illustrate the definition with a further example.

Example 2.11. Consider the free monoid ring $\mathbb{F}_2\langle x_1, \dots, x_4\rangle$, equipped with the word ordering σ = LLex. Given two polynomials g_1, g_2 with $\mathrm{Lw}_\sigma(g_1) = x_3(x_1x_2)^3$ and $\mathrm{Lw}_\sigma(g_2) = (x_2x_1)^3x_4$, it is easy to check that the obstructions in $\mathrm{Obs}(1,1)$ and $\mathrm{Obs}(2,2)$ have no overlaps. However, the set $\mathrm{Obs}(1,2)$ contains three obstructions with overlaps, namely:

(1) The obstruction $\mathbf{o}_{12}(1, x_1x_4; x_3x_1, 1)$ has the overlap $x_2x_1x_2x_1x_2$.
(2) The obstruction $\mathbf{o}_{12}(1, x_1x_2x_1x_4; x_3x_1x_2x_1)$ has the overlap $x_2x_1x_2$.
(3) The obstruction $\mathbf{o}_{12}(1, (x_1x_2)^2x_1x_4; x_3(x_1x_2)^2x_1)$ has the overlap x_2.

As shown in (T2), there are infinitely many obstructions without overlaps in each set $\mathrm{Obs}(i, j)$. The following lemma gets rid of these trivial obstructions.

Lemma 2.12. *If $\mathbf{o}_{i,j}(w_i, w_i'; w_j, w_j') \in \mathrm{Obs}(i, j)$ has no overlap, then $S_{i,j}(w_i, w_i'; w_j, w_j')$ has a Gröbner representation in terms of G.*

Proof. See [20], Lemma 5.4. □

Observe that Lemma 2.12 is indeed a noncommutative version of the *product criterion* (or *criterion 2*) of Buchberger (cf. [6]).

Definition 2.13. Let $G = \{g_1, \ldots, g_s\} \subseteq K\langle X\rangle \setminus \{0\}$ with $s \geq 1$.
(a) Let $i, j \in \{1, \ldots, s\}$ and $i < j$. An obstruction in $\mathrm{Obs}(i, j)$ is called **nontrivial** if it has an overlap and is of the form $\mathbf{o}_{i,j}(w_i, 1; 1, w_j')$, or $\mathbf{o}_{i,j}(1, w_i'; w_j, 1)$, or $\mathbf{o}_{i,j}(w_i, w_i'; 1, 1)$, or $\mathbf{o}_{i,j}(1, 1; w_j, w_j')$ with $w_i, w_i', w_j, w_j' \in \langle X\rangle$.
(b) Let $i \in \{1, \ldots, s\}$. A self obstruction in $\mathrm{Obs}(i, i)$ is called **nontrivial** if it has an overlap and is of the form $\mathbf{o}_{i,i}(1, w_i'; w_i, 1)$ with $w_i, w_i' \in \langle X\rangle \setminus \{1\}$.
(c) Let $i, j \in \{1, \ldots, s\}$ and $i \leq j$. The set of all nontrivial obstructions of g_i and g_j will be denoted by $\mathrm{NTO}(i, j)$.

In the literature, a nontrivial obstruction of the form $\mathbf{o}_{i,j}(w_i, 1; 1, w_j')$ is called a **left obstruction**, a nontrivial obstruction of the form $\mathbf{o}_{i,j}(1, w_i'; w_j, 1)$ is called a **right obstruction**, and a nontrivial obstruction which is of the form $\mathbf{o}_{i,j}(w_i, w_i'; 1, 1)$ or $\mathbf{o}_{i,j}(1, 1; w_j, w_j')$ is called a **center obstruction**. We picture these four types of obstructions as follows:

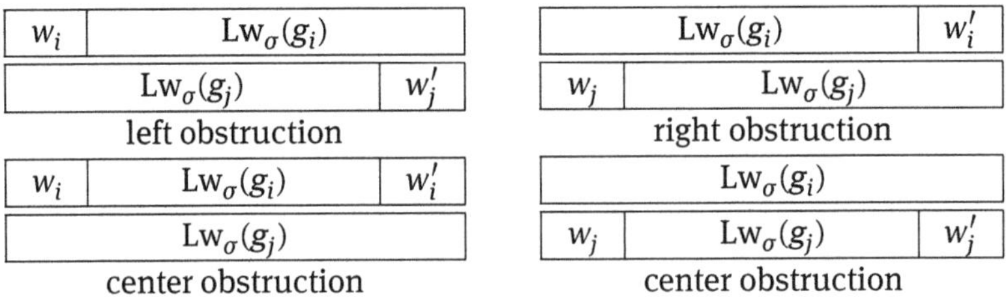

At this point, we can refine the characterization of Gröbner bases given in Proposition 2.7 in the following way.

Proposition 2.14 (The Buchberger criterion). *Let $G = \{g_1, \ldots, g_s\} \subseteq K\langle X\rangle$ be a set of nonzero polynomials which generate a two-sided ideal $I = \langle G\rangle \subseteq K\langle X\rangle$. Then the set G is a σ-Gröbner basis of I if and only if, for each nontrivial obstruction $\mathbf{o}_{i,j}(w_i, w_i'; w_j, w_j') \in$*

$\bigcup_{1\le i\le j\le s} \mathrm{NTO}(i,j)$, *its S-polynomial* $S_{i,j}(w_i, w_i'; w_j, w_j')$ *has a Gröbner representation in terms of* G.

Proof. This follows directly from Proposition 2.7 and Lemmas 2.8 and 2.12. In view of Lemma 2.12, it suffices to consider each obstruction with overlap, which is a nontrivial obstruction or a proper multiple of a nontrivial obstruction. Further, Lemma 2.8 treats a proper multiple of a nontrivial obstruction via the corresponding nontrivial obstruction. Therefore, it is sufficient to consider only nontrivial obstructions. □

The Buchberger criterion enables us to formulate the following procedure for computing Gröbner bases of two-sided ideals. Note that, in the procedure, by a **fair strategy** we mean a selection strategy which ensures that every obstruction is selected eventually. Since these Gröbner bases need not be finite, we have to content ourselves with an enumerating procedure.

Theorem 2.15 (The Buchberger procedure). *Let* $s \ge 1$, *and let* $G = \{g_1, \dots, g_s\} \subseteq K\langle X\rangle$ *be a set of nonzero polynomials which generate a two-sided ideal* $I = \langle G\rangle \subseteq K\langle X\rangle$. *Consider the following sequence of instructions:*

(B1) *Let* $B = \bigcup_{1\le i\le j\le s} \mathrm{NTO}(i,j)$.

(B2) *If* $B = \emptyset$, *return the result* G. *Otherwise, select an obstruction* $\mathbf{o}_{i,j}(w_i, w_i'; w_j, w_j') \in B$ *using a fair strategy and delete it from* B.

(B3) *Compute the S-polynomial* $S = S_{i,j}(w_i, w_i'; w_j, w_j')$ *and its normal remainder* $S' = \mathrm{NR}_{\sigma,G}(S)$. *If* $S' = 0$, *continue with step* (B2).

(B4) *Increase* s *by one, append* $g_s = S'$ *to the set* G, *and append the set of obstructions* $\bigcup_{1\le i\le s} \mathrm{NTO}(i,s)$ *to the set* B. *Then continue with step* (B2).

This is a procedure that enumerates a σ*-Gröbner basis* G *of* I. *If* I *has a finite* σ*-Gröbner basis, the procedure stops after finitely many steps, and the resulting set* G *is a finite* σ*-Gröbner basis of* I.

Proof. Note that this is a straightforward generalization of the commutative version of Buchberger's algorithm to the noncommutative case. We refer to [20] for the original form of this procedure and to [25], Theorem 4.1.14 for a detailed proof. □

3 Noncommutative Gebauer–Möller criteria

In this section, we present noncommutative Gebauer–Möller criteria. They check whether an obstruction can be represented by "smaller" obstructions. If so, we declare such obstructions to be *unnecessary*. Before going into details, for $s \ge 1$, we define a certain well-ordering τ on $\mathbb{T}(F_s) = \{we_iw' \mid i \in \{1,\dots,s\}, w, w' \in \langle X\rangle\}$ and use it to order obstructions. In the following, let $s \ge 1$, and let $G = \{g_1, \dots, g_s\} \subseteq K\langle X\rangle \setminus \{0\}$ be a finite set of noncommutative polynomials.

Definition 3.1. Let us define a relation τ on $\mathbb{T}(F_s)$ as follows. For two terms $w_1 e_i w_1'$, $w_2 e_j w_2' \in \mathbb{T}(F_s)$, we let $w_1 e_i w_1' \geq_\tau w_2 e_j w_2'$ if:
(a) $w_1 \operatorname{Lw}_\sigma(g_i) w_1' >_\sigma w_2 \operatorname{Lw}_\sigma(g_j) w_2'$, or
(b) $w_1 \operatorname{Lw}_\sigma(g_i) w_1' = w_2 \operatorname{Lw}_\sigma(g_j) w_2'$ and $i > j$, or
(c) $w_1 \operatorname{Lw}_\sigma(g_i) w_1' = w_2 \operatorname{Lw}_\sigma(g_j) w_2'$ and $i = j$ and $w_1 \geq_\sigma w_2$.

One can check that τ is a well-ordering and is compatible with scalar multiplication. The relation τ is called the module term ordering **induced by** (σ, G) on $\mathbb{T}(F_s)$.

By definition, for every obstruction $\mathbf{o}_{i,j}(w_i, w_i'; w_j, w_j') \in \bigcup_{1\le i\le j\le s} \operatorname{Obs}(i,j)$, we have $w_i e_i w_i' <_\tau w_j e_j w_j'$. We extend the ordering τ to the set of obstructions $\bigcup_{1\le i'\le j'\le s} \operatorname{Obs}(i',j')$ by committing the following slight abuse of notation.

Definition 3.2. Let τ be the module term ordering induced by (σ, G) on $\mathbb{T}(F_s)$. Let $\mathbf{o}_{i,j}(w_i, w_i'; w_j, w_j')$, $\mathbf{o}_{k,l}(w_k, w_k'; w_l, w_l')$ be two obstructions in the set $\bigcup_{1\le i'\le j'\le s} \operatorname{Obs}(i',j')$. If we have $w_j e_j w_j' >_\tau w_l e_l w_l'$, or if we have $w_j e_j w_j' = w_l e_l w_l'$ and $w_i e_i w_i' \geq_\tau w_k e_k w_k'$, then we let $\mathbf{o}_{i,j}(w_i, w_i'; w_j, w_j') \geq_\tau \mathbf{o}_{k,l}(w_k, w_k'; w_l, w_l')$. The ordering τ is called the ordering **induced by** (σ, G) on the set of obstructions.

One can verify that τ is also a well-ordering on $\bigcup_{1\le i\le j\le s} \operatorname{Obs}(i,j)$ and compatible with scalar multiplication.

Now we are ready to generalize the commutative Gebauer–Möller criteria (see [7] and [11]) to the noncommutative case. Recall that, in step (B4) of the Buchberger procedure, when a new generator g_s is added, we immediately construct new obstructions $\bigcup_{1\le i\le s} \operatorname{NTO}(i,s)$. We want to detect unnecessary obstructions in the set $\bigcup_{1\le i\le s} \operatorname{NTO}(i,s)$ of newly constructed obstructions as well as in the set $\bigcup_{1\le i\le j\le s-1} \operatorname{NTO}(i,j)$ of previously constructed obstructions. We achieve this goal via the following three steps.

First, we detect unnecessary obstructions in the set $\bigcup_{1\le i\le s} \operatorname{NTO}(i,s)$ with the aid of other obstructions also in this set. This step is called a *head reduction step* in [7]. Second, we detect unnecessary obstructions in the set $\bigcup_{1\le i\le s} \operatorname{NTO}(i,s)$ with the aid of obstructions in $\bigcup_{1\le i\le j\le s-1} \operatorname{NTO}(i,j)$. This step is called a *tail reduction step* in [7]. Third, we detect unnecessary obstructions in the set $\bigcup_{1\le i\le j\le s-1} \operatorname{NTO}(i,j)$ using the new generator g_s. Indeed, the first step corresponds to the commutative Gebauer–Möller criterion M, the second one to F, and the last step corresponds to criterion B_k (cf. [11], Subsection 3.4). In the following, they are the contents of Propositions 3.5, 3.6, and 3.13, respectively.

The next lemma helps us to implement the first step, that is, to detect unnecessary obstructions in the set $\bigcup_{1\le i\le s} \operatorname{NTO}(i,s)$ of newly constructed obstructions via other obstructions in this set.

Lemma 3.3. *Let $\mathbf{o}_{i,s}(w_i, w_i'; u_s, u_s')$ and $\mathbf{o}_{j,s}(w_j, w_j'; v_s, v_s')$ be two distinct non-trivial obstructions in $\bigcup_{1\le i'\le s} \operatorname{NTO}(i',s)$ with two words $w, w' \in \langle X\rangle$ satisfying $u_s = w v_s$ and $u_s' = v_s' w'$.*

(a) *If $i < j$ and $ww' \neq 1$, then we have*

$$\mathbf{o}_{i,s}(w_i, w_i'; u_s, u_s') = w\,\mathbf{o}_{j,s}(w_j, w_j'; v_s, v_s')w' + \mathbf{o}_{i,j}(w_i, w_i'; ww_j, w_j'w')$$

with $\mathbf{o}_{i,s}(w_i, w_i'; u_s, u_s') >_\tau \mathbf{o}_{j,s}(w_j, w_j'; v_s, v_s')$ *and* $\mathbf{o}_{i,s}(w_i, w_i'; u_s, u_s') >_\tau \mathbf{o}_{i,j}(w_i, w_i'; ww_j, w_j'w')$*. Further, if the S-polynomials* $S_{j,s}(w_j, w_j'; v_s, v_s')$ *and* $S_{i,j}(w_i, w_i'; ww_j, w_j'w')$ *have Gröbner representations in terms of G, then so does* $S_{i,s}(w_i, w_i'; u_s, u_s')$.

(b) *If $i > j$, then we have*

$$\mathbf{o}_{i,s}(w_i, w_i'; u_s, u_s') = w\,\mathbf{o}_{j,s}(w_j, w_j'; v_s, v_s')w' - \mathbf{o}_{j,i}(ww_j, w_j'w'; w_i, w_i')$$

with $\mathbf{o}_{i,s}(w_i, w_i'; u_s, u_s') >_\tau \mathbf{o}_{j,s}(w_j, w_j'; v_s, v_s')$ *and* $\mathbf{o}_{i,s}(w_i, w_i'; u_s, u_s') >_\tau \mathbf{o}_{j,i}(ww_j, w_j'w'; w_i, w_i')$*. Further, if the S-polynomials* $S_{j,s}(w_j, w_j'; v_s, v_s')$ *and* $S_{j,i}(ww_j, w_j'w'; w_i, w_i')$ *have Gröbner representations in terms of G, then so does* $S_{i,s}(w_i, w_i'; u_s, u_s')$.

(c) *If $i = j$ and $ww' \neq 1$ or if $i = j$ and $ww' = 1$ and $w_i >_\sigma w_j$, then we have*

$$\mathbf{o}_{i,s}(w_i, w_i'; u_s, u_s') = w\,\mathbf{o}_{j,s}(w_j, w_j'; v_s, v_s')w' + \mathbf{o}_{i,j}(w_i, w_i'; ww_j, w_j'w')$$

with $\mathbf{o}_{i,s}(w_i, w_i'; u_s, u_s') >_\tau \mathbf{o}_{j,s}(w_j, w_j'; v_s, v_s')$ *and* $\mathbf{o}_{i,s}(w_i, w_i'; u_s, u_s') >_\tau \mathbf{o}_{i,j}(w_i, w_i'; ww_j, w_j'w')$*. Further, if the S-polynomials* $S_{j,s}(w_j, w_j'; v_s, v_s')$ *and* $S_{i,j}(w_i, w_i'; ww_j, w_j'w')$ *have Gröbner representations in terms of G, then so does* ${}_{i,s}(w_i, w_i'; u_s, u_s')$.

Proof. We prove case (a). Cases (b) and (c) can be proved similarly. The equation in case (a) follows from Definition 2.5.a and from the conditions $u_s = wv_s$, $u_s' = v_s'w'$ and $i < j$. Because of $ww' > 1$, we have $u_s \operatorname{Lw}(g_s)u_s' = wv_s \operatorname{Lw}(g_s)v_s'w' >_\sigma v_s \operatorname{Lw}(g_s)v_s'$. Consequently, we get $u_s e_s u_s' >_\tau v_s e_s v_s'$ and $\mathbf{o}_{i,s}(w_i, w_i'; u_s, u_s') >_\tau \mathbf{o}_{j,s}(w_j, w_j'; v_s, v_s')$. From $u_s \operatorname{Lw}(g_s)u_s' = w_i \operatorname{Lw}(g_i)w_i' = ww_j \operatorname{Lw}(g_j)w_j'w'$ and $s > j$, we get the inequalities $u_s e_s u_s' >_\tau ww_j e_j w_j'w'$ and $\mathbf{o}_{i,s}(w_i, w_i'; u_s, u_s') >_\tau \mathbf{o}_{i,j}(w_i, w_i'; ww_j, w_j'w')$.

Next, we show that, if $S_{j,s}(w_j, w_j'; v_s, v_s')$ and $S_{i,j}(w_i, w_i'; ww_j, w_j'w')$ have Gröbner representations in terms of G, then so does $S_{i,s}(w_i, w_i'; u_s, u_s')$. Clearly, we have

$$S_{i,s}(w_i, w_i'; u_s, u_s') = wS_{j,s}(w_j, w_j'; v_s, v_s')w' + S_{i,j}(w_i, w_i'; ww_j, w_j'w').$$

Without loss of generality, we assume that $S_{i,s}(w_i, w_i'; u_s, u_s')$, $S_{j,s}(w_j, w_j'; v_s, v_s')$ and $S_{i,j}(w_i, w_i'; ww_j, w_j'w')$ are nonzero. Since there is a Gröbner representation for $S_{j,s}(w_j, w_j'; v_s, v_s')$, we have

$$S_{j,s}(w_j, w_j'; v_s, v_s') = \sum_{k=1}^{\mu} a_k w_k g_{i_k} w_k'$$

with $a_k \in K \setminus \{0\}$, $w_k, w_k' \in \langle X \rangle$, $g_{i_k} \in G$ for all $k \in \{1, \dots, \mu\}$, such that $\operatorname{Lw}_\sigma(v_s g_s v_s') >_\sigma \operatorname{Lw}_\sigma(a_k w_k g_{i_k} w_k')$. Similarly, for $S_{i,j}(w_i, w_i'; ww_j, w_j'w')$ we have

$$S_{i,j}(w_i, w_i'; ww_j, w_j'w') = \sum_{l=1}^{\nu} b_l w_l g_{i_l} w_l'$$

with $b_l \in K \setminus \{0\}$, $w_l, w_l' \in \langle X \rangle$, $g_{i_l} \in G$ for all $l \in \{1, \dots, \nu\}$, such that $\mathrm{Lw}_\sigma(ww_jg_jw_j'w') >_\sigma \mathrm{Lw}_\sigma(b_lw_lg_{i_l}w_l')$. Therefore, we have

$$\begin{aligned} S_{i,s}(w_i, w_i'; u_s, u_s') &= w\left(\sum_{k=1}^{\mu} a_k w_k g_{i_k} w_k'\right)w' + \sum_{l=1}^{\nu} b_l w_l g_{i_l} w_l' \\ &= \sum_{k=1}^{\mu} a_k ww_k g_{i_k} w_k' w' + \sum_{l=1}^{\nu} b_l w_l g_{i_l} w_l'. \end{aligned}$$

As $u_s \,\mathrm{Lw}_\sigma(g_s) u_s' = wv_s \,\mathrm{Lw}_\sigma(g_s) v_s' w'$, we have $\mathrm{Lw}_\sigma(u_sg_su_s') = \mathrm{Lw}_\sigma(wv_sg_sv_s'w') >_\sigma \mathrm{Lw}_\sigma(ww_kg_{i_k}w_k'w')$ for all $k \in \{1, \dots, \mu\}$. By Definition 2.5, we have $\mathrm{Lw}_\sigma(u_sg_su_s') = \mathrm{Lw}_\sigma(w_ig_iw_i') = \mathrm{Lw}_\sigma(ww_jg_jw_j'w') >_\sigma \mathrm{Lw}_\sigma(b_lw_lg_{i_l}w_l')$ for all $l \in \{1, \dots, \nu\}$. Therefore,

$$S_{i,s}(w_i, w_i'; u_s, u_s') = \sum_{k=1}^{\mu} a_k ww_k g_{i_k} w_k' w' + \sum_{l=1}^{\nu} b_l w_l g_{i_l} w_l'$$

is a Gröbner representation of $S_{i,s}(w_i, w_i'; u_s, u_s')$. □

The following example shows that the obstruction $\mathbf{o}_{i,j}(w_i, w_i'; ww_j, w_j'w')$ in case (a) of Lemma 3.3 can be a proper multiple of a nontrivial obstruction, or an obstruction without overlap. Similar phenomena occur in cases (b) and (c) of Lemma 3.3, as well as in Lemmas 3.7 and 3.11.

Example 3.4. Consider polynomials $G = \{g_1, g_2, g_3\}$ in the noncommutative polynomial ring $K\langle x, y\rangle$.

(a) Assume that $\mathrm{Lm}_\sigma(g_1) = y^3$, $\mathrm{Lm}_\sigma(g_2) = x^2y^2$ and $\mathrm{Lm}_\sigma(g_3) = xyx^2y$. Then we have $\mathbf{o}_{1,3}(xyx^2, 1; 1, y^2), \mathbf{o}_{2,3}(xy, 1; 1, y) \in \bigcup_{1 \le i \le 3} \mathrm{NTO}(i, 3)$, and

$$\mathbf{o}_{1,3}(xyx^2, 1; 1, y^2) = \mathbf{o}_{2,3}(xy, 1; 1, y)y + \mathbf{o}_{1,2}(xyx^2, 1; xy, y).$$

Observe that $\mathbf{o}_{1,2}(xyx^2, xy; y) = xy\, \mathbf{o}_{1,2}(x^2, 1; 1, y)$ is a proper multiple of the nontrivial obstruction $\mathbf{o}_{1,2}(x^2, 1; 1, y)$.

(b) Now assume that $\mathrm{Lm}_\sigma(g_1) = (xy)^2$, $\mathrm{Lm}_\sigma(g_2) = y$ and $\mathrm{Lm}_\sigma(g_3) = xyx^2y$. Then we have $\mathbf{o}_{1,3}(xyx, 1; 1, xy), \mathbf{o}_{2,3}(x, x^2y; 1, 1) \in \bigcup_{1 \le i \le 3} \mathrm{NTO}(i, 3)$, and

$$\mathbf{o}_{1,3}(xyx, 1; 1, xy) = \mathbf{o}_{2,3}(x, x^2y; 1, 1)xy + \mathbf{o}_{1,2}(xyx, 1; x, x^2yxy).$$

One can check that $\mathbf{o}_{1,2}(xyx, 1; x, x^2yxy)$ is an obstruction without overlap.

In the following, we present the noncommutative multiply criterion. It is a noncommutative analog of the Gebauer–Möller criterion M.

Proposition 3.5 (Noncommutative multiply criterion). *Suppose that $\mathbf{o}_{i,s}(w_i, w_i'; u_s, u_s')$ and $\mathbf{o}_{j,s}(w_j, w_j'; v_s, v_s')$ are two distinct non-trivial obstructions in $\bigcup_{1 \le i' \le s} \mathrm{NTO}(i', s)$ such that there exist two words w, w' in $\langle X \rangle$ satisfying $u_s = wv_s$ and $u_s' = v_s'w'$ and $ww' \ne 1$. Then we can remove the obstruction $\mathbf{o}_{i,s}(w_i, w_i'; u_s, u_s')$ from $\bigcup_{1 \le i' \le s} \mathrm{NTO}(i', s)$ in the execution of the Buchberger procedure.*

Proof. By the previous lemma, the obstruction $\mathbf{o}_{i,s}(w_i, w_i'; u_s, u_s')$ can be represented as

$$\mathbf{o}_{i,s}(w_i, w_i'; u_s, u_s') = w\,\mathbf{o}_{j,s}(w_j, w_j'; v_s, v_s')w' + a\,\mathbf{o}_{k,l}(w_k, w_k'; w_l, w_l')$$

with $a \in \{1, -1\}$ and $k = \min\{i, j\}$, $l = \max\{i, j\}$. To prove that $\mathbf{o}_{i,s}(w_i, w_i'; u_s, u_s')$ is strictly larger than $\mathbf{o}_{j,s}(w_j, w_j'; v_s, v_s')$ and $\mathbf{o}_{k,l}(w_k, w_k'; w_l, w_l')$, we consider two cases. If $i > j$, then the result follows from Lemma 3.3.b; if $i \le j$, then the result follows from Lemma 3.3.a and 3.3.c and the condition $ww' \neq 1$. Moreover, the S-polynomial $S_{i,s}(w_i, w_i'; u_s, u_s')$ has a Gröbner representation in terms of G if $S_{j,s}(w_j, w_j'; v_s, v_s')$ and $S_{k,l}(w_k, w_k'; w_l, w_l')$ have Gröbner representations in terms of G. Theorem 2.15 ensures that $S_{j,s}(w_j, w_j'; v_s, v_s')$ has a Gröbner representation in terms of G. Note that the obstruction $\mathbf{o}_{k,l}(w_k, w_k'; w_l, w_l')$ can be a proper multiple of a nontrivial obstruction or an obstruction without overlap (for instance, see Example 3.4). If $\mathbf{o}_{k,l}(w_k, w_k'; w_l, w_l')$ is a proper multiple of a nontrivial obstruction, then Lemma 2.8 and Theorem 2.15 guarantee that $S_{k,l}(w_k, w_k'; w_l, w_l')$ has a Gröbner representation in terms of G. If $\mathbf{o}_{k,l}(w_k, w_k'; w_l, w_l')$ is an obstruction without overlap, then by Lemma 2.12, its S-polynomial has a Gröbner representation in terms of G. Now the conclusion follows from Proposition 2.14 and Theorem 2.15. □

The next result is the noncommutative version of the Gebauer–Möller criterion F.

Proposition 3.6 (Noncommutative leading word criterion). *Suppose that* $\mathbf{o}_{i,s}(w_i, w_i'; u_s, u_s')$ *and* $\mathbf{o}_{j,s}(w_j, w_j'; v_s, v_s')$ *are two distinct non-trivial obstructions in* $\bigcup_{1\le i'\le s} \mathrm{NTO}(i', s)$ *such that there exist two words* w, w' *in* $\langle X\rangle$ *satisfying* $u_s = wv_s$ *and* $u_s' = v_s'w'$. *Then* $\mathbf{o}_{i,s}(w_i, w_i'; u_s, u_s')$ *can be removed from* $\bigcup_{1\le i'\le s} \mathrm{NTO}(i', s)$ *in the execution of the Buchberger procedure if one of the following conditions is satisfied:*
(a) $i > j$.
(b) $i = j$ *and* $ww' = 1$ *and* $w_i >_\sigma w_j$.

Proof. Observe that condition (a) corresponds to Lemma 3.3.b, while condition (b) corresponds to Lemma 3.3.c. We represent $\mathbf{o}_{i,s}(w_i, w_i'; u_s, u_s')$ as

$$\mathbf{o}_{i,s}(w_i, w_i'; u_s, u_s') = w\,\mathbf{o}_{j,s}(w_j, w_j'; v_s, v_s')w' - \mathbf{o}_{j,i}(ww_j, w_j'w'; w_i, w_i').$$

By Lemma 3.3.b and 3.3.c, we have $\mathbf{o}_{i,s}(w_i, w_i'; u_s, u_s')$ is strictly larger than $\mathbf{o}_{j,s}(w_j, w_j'; v_s, v_s')$ and $\mathbf{o}_{j,i}(ww_j, w_j'w'; w_i, w_i')$. Moreover, if the S-polynomials $S_{j,s}(w_j, w_j'; v_s, v_s')$ and $S_{j,i}(ww_j, w_j'w'; w_i, w_i')$ have Gröbner representations in terms of G, then so does $S_{i,s}(w_i, w_i'; u_s, u_s')$. Theorem 2.15 ensures $S_{j,s}(w_j, w_j'; v_s, v_s')$ has a Gröbner representation in terms of G. Note that the obstruction $\mathbf{o}_{j,i}(ww_j, w_j'w'; w_i, w_i')$ can be a proper multiple of a nontrivial obstruction or an obstruction without overlap (for instance, see Example 3.4). If the obstruction $\mathbf{o}_{j,i}(ww_j, w_j'w'; w_i, w_i')$ is a proper multiple of a nontrivial obstruction, then Lemma 2.8 and Theorem 2.15 guarantee that $S_{j,i}(ww_j, w_j'w'; w_i, w_i')$ has a Gröbner representation in terms of G. If $\mathbf{o}_{j,i}(ww_j, w_j'w'; w_i, w_i')$ is an obstruction without overlap, then by Lemma 2.12, its S-polynomial has a Gröbner representation in terms of G. Now the conclusion follows from Proposition 2.14 and Theorem 2.15. □

Next, we work on detecting unnecessary obstructions in $\bigcup_{1\le i\le s}\mathrm{NTO}(i,s)$ via obstructions in the set $\bigcup_{1\le i\le j\le s-1}\mathrm{NTO}(i,j)$ of previously constructed obstructions.

Lemma 3.7. *Let $\mathbf{o}_{j,s}(u_j,u_j';w_s,w_s')$ and $\mathbf{o}_{i,j}(w_i,w_i';v_j,v_j')$ be nontrivial obstructions in $\bigcup_{1\le i'\le s}\mathrm{NTO}(i',s)$ and $\bigcup_{1\le i'\le j'\le s-1}\mathrm{NTO}(i',j')$, respectively. If there exist two words $w,w'\in\langle X\rangle$ such that $u_j=wv_j$ and $u_j'=v_j'w'$, then we have*

$$\mathbf{o}_{j,s}(u_j,u_j';w_s,w_s')=-w\,\mathbf{o}_{i,j}(w_i,w_i';v_j,v_j')w'+\mathbf{o}_{i,s}(ww_i,w_i'w';w_s,w_s')$$

where the inequalities $\mathbf{o}_{j,s}(u_j,u_j';w_s,w_s')>_\tau\mathbf{o}_{i,j}(w_i,w_i';v_j,v_j')$ and $\mathbf{o}_{j,s}(u_j,u_j';w_s,w_s')>_\tau\mathbf{o}_{i,s}(ww_i,w_i'w';w_s,w_s')$ hold. Furthermore, if the two S-polynomials $S_{i,j}(w_i,w_i';v_j,v_j')$ and $S_{i,s}(ww_i,w_i'w';w_s,w_s')$ have Gröbner representations in terms of G, then so does $S_{j,s}(u_j,u_j';w_s,w_s')$.

Proof. The claimed equality follows from Definition 2.5.a and from the conditions $u_j=wv_j$ and $u_j'=v_j'w'$. We have $\mathbf{o}_{j,s}(u_j,u_j';w_s,w_s')>_\tau\mathbf{o}_{i,j}(w_i,w_i';v_j,v_j')$ for $w_se_sw_s'>_\tau u_je_ju_j'=wv_je_jv_j'w\ge_\tau v_je_jv_j'$. From the inequality $u_je_ju_j'=wv_je_jv_j'w>_\tau ww_ie_iw_i'w'$, it follows that $\mathbf{o}_{j,s}(u_j,u_j';w_s,w_s')>_\tau\mathbf{o}_{i,s}(ww_i,w_i'w';w_s,w_s')$. Again, we can prove the second part by following the same argument as in the proof of Lemma 3.3.a. □

Note that the obstruction $\mathbf{o}_{i,s}(ww_i,w_i'w';w_s,w_s')$ in Lemma 3.7 can be a proper multiple of a nontrivial obstruction or an obstruction without overlap. However, it suffices for us to consider only the latter case, since the former case has been considered in Proposition 3.5, and, more precisely, in Lemma 3.3.b.

Proposition 3.8 (Noncommutative tail reduction). *Suppose that $\mathbf{o}_{j,s}(u_j,u_j';w_s,w_s')$ and $\mathbf{o}_{i,j}(w_i,w_i';v_j,v_j')$ are nontrivial obstructions in $\bigcup_{1\le i'\le s}\mathrm{NTO}(i',s)$ and $\bigcup_{1\le i'\le j'\le s-1}\mathrm{NTO}(i',j')$, respectively, such that there exist two words $w,w'\in\langle X\rangle$ satisfying $u_j=wv_j$ and $u_j'=v_j'w'$. If the word ww_i is a multiple of $w_s\,\mathrm{Lw}_\sigma(g_s)$, or if the word $w_i'w'$ is a multiple of $\mathrm{Lw}_\sigma(g_s)w_s'$, then $\mathbf{o}_{j,s}(u_j,u_j';w_s,w_s')$ can be removed from $\bigcup_{1\le i'\le s}\mathrm{NTO}(i',s)$ in the execution of the Buchberger procedure.*

Proof. By Lemma 3.7, the obstruction $\mathbf{o}_{j,s}(u_j,u_j';w_s,w_s')$ can be represented as

$$\mathbf{o}_{j,s}(u_j,u_j';w_s,w_s')=-w\,\mathbf{o}_{i,j}(w_i,w_i';v_j,v_j')w'+\mathbf{o}_{i,s}(ww_i,w_i'w';w_s,w_s')$$

where the inequalities $\mathbf{o}_{j,s}(u_j,u_j';w_s,w_s')>_\tau\mathbf{o}_{i,j}(w_i,w_i';v_j,v_j')$ and $\mathbf{o}_{j,s}(u_j,u_j';w_s,w_s')>_\tau\mathbf{o}_{i,s}(ww_i,w_i'w';w_s,w_s')$ hold. Furthermore, if the two S-polynomials $S_{i,j}(w_i,w_i';v_j,v_j')$ and $S_{i,s}(ww_i,w_i'w';w_s,w_s')$ have Gröbner representations in terms of G, then so does $S_{j,s}(u_j,u_j';w_s,w_s')$. Consequently, Theorem 2.15 ensures that $S_{i,j}(w_i,w_i';v_j,v_j')$ has a Gröbner representation in terms of G. Note that ww_i is a multiple of $w_s\,\mathrm{Lw}_\sigma(g_s)$ or $w_i'w'$ is a multiple of $\mathrm{Lw}_\sigma(g_s)w_s'$. This implies that $\mathbf{o}_{i,s}(ww_i,w_i'w';w_s,w_s')$ has no overlap. Therefore, Lemma 2.12 shows that $S_{i,s}(ww_i,w_i'w';w_s,w_s')$ has a Gröbner representation in terms of G. Now the conclusion follows from Proposition 2.14 and Theorem 2.15. □

Let us see a concrete example where the noncommutative tail reduction applies and detects a redundant obstruction.

Example 3.9. Let K be a field and let $K\langle x,y\rangle$ be equipped with the word ordering $\sigma =$ LLex. For the set of polynomials $G = \{g_1, g_2, g_3\}$ such that $g_1 = xy + x$, $g_2 = xyxy + 1$, and $g_3 = xyx + y$, we construct the sets of all nontrivial obstructions and find $\mathrm{NTO}(1,1) = \emptyset$, $\mathrm{NTO}(1,2) \cup \mathrm{NTO}(2,2) = \{\mathbf{o}_{1,2}(xy,1;1,1), \mathbf{o}_{1,2}(1,xy;1,1), \mathbf{o}_{2,2}(1,xy;xy,1)\}$ as well as

$$\begin{aligned}&\mathrm{NTO}(1,3) \cup \mathrm{NTO}(2,3) \cup \mathrm{NTO}(3,3)\\&\quad = \{\mathbf{o}_{1,3}(1,x;1,1), \mathbf{o}_{1,3}(xy,1;1,y), \mathbf{o}_{2,3}(1,1;1,y), \mathbf{o}_{2,3}(xy,1;1,yxy), \mathbf{o}_{3,3}(1,yx;xy,1)\}\end{aligned}$$

When we apply the noncommutative tail reduction to these obstructions, we see that $\mathbf{o}_{2,3}(xy,1;1,yxy)$ and $\mathbf{o}_{1,2}(xy,1;1,1)$ satisfy the hypothesis of the proposition with $w = xy$ and $w' = 1$, since $ww_1 = xyxy$ is a multiple of $w_3\,\mathrm{Lw}_\sigma(g_3) = xyx$. Hence the obstruction $\mathbf{o}_{2,3}(xy,1;1,yxy)$ is detected as redundant and can be discarded.

In terms of the syzygies of the corresponding leading words, this result can be expressed by the equality

$$xye_2 - e_3yxy = -xy(xye_1 - e_2) + (xyxye_1 - e_3xyx)$$

where the syzygy $xyxye_1 - e_3xyx$ corresponds to a trivial obstruction without overlap.

Remark 3.10. Our experiments in the final section show that, after applying the previous two criteria, the noncommutative tail reduction is unlikely to apply in the Buchberger procedure. This is due to the fact that frequently the noncommutative multiply criterion and the noncommutative leading word criterion have already detected all unnecessary obstructions in the set $\bigcup_{1\le i\le s}\mathrm{NTO}(i,s)$ of newly constructed obstructions.

In the Buchberger procedure, we usually use the selection strategy which chooses the least **σ-degree** of an obstruction in B, that is, the smallest leading term of the corresponding S-polynomial. In the above example, this implies that we probably never construct a set B which contains $\mathbf{o}_{2,3}(xy,1;1,yxy)$ and $\mathbf{o}_{1,2}(xy,1;1,1)$. Namely, the latter obstruction will be chosen early on and removed as redundant using the noncommutative leading word criterion and the obstruction $\mathbf{o}_{1,2}(1,xy;1,1)$.

This observation indicates that it may be an interesting attempt to improve the procedure further by postponing the removal of redundant obstructions in a suitable way.

So far, we have detected unnecessary obstructions in the set $\bigcup_{1\le i\le s}\mathrm{Obs}(i,s)$ of newly constructed obstructions. Intuitively, we should also be able to detect unnecessary obstructions in the set $\bigcup_{1\le i\le j\le s-1}\mathrm{Obs}(i,j)$ of previously constructed obstructions. Thus, in the last step, we detect unnecessary obstructions in this set by using the new generator g_s.

Lemma 3.11. *Let* $\mathbf{o}_{i,j}(w_i, w_i'; w_j, w_j') \in \bigcup_{1\le i'\le j'\le s-1} \mathrm{NTO}(i', j')$ *be a non-trivial obstruction. If there are two words* $w, w' \in \langle X\rangle$ *satisfying* $w_j \,\mathrm{Lw}_\sigma(g_j) w_j' = w \,\mathrm{Lw}_\sigma(g_s) w'$*, then we can represent* $\mathbf{o}_{i,j}(w_i, w_i'; w_j, w_j')$ *as*

$$\mathbf{o}_{i,j}(w_i, w_i'; w_j, w_j') = \mathbf{o}_{i,s}(w_i, w_i'; w, w') - \mathbf{o}_{j,s}(w_j, w_j'; w, w').$$

Moreover, if $S_{i,s}(w_i, w_i'; w, w')$ *and* $S_{j,s}(w_j, w_j'; w, w')$ *have Gröbner representations in terms of G, then so does* $S_{i,j}(w_i, w_i'; w_j, w_j')$.

Proof. The claimed equality follows from Definition 2.5.a and the condition $w_j \,\mathrm{Lw}_\sigma(g_j) w_j' = w \,\mathrm{Lw}_\sigma(g_s) w'$. The proof of the second part is analogous to the proof of the second part of Lemma 3.3.a. □

The following example shows that the obstruction $\mathbf{o}_{i,s}(w_i, w_i'; w, w')$ in the equation of Lemma 3.11 can be an obstruction without overlap or a proper multiple of a nontrivial obstruction. In the case that $\mathbf{o}_{i,s}(w_i, w_i'; w, w')$ is a proper multiple of a nontrivial obstruction, say $\mathbf{o}_{i,s}(\tilde{w}_i, \tilde{w}_i'; \tilde{w}, \tilde{w}')$, the example shows that it does not necessarily follow that $\mathbf{o}_{i,s}(w_i, w_i'; w, w') >_\tau \mathbf{o}_{i,s}(\tilde{w}_i, \tilde{w}_i'; \tilde{w}, \tilde{w}')$ (compare Lemmas 3.3 and 3.7). The same statement holds for the obstruction $\mathbf{o}_{j,s}(w_j, w_j'; w, w')$ in the equality of Lemma 3.11.

Example 3.12. Consider the set of polynomials $G = \{g_1, g_2, g_3\}$ in the noncommutative polynomial ring $K\langle x, y\rangle$ with $\mathrm{Lm}_\sigma(g_1) = x^3yx$, $\mathrm{Lm}_\sigma(g_2) = x^2$ and $\mathrm{Lm}_\sigma(g_3) = x$. We have $\mathbf{o}_{1,2}(1, 1; x, yx) \in \bigcup_{1\le i\le j\le 2} \mathrm{NTO}(i, j)$ and $x\,\mathrm{Lw}_\sigma(g_2) yx = x^3yx = x^3y\,\mathrm{Lw}_\sigma(g_3)$ and

$$\mathbf{o}_{1,2}(1, 1; x, yx) = \mathbf{o}_{1,3}(1, 1; x^3y, 1) - \mathbf{o}_{2,3}(x, yx; x^3y, 1).$$

One can check that $\mathbf{o}_{1,3}(1, 1; x^3y, 1)$ is a nontrivial obstruction in $\mathrm{NTO}(1, 3)$ and $\mathbf{o}_{1,2}(1, 1; x, yx) <_\tau \mathbf{o}_{1,3}(1, 1; x^3y, 1)$. Moreover, $\mathbf{o}_{2,3}(x, yx; x^3y, 1)$ is an obstruction without overlap.

The following is a noncommutative analogue of the Gebauer–Möller criterion B_k, which is also known as the *chain criterion* (or *criterion 1*) of Buchberger (cf. [6]).

Proposition 3.13 (Noncommutative backward criterion). *Suppose that* $\mathbf{o}_{i,j}(w_i, w_i'; w_j, w_j') \in \bigcup_{1\le i'\le j'\le s-1} \mathrm{NTO}(i', j')$ *is a nontrivial obstruction. Then, in the execution of the Buchberger procedure, the obstruction* $\mathbf{o}_{i,j}(w_i, w_i'; w_j, w_j')$ *can be removed from* $\bigcup_{1\le i'\le j'\le s-1} \mathrm{NTO}(i', j')$ *if the following three conditions are satisfied:*
(a) *There are two words* $w, w' \in \langle X\rangle$ *such that* $w_j \,\mathrm{Lw}_\sigma(g_j) w_j' = w \,\mathrm{Lw}_\sigma(g_s) w'$.
(b) *The obstruction* $\mathbf{o}_{i,s}(w_i, w_i'; w, w')$ *is an obstruction without overlap or a multiple of an obstruction in* $\bigcup_{1\le i'\le s} \mathrm{NTO}(i', s)$.
(c) *The obstruction* $\mathbf{o}_{j,s}(w_j, w_j'; w, w')$ *is an obstruction without overlap or a multiple of an obstruction in* $\bigcup_{1\le i'\le s} \mathrm{NTO}(i', s)$.

Proof. By Lemma 3.11, we can represent $\mathbf{o}_{i,j}(w_i, w_i'; w_j, w_j')$ as

$$\mathbf{o}_{i,j}(w_i, w_i'; w_j, w_j') = \mathbf{o}_{i,s}(w_i, w_i'; w, w') - \mathbf{o}_{j,s}(w_j, w_j'; w, w').$$

Moreover, $S_{i,j}(w_i, w_i'; w_j, w_j')$ has a Gröbner representation in terms of G if $S_{i,s}(w_i, w_i'; w, w')$ and $S_{j,s}(w_j, w_j'; w, w')$ have Gröbner representations in terms of G. If $\mathbf{o}_{i,s}(w_i, w_i'; w, w')$ is an obstruction without overlap, then Lemma 2.12 shows that its S-polynomial has a Gröbner representations in terms of G. If it is a multiple of an obstruction in $\bigcup_{1\le i'\le s} \mathrm{NTO}(i', s)$, then Lemma 2.12 and Theorem 2.15 ensure that $S_{i,s}(w_i, w_i'; w, w')$ has a Gröbner representation in terms of G. By the same argument, one can show that $S_{j,s}(w_j, w_j'; w, w')$ has a Gröbner representation in terms of G. Now the conclusion follows from Proposition 2.14 and Theorem 2.15. □

Let us mention that the noncommutative backward criterion given in Proposition 3.13 covers, in particular, all useless obstructions presented by T. Mora in [20], Lemma 5.11, and that this result (called *Möller's lemma*) is actually a proper subset of the cases covered by the proposition, as we show in Section 5.1.

Remark 3.14. In order to apply Propositions 3.5, 3.6, 3.8, and 3.13 to remove unnecessary obstructions during the execution of the Buchberger procedure, it is crucial to make sure that the S-polynomials of those removed obstructions have Gröbner representations.

(a) Propositions 3.5, 3.6, and 3.8 remove unnecessary nontrivial obstructions, say $\mathbf{o}_{i,s}(w_i, w_i'; w_s, w_s')$, from the set $\bigcup_{1\le i'\le s} \mathrm{NTO}(i', s)$ of newly constructed obstructions. The Gröbner representation of the S-polynomial $S_{i,s}(w_i, w_i'; w_s, w_s')$ depends on the Gröbner representations of the S-polynomials of two smaller obstructions in the set $\bigcup_{1\le i'\le j'\le s-1} \mathrm{Obs}(i', j')$ and the set $\bigcup_{1\le i'\le s} \mathrm{Obs}(i', s)$.

(b) Proposition 3.13 removes unnecessary obstructions, say $\mathbf{o}_{i,j}(w_i, w_i'; w_j, w_j')$, from the set $\bigcup_{1\le i'\le j'\le s-1} \mathrm{Obs}(i', j')$ of previously constructed obstructions. The Gröbner representation of $S_{i,j}(w_i, w_i'; w_j, w_j')$ depends on the Gröbner representations of the S-polynomials of two obstructions, say $\mathbf{o}_{k,s}(w_k, w_k'; u_s, u_s')$ and $\mathbf{o}_{l,s}(w_l, w_l'; v_s, v_s')$, in $\bigcup_{1\le i'\le s} \mathrm{Obs}(i', s)$, which are not necessarily smaller than $\mathbf{o}_{i,j}(w_i, w_i'; w_j, w_j')$. If $\mathbf{o}_{k,s}(w_k, w_k'; u_s, u_s')$ is a proper multiple of a nontrivial obstruction, say $\mathbf{o}_{k,s}(\tilde{w}_k, \tilde{w}_k'; \tilde{u}_s, \tilde{u}_s')$, in the set $\bigcup_{1\le i'\le s} \mathrm{NTO}(i', s)$, then, before removing $\mathbf{o}_{i,j}(w_i, w_i'; w_j, w_j')$, it is important to ensure that $\mathbf{o}_{k,s}(\tilde{w}_k, \tilde{w}_k'; \tilde{u}_s, \tilde{u}_s')$ is in $\bigcup_{1\le i'\le s} \mathrm{NTO}(i', s)$. The same check should be applied to $\mathbf{o}_{l,s}(w_l, w_l'; v_s, v_s')$.

Observe that Propositions 3.5, 3.6, and 3.13 are actually generalizations of the well-known Gebauer–Möller criteria (see [7] and [11]) in commutative polynomial rings. More precisely, Propositions 3.5, 3.6, and 3.13 correspond to criterion M, criterion F, and criterion B_k, respectively (see [11], Subsection 3.4).

4 The improved Buchberger procedure

Let us continue to use the setting of the previous sections. With the help of the Gebauer–Möller criteria, we can improve the Buchberger procedure as follows.

Theorem 4.1 (The improved Buchberger procedure). *In the setting of Theorem 2.15, we replace step* (B4) *by the following sequence of instructions:*

(4a) *Increase s by one. Append* $g_s = S'$ *to the set G, and form the set of nontrivial obstructions* $\mathrm{NTO}(s) = \bigcup_{1\le i\le s} \mathrm{NTO}(i,s)$.

(4b) *Remove from* NTO(s) *all obstructions* $\mathbf{o}_{i,s}(w_i, w_i'; u_s, u_s')$ *such that there exists an obstruction* $\mathbf{o}_{j,s}(w_j, w_j'; v_s, v_s') \in \mathrm{NTO}(s)$ *with the properties that there exist two words* $w, w' \in \langle X\rangle$ *satisfying* $u_s = wv_s$, $u_s' = v_s'w'$, *and* $ww' \neq 1$.

(4c) *Remove from* NTO(s) *all obstructions* $\mathbf{o}_{i,s}(w_i, w_i'; u_s, u_s')$ *such that there exists an obstruction* $\mathbf{o}_{j,s}(w_j, w_j'; v_s, v_s') \in \mathrm{NTO}(s)$ *with the properties that there exist two words* $w, w' \in \langle X\rangle$ *satisfying* $u_s = wv_s$, $u_s' = v_s'w'$, *and such that* $i > j$, *or* $i = j$ *and* $ww' = 1$ *and* $w_i >_\sigma w_j$.

(4d) *Remove from* NTO(s) *all obstructions* $\mathbf{o}_{j,s}(u_j, u_j'; w_s, w_s')$ *such that there exists an obstruction* $\mathbf{o}_{i,j}(w_i, w_i'; v_j, v_j') \in B$ *with the properties that there exist two words* $w, w' \in \langle X\rangle$ *satisfying* $u_j = wv_j$, $u_j' = v_j'w'$, *and such that* $\mathbf{o}_{i,s}(ww_i, w_i'w'; w_s, w_s')$ *has no overlap.*

(4e) *Remove from B all obstructions* $\mathbf{o}_{i,j}(w_i, w_i'; w_j, w_j')$ *such that there exist two words* $w, w' \in \langle X\rangle$ *satisfying* $w\,\mathrm{Lw}_\sigma(g_s)w' = w_j\,\mathrm{Lw}_\sigma(g_j)w_j'$, *and such that the following conditions are satisfied.*
   (i) $\mathbf{o}_{i,s}(w_i, w_i'; w, w')$ *is an obstruction without overlap or a multiple of an obstruction in* NTO(s).
   (ii) $\mathbf{o}_{j,s}(w_j, w_j'; w, w')$ *is an obstruction without overlap or a multiple of an obstruction in* NTO(s).

(4f) *Replace B by* $B \cup \mathrm{NTO}(s)$ *and continue with step* (B2).

Then the resulting set of instructions is a procedure that enumerates a σ-Gröbner basis G of I. If I has a finite σ-Gröber basis, it stops after finitely many steps and the resulting set G is a finite σ-Gröbner basis of I.

Proof. This follows from Theorem 2.15 and Propositions 3.5, 3.6, 3.8, and 3.13. □

To illustrate the improved Buchberger procedure, we apply it to the following example which is taken from [20], Example 6.1.

Example 4.2. Consider the free monoid ring $K\langle a,b,c,d,e,f\rangle$ equipped with the weighted lexicographic word ordering σ using the weights $(3,1,1,1,1,1)$ and $a <_\sigma b <_\sigma c <_\sigma d <_\sigma e <_\sigma f$. We want to enumerate a σ-Gröbner basis of the ideal $I = \langle f_1,\dots,f_7\rangle$, where $f_1 = ca - ac$, $f_2 = da - ad$, $f_3 = ba - b^2c$, $f_4 = be - b$, $f_5 = bf - b$, $f_6 = ef - b$, and $f_7 = bcd$.

For the selection strategy in step (B2) of the Buchberger procedure, we shall use two different strategies based on the **σ-degree** of an obstruction: this is the leading term of the corresponding S-polynomial and can be compared using σ (see also [13], Section 2.3). The algorithm starts by initialising $G = \{f_1,\dots,f_7\}$ and $B =$

$\{\mathbf{o}_{4,6}(1,f;b,1), \mathbf{o}_{2,7}(bc,1;1,a)\}$. Then it proceeds differently according to the chosen selection strategy.

(a) Let us always choose the last element of B, that is, the one with the largest σ-degree.
   (1) Select $\mathbf{o}_{2,7}(bc,1;1,a)$ whose σ-degree is $bcda$. Then $S_{2,7}(bc,1;1,a) = -bcad$ and $\mathrm{NR}_{\sigma,G}(-bcad) = -b^2c^2d$. Hence $f_8 = b^2c^2d$ is appended to G and $\mathbf{o}_{2,8}(b^2c^2,1;1,a)$ to B.
   (2) Select $\mathbf{o}_{2,8}(b^2c^2,1;1,a)$ whose σ-degree is b^2c^2da. Then we obtain $S_{2,8}(b^2c^2,1;1,a) = -b^2c^2ad$ and $\mathrm{NR}_{\sigma,G}(-b^2c^2ad) = -b^3c^3d$. Hence $f_9 = b^3c^3d$ is appended to G and $\mathbf{o}_{2,9}(b^3c^3,1;1,a)$ to B.
   (3) The procedure continues in this way forever. At stage k, we have $G = \{f_1, \dots, f_{7+k}\}$ with $f_{7+k} = b^{k+1}c^{k+1}d$ and $B = \{\mathbf{o}_{4,6}(1,f;b,1), \mathbf{o}_{2,k+7}(b^{k+1}c^{k+1},1;1,a)\}$.

   Clearly, this selection strategy is not fair, since $o_{4,6}(1,f;b,1)$ is never chosen. It enumerates an infinite set G.

(b) Now we always choose the element of B with the smallest σ-degree.
   (1) Select $\mathbf{o}_{4,6}(1,f;b,1)$. Here, we obtain $S_{4,6}(1,f;b,1) = -bf + b^2$ and $\mathrm{NR}_{\sigma,G}(-bf + b^2) = b^2 - b$. Hence $f_8 = b^2 - b$ is appended to G. Using the new step (B4) of the improved Buchberger procedure, we find that the five obstructions $\mathbf{o}_{3,8}(b,1;1,a)$, $\mathbf{o}_{4,8}(b,1;1,e)$, $\mathbf{o}_{5,8}(b,1;1,f)$, $\mathbf{o}_{7,8}(b,1;1,cd)$, and $\mathbf{o}_{8,8}(1,b;b,1)$ have to be appended to B.
   (2) Now the σ-degrees of the obstructions in B are $b^3 <_\sigma b^2e <_\sigma b^2f <_\sigma b^2cd <_\sigma b^2a <_\sigma bcda$. Using $\mathbf{o}_{8,8}(1,b;b,1)$, we find $S_{8,8}(1,b;b,1) = 0$.
   (3) In the next four iterations, we choose $\mathbf{o}_{4,8}(b,1;1,e)$, $\mathbf{o}_{5,8}(b,1;1,f)$, $\mathbf{o}_{7,8}(b,1;1,cd)$, and $\mathbf{o}_{3,8}(b,1;1,a)$. We find that every time the normal remainder of the corresponding S-polynomial is zero.
   (4) Finally, we choose $\mathbf{o}_{2,7}(bc,1;1,a)$. We obtain $S_{2,7}(bc,1;1,a) = -bcad$ and $\mathrm{NR}_{\sigma,G}(-bcad) = -bc^2d$. Hence we append $f_9 = bc^2d$ to G and $\mathbf{o}_{2,9}(bc^2,1;1,a)$ to B.
   (5) The σ-degrees of the elements of B are $b^2c^2d <_\sigma bc^2da$. Thus we choose $\mathbf{o}_{8,9}(1,c^2d;b,1)$ and get $\mathrm{NR}_{\sigma,G}(S_{8,9}(1,c^2d;b,1)) = 0$.
   (6) It remains to choose $\mathbf{o}_{2,9}(bc^2,1;1,a)$. We compute $S_{2,9}(bc^2,1;1,a) = -bc^2ad$ and $\mathrm{NR}_{\sigma,G}(-bc^2ad) = -bc^3d$. Consequently, we append the element $f_{10} = bc^3d$ to G and the two obstructions $\mathbf{o}_{2,10}(bc^3,1;1,a)$ and $\mathbf{o}_{8,10}(1,c^3d;b,1)$ to B.
   (7) Clearly, the procedure continues in the same way forever. At every iteration, one S-polynomial reduces to zero and the other one yields a new Gröbner basis element and two new obstructions.

   Again the procedure continues indefinitely and enumerates the set $G = \{f_1, f_2, \dots\}$, where $f_8 = b^2 - b$ and $f_{i+7} = bc^id$ for $i \geq 2$. Here, we know that G is a σ-Gröbner basis of I, since the chosen selection strategy is fair.

In both cases, we enumerate an infinite set. Unfortunately, the claim in [20] that the second selection strategy yields a finite Gröbner basis of I is not correct, and the example cannot be used to support the claims made there. In particular, the set $\{f_1, \ldots, f_7, b^2 - b\}$ is apparently not a σ-Gröbner basis of I.

5 Mora's lemma

In this section, we discuss the relation between the versions of the noncommutative Gebauer–Möller criteria given in Section 3 and some previous versions, in particular, the ones by Teo Mora in [20] (which are recalled and elaborated in [22]). The latter are proven using a noncommutative version of a result called *Möller's lemma* in [21], Lemma 25.1.4. This noncommutative version, which we call Mora's lemma, is the content of [20], Lemma 5.11. (It is reproduced in [22], Remark 47.6.14.) In this section, we show that Mora's lemma covers only some cases of our noncommutative backward criterion 3.13, and that therefore the criteria presented in Section 3 are, in general, more powerful than any criteria derived from Mora's lemma alone.

In the following, we use the notation and the setting of the preceding sections. To begin with, let us formulate Mora's lemma in our notation.

Proposition 5.1 (Mora's lemma). *Let $G = \{g_1, \ldots, g_s\}$ be a set of nonzero polynomials in $K\langle x_1, \ldots, x_n\rangle$. Assume that there exist $i, j \in \{1, \ldots, s-1\}$ and $v_\ell, v_r, w_\ell, w_r \in \langle X\rangle$ such that $v_\ell \operatorname{Lw}_\sigma(g_i) = \operatorname{Lw}_\sigma(g_j) v_r = w_\ell \operatorname{Lw}_\sigma(g_s) w_r$. Then the following statements hold:*

(a) *If there exist words $u_\ell, u_r \in \langle X\rangle$ such that $v_\ell = w_\ell u_\ell$ and $v_r = u_r w_r$, as well as $u_\ell \operatorname{Lw}_\sigma(g_i) = \operatorname{Lw}_\sigma(g_s) w_r$ and $\operatorname{Lw}_\sigma(g_j) u_r = w_\ell \operatorname{Lw}_\sigma(g_s)$, then we have*

$$S_{i,j}(v_\ell, 1; 1, v_r) = w_\ell S_{i,s}(u_\ell, 1; 1, w_r) - S_{j,s}(1, u_r; w_\ell, 1) w_r.$$

In particular, if $S_{i,s}(u_\ell, 1; 1, w_r)$ and $S_{j,s}(1, u_r; w_\ell, 1)$ have weak Gröbner representations, then so does $S_{i,j}(v_\ell, 1; 1, w_r)$.

(b) *If there exist words $u_\ell, u_r \in \langle X\rangle$ such that $w_\ell = v_\ell u_\ell$ and $v_r = u_r w_r$, as well as $\operatorname{Lw}_\sigma(g) = u_\ell \operatorname{Lw}_\sigma(g_j) w_r$ and $\operatorname{Lw}_\sigma(g_j) u_r = w_\ell \operatorname{Lw}_\sigma(g_s)$, then we have*

$$S_{i,j}(v_\ell, 1; 1, v_r) = u_\ell S_{i,s}(1, 1; u_\ell, w_r) - S_{j,s}(1, u_r; w_\ell, 1).$$

In particular, if $S_{i,s}(1, 1; u_\ell, w_r)$ and $S_{j,s}(1, u_r; w_\ell, 1)$ have weak Gröbner representations, then so does $S_{i,j}(v_\ell, 1; 1, v_r)$.

(c) *If there exist words $u_\ell, u_r \in \langle X\rangle$ such that $v_\ell = w_\ell u_\ell$ and $w_r = u_r v_r$, as well as $u_\ell \operatorname{Lw}_\sigma(g_i) = \operatorname{Lw}_\sigma(g_s) w_r$ and $\operatorname{Lw}_\sigma(g_j) = w_\ell \operatorname{Lw}_\sigma(g_s) u_r$, then we have*

$$S_{i,j}(v_\ell, 1; 1, v_r) = w_\ell S_{i,s}(u_\ell, 1; 1, w_r) - S_{j,s}(1, 1; w_\ell, u_r) v_r.$$

In particular, if $S_{i,s}(u_\ell, 1; 1, w_r)$ and $S_{j,s}(1, 1; w_\ell, u_r)$ have weak Gröbner representations, then so does $S_{i,j}(v_\ell, 1; 1, v_r)$.

(d) *If there exist words* $u_\ell, u_r \in \langle X \rangle$ *such that* $w_\ell = v_\ell u_\ell$ *and* $w_r = u_r v_r$, *as well as* $\mathrm{Lw}_\sigma(g_i) = u_\ell \, \mathrm{Lw}_\sigma(g_s)$ *and* $\mathrm{Lw}_\sigma(g_j) = w_\ell \, \mathrm{Lw}_\sigma(g_s) u_r$, *then we have*

$$S_{i,j}(v_\ell, 1; 1, v_r) = v_\ell S_{i,s}(1, 1; u_\ell, w_r) - S_{j,s}(1, 1; w_\ell, u_r) v_r.$$

In particular, if $S_{i,s}(1, 1; u_\ell, w_r)$, *and* $S_{j,s}(1, 1; w_\ell, u_r)$ *have weak Gröbner representations, then so does* $S_{i,j}(v_\ell, 1; 1, v_r)$.

Proof. This is stated in [20], Lemma 5.11, and in [22], Remark 47.6.14 as a consequence of [22], Lemma 47.6.12. □

In order to understand the relation of Mora's lemma to the noncommutative backward criterion 3.13, we elaborate the different cases in which Proposition 3.13 applies in the following remark.

Remark 5.2 (Types of left obstructions). Assume that we are in the setting of the proposition. In particular, suppose that $g_1, \ldots, g_s \in K\langle X \rangle$ and that $v_\ell, v_r, w_\ell, w_r \in \langle X \rangle$ satisfy $v_\ell \, \mathrm{Lw}_\sigma(g_i) = \mathrm{Lw}_\sigma(g_j) v_r = w_\ell \, \mathrm{Lw}_\sigma(g_s) w_r$. In particular, we assume that $\mathbf{o}_{i,j}(w_\ell, 1; 1, w_r)$ is a left obstruction. Depending on whether w_ℓ and w_r are trivial or nontrivial, we obtain the following four cases:

(a) If $w_\ell = w_r = 1$, we have the following picture:

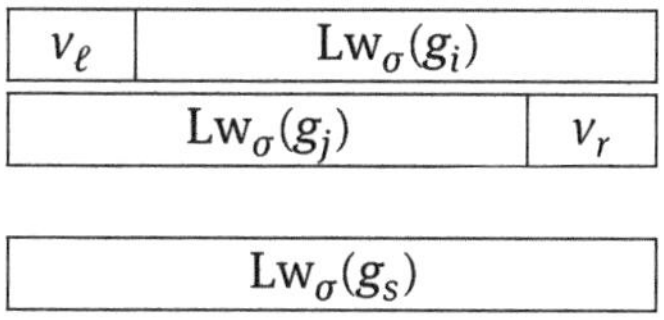

Clearly, the corresponding leading word syzygies satisfy

$$v_\ell e_i - e_j v_r = (v_\ell e_i - e_s) - (e_j v_r - e_s).$$

(b) In the case $w_\ell \neq 1$ and $w_r = 1$, we have three subcases which depend on the position of the word u_ℓ.

(b.1) If $v_\ell = w_\ell u_\ell$ for some $u_\ell \in \langle X \rangle$, we have the following situation:

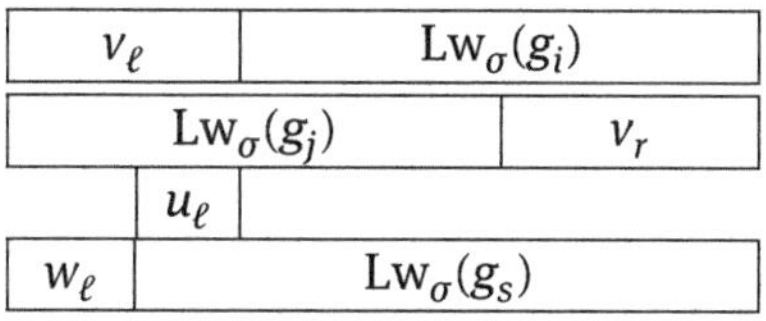

The corresponding leading word syzygies satisfy

$$v_\ell e_i - e_j v_r = w_\ell (u_\ell e_i - e_s) - (e_j v_r - w_\ell e_s).$$

(b.2) If $w_\ell = v_\ell u_\ell$ for some $u_\ell \in \langle X \rangle$ with $\mathrm{len}(w_\ell) < \mathrm{len}(\mathrm{Lw}_\sigma(g_j))$, then we have the following situation:

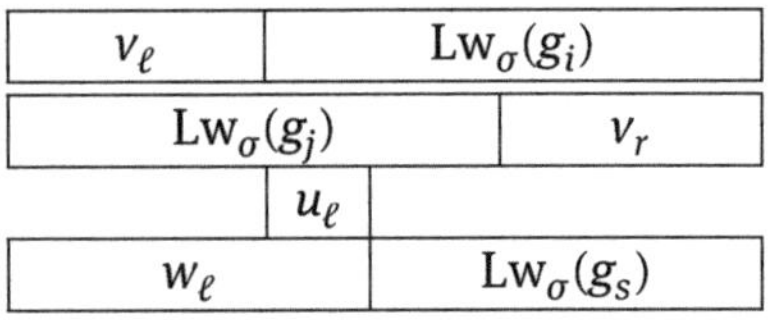

The corresponding leading word syzygies satisfy

$$v_\ell e_i - e_j v_r = w_\ell(e_i - u_\ell e_s) - (e_j v_r - w_\ell e_s).$$

(b.3) If $w_\ell = v_\ell u_\ell$ for some $u_\ell \in \langle X\rangle$ with $\mathrm{len}(w_\ell) \geq \mathrm{len}(\mathrm{Lw}_\sigma(g_j))$, then we have the following situation:

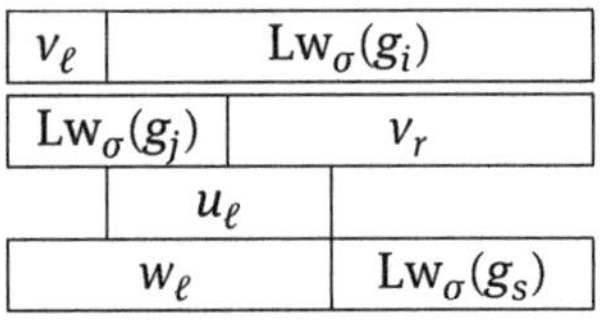

The corresponding leading word syzygies satisfy

$$v_\ell e_i - e_j v_r = w_\ell(e_i - u_\ell e_s) - (e_j v_r - w_\ell e_s).$$

(c) The case $w_\ell = 1$ and $w_r \neq 1$ is similar to (b) and left to the reader.

(d) Finally, we consider the case $w_\ell \neq 1$ and $w_r \neq 1$. Here, we have six subcases which depend on the positions of the words u_ℓ and u_r.

(d.1) Assume that $v_\ell = w_\ell \leq_\sigma \mathrm{Lw}_\sigma(g_s)u_\ell$ for some nontrivial word $u_\ell \in \langle X\rangle$. Then $\mathrm{Lw}_\sigma(g_i)$ and $\mathrm{Lw}_\sigma(g_s)$ have no overlap, but $\mathrm{Lw}_\sigma(g_i)$ and $\mathrm{Lw}_\sigma(g_j)$ have an overlap at $u_r \in \langle X\rangle$. Thus we have the following picture:

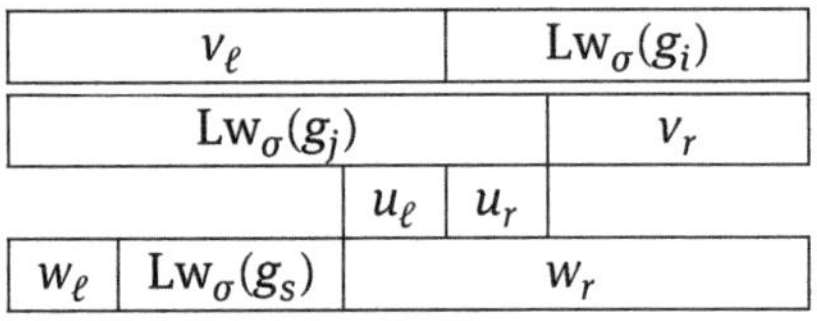

The corresponding leading word syzygies satisfy

$$v_\ell e_i - e_j v_r = w_\ell(\mathrm{Lw}_\sigma(g_i)u_\ell e_i - e_s w_r) - (e_j - w_\ell e_s u_\ell u_r)v_r.$$

(d.2) If $v_r = u_\ell \,\mathrm{Lw}_\sigma(g_s)w_r$ with $u_\ell \in \langle X\rangle$, then $\mathrm{Lw}_\sigma(g_j)$ and $\mathrm{Lw}_\sigma(g_s)$ have no overlap and the situation is analogous to case (d.1).

(d.3) Now suppose that $v_\ell = w_\ell u_\ell$ and $w_r = u_r v_r$ for some nontrivial words $u_\ell, u_r \in \langle X\rangle$. Then we are in the following situation:

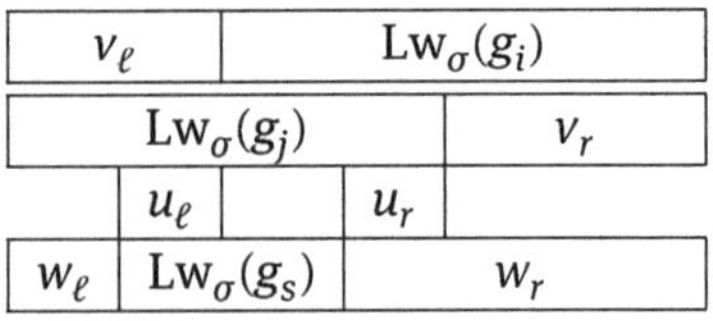

The corresponding leading word syzygies satisfy

$$v_\ell e_i - e_j v_r = w_\ell(u_\ell e_i - e_s w_r) - (e_j - w_\ell e_s u_r)v_r.$$

(d.4) Next we consider the case $w_\ell = v_\ell u_\ell$ and $v_r = u_r w_r$ with $u_\ell, u_r \in \langle X\rangle$, that is, the following situation:

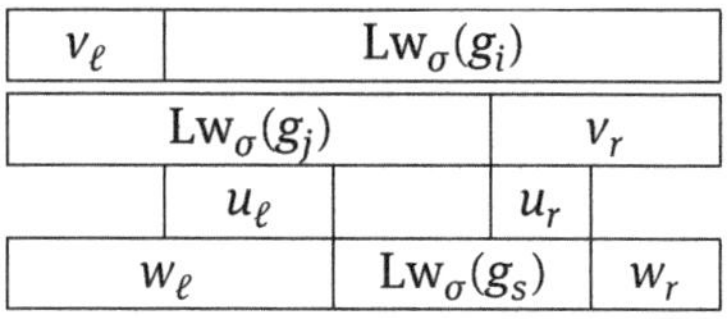

The corresponding leading word syzygies satisfy

$$v_\ell e_i - e_j v_r = v_\ell(e_i - u_\ell e_s w_r) - (e_j u_r - w_\ell e_s)w_r.$$

(d.5) In the case $v_\ell = w_\ell u_\ell$ and $v_r = u_r w_r$ with $u_\ell, u_r \in \langle X\rangle$, we have the situation

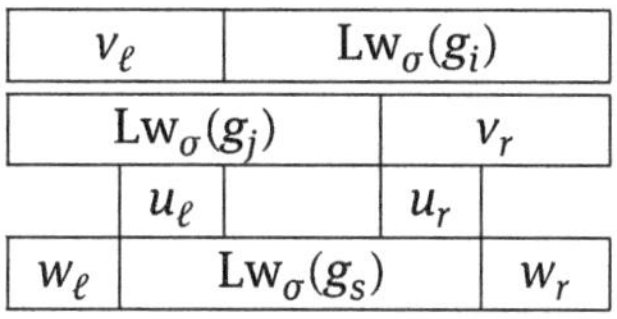

The corresponding leading word syzygies satisfy

$$v_\ell e_i - e_j v_r = w_\ell(u_\ell e_i - e_s w_r) - (e_j u_r - w_\ell e_s)w_r.$$

(d.6) Finally, if $w_\ell = v_\ell u_\ell$ and $w_r = u_r v_r$ with $u_\ell, u_r \in \langle X\rangle$, then we have the following picture:

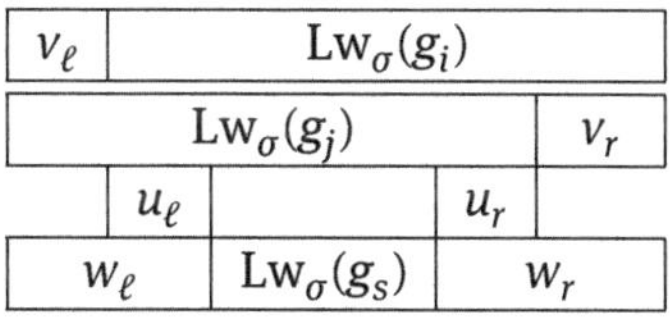

The corresponding leading word syzygies satisfy

$$v_\ell e_i - e_j v_r = v_\ell(e_i - u_\ell e_s w_r) - (e_j - w_\ell e_s u_r)v_r.$$

In all cases, we see from the leading word syzygies that the noncommutative backward criterion applies and the obstruction $\mathbf{o}_{i,j}(v_\ell, 1; 1, v_r)$ can be discarded.

A similar case-by-case analysis can be made for center obstructions (see [25], Remark 4.2.20). At this point, we can deduce that Mora's lemma treats the following subcases of Remark 5.2:

(a) Proposition 5.1.a corresponds to Case (d.5).
(b) Proposition 5.1.b corresponds to Case (d.4).
(c) Proposition 5.1.c corresponds to Case (d.3).
(d) Proposition 5.1.b corresponds to Case (d.6).

Thus we conclude that Proposition 3.13 allows us to discard, in general, more obstructions. One reason for this is that we are also using obstructions without overlap to find superfluous obstructions, for instance as in Case (b.3). Next, we check how the noncommutative Gebauer–Möller perform in practice.

6 Some experiments

In this section, we present some experimental data which illustrate the performance of the Gebauer–Möller criteria presented in Propositions 3.5, 3.6, 3.8, and 3.13. The computations are based on a C++ implementation in the ApCoCoA library (see [1]) by the second author. The examples are taken from some problems in group theory and may be considered to be of medium difficulty.

Example 6.1. Consider the noncommutative polynomial ring $\mathbb{Q}\langle a,b\rangle$ equipped with the word ordering LLex on $\langle a,b\rangle$ such that $a >_{\mathrm{LLex}} b$. We take the list of *finite generalized triangle groups* from [23], Theorem 2.12, and construct a list of ideals in $\mathbb{Q}\langle a,b\rangle$. For $k = 1,\dots,13$, let $I_k = \langle G_k\rangle \subseteq \mathbb{Q}\langle a,b\rangle$ be the ideal generated by the following set of polynomials G_k in $\mathbb{Q}\langle a,b\rangle$:

$$\begin{aligned}
G_1 &= \{a^2-1, b^3-1, (ababab^2ab^2)^2-1\},\\
G_2 &= \{a^2-1, b^3-1, (ababab^2)^3-1\},\\
G_3 &= \{a^3-1, b^3-1, (abab^2)^2-1\},\\
G_4 &= \{a^3-1, b^3-1, (aba^2b^2)^2-1\},\\
G_5 &= \{a^2-1, b^5-1, (abab^2)^2-1\},\\
G_6 &= \{a^2-1, b^5-1, (ababab^4)^2-1\},\\
G_7 &= \{a^2-1, b^5-1, (abab^2ab^4)^2-1\},\\
G_8 &= \{a^2-1, b^4-1, (ababab^3)^2-1\},\\
G_9 &= \{a^2-1, b^3-1, (abab^2)^2-1\},\\
G_{10} &= \{a^2-1, b^3-1, (ababab^2)^2-1\},\\
G_{11} &= \{a^2-1, b^3-1, (abababab^2)^2-1\},\\
G_{12} &= \{a^2-1, b^3-1, (ababab^2abab^2)^2-1\},\\
G_{13} &= \{a^2-1, b^3-1, (ababababab^2ab^2)^2-1\}.
\end{aligned}$$

The following table lists some numbers of polynomials and obstructions treated by the improved Buchberger procedure given in Theorem 4.1.

k	#(Gb)	#(RGb)	#(Tot)	#(Sel)	#(M)	#(F)	#(B_k)	ρ
1	62	35	7,032	248	6,512	48	224	0.0353
2	133	96	31,700	533	30,571	70	526	0.0168
3	50	40	2,828	197	2,489	11	131	0.0697
4	64	28	4,702	253	4,185	46	218	0.0538
5	35	21	1,580	115	1,348	24	93	0.0728
6	199	164	51,175	882	49,126	26	1,141	0.0172
7	200	164	51,864	886	49,818	17	1,143	0.0170
8	53	37	3,756	192	3,357	19	188	0.0511
9	11	5	150	31	98	8	13	0.2067
10	22	15	741	74	605	18	44	0.0999
11	30	21	1,573	116	1,324	50	83	0.0737
12	97	70	16,841	365	15,989	97	390	0.0217
13	220	194	87,673	1021	85,136	153	1,363	0.0116

Here, we used the following abbreviations:

- #(Gb) is the number of elements of the Gröbner basis returned by the procedure.
- #(RGb) is the cardinality of the reduced Gröbner basis of the corresponding ideal.
- #(Tot) is the total number of nontrivial obstructions constructed during the Buchberger procedure.
- #(Sel) is the number of actually selected and analyzed nontrivial obstructions.
- #(M) is the number of unnecessary nontrivial obstructions detected by the noncommutative multiply criterion given in Proposition 3.5.
- #(F) is the number of unnecessary nontrivial obstructions detected by the noncommutative leading word criterion given in Proposition 3.6.
- #(B_k) is the number of unnecessary nontrivial obstructions detected by the noncommutative backward criterion given in Proposition 3.13.
- $\rho = \#(\text{Sel})/\#(\text{Tot})$.

Note that #(RGb) is an invariant of the ideal which only depends on the chosen word ordering. Other numbers in the table rely also on the selection strategy. In our experiments, we used the *normal strategy* which first chooses the obstruction whose S-polynomial has the lowest degree and then breaks ties by choosing the obstruction whose S-polynomial has the smallest leading word with respect to the word ordering. In these experiments, the noncommutative tail reduction given in Proposition 3.8 detected no unnecessary nontrivial obstruction (see Remark 3.10). Apparently, the noncommutative multiply criterion and the noncommutative leading word criterion had already detected all unnecessary obstructions in the set $\bigcup_{1\le i\le s} \mathrm{NTO}(i,s)$ of newly constructed obstructions. The low ratios ρ in the table indicate that the noncommutative Gebauer–Möller criteria we obtained can detect most unnecessary obstructions during the procedure.

Example 6.2. The following ideals braid3 and braid4 in the noncommutative polynomial ring $\mathbb{Q}\langle x_1, x_2, x_3\rangle$ are taken from [24], Section 5. More precisely, braid3 is the ideal generated by the set $\{-x_2x_3x_1+x_3x_1x_3, x_2x_1x_2-x_3x_2x_3, x_1x_2x_1-x_3x_1x_2, x_1^3+x_1x_2x_3+x_2^3+x_3^3\}$, and braid4 is the ideal generated by the set $\{-x_2x_3x_1 + x_3x_1x_3, x_2x_1x_2 - x_3x_2x_3, x_1x_2x_3 - x_3x_1x_2, x_1^3 + x_1x_2x_3 + x_2^3 + x_3^3\}$. These ideals are generated by sets of homogeneous generators. The following table lists the results of the computations of Gröbner bases truncated at degree 11 with respect to LLex via the improved Buchberger procedure.

	#(Gb)	#(Tot)	#(Sel)	ρ
braid3-11	726	289,642	1,663	0.0057
braid4-11	416	93,252	1,150	0.0123

The meaning of the symbols is the same as in Example 6.1. In this experiment, we also used the normal strategy. Moreover, since we compute truncated Gröbner bases, we discard those obstructions whose S-polynomial have degrees larger than the degree of truncation. Thus the ratios ρ in the table are lower than the ratios in the table of Example 6.1. Again, the noncommutative Gebauer–Möller criteria detect most unnecessary obstructions during the procedure.

The experimental data in Examples 6.1 and 6.2 show that the generalizations of the Gebauer–Möller criteria presented in Section 3 can successfully detect a large number of unnecessary obstructions. In fact, they apparently detect almost all unnecessary obstructions during the Buchberger procedure.

Bibliography

[1] ApCoCoA. Applied Computations in Commutative Algebra. Available at http://apcocoa.uni-passau.de.

[2] Bergman GH. The diamond lemma for ring theory. Adv Math. 1978;29:178–218.

[3] Bosma W, Cannon J, Playoust C. The Magma algebra system I: the user language. J Symb Comput. 1997;24:235–65.

[4] Buchberger B. Ein Algorithmus zum Auffinden der Basiselemente des Restklassenrings nach einem nulldimensionalen Polynomideal (in German), dissertation, Universität Innsbruck, 1965.

[5] Buchberger B. A criterion for detecting unnecessary reductions in the construction of Groebner bases. In: Ng EW, editor. Proc. Int. Symp. Symbolic and Algebraic Manipulation (EUROSAM '79). Lect Notes Comp Sci. vol. 72. Heidelberg: Springer; 1979. p. 3–21.

[6] Buchberger B. Gröbner bases: an algorithmic method in polynomial ideal theory. In: Bose NK, editor. Multidimensional Systems Theory. Dordrecht: D. Reidel Publ.; 1985. p. 184–232.

[7] Caboara M, Kreuzer M, Robbiano L. Efficiently computing minimal sets of critical pairs. J Symb Comput. 2004;38:1169–90.

[8] Cohen AM. Non-commutative polynomial computations. 2007. Available at www.win.tue.nl/_amc/pub/gbnpaangepast.pdf.

[9] Fine B, Kreuzer M, Rosenberger G. Free Polynomial Algebras and Magnus's Freiheitssatz. Int J Group Theory. 2015;4:13–9.

[10] Green EL, Heath S, Keller BJ. Opal: a system for computing noncommutative Gröbner bases. In: Comon H, editor. Proc. Int. Conf. Rewriting Techniques and Applications (RTA '97). Lect Notes Comp Sci. vol. 1232. Heidelberg: Springer; 1997. p. 331–4.
[11] Gebauer R, Möller HM. On an installation of Buchberger's algorithm. J Symb Comput. 1988;6:275–86.
[12] Kreuzer M, Myasnikov A, Rosenberger G, Ushakov A. Quotient tests and Gröbner bases. Contemp Math. 2007;421:187–200.
[13] Kreuzer M, Robbiano L. Computational Commutative Algebra 1. Heidelberg: Springer; 2000.
[14] Kreuzer M, Robbiano L. Computational Commutative Algebra 2. Heidelberg: Springer; 2005.
[15] Kreuzer M, Rosenberger G. Growth in Hecke groups. In: Izquierdo M, editor. Riemann and Klein Surfaces, Symmetries and Moduli Spaces. Contemp Math. vol. 629. 2014. p. 261–81.
[16] Levandovskyy V. Plural, a non-commutative extension of Singular: past, present, future. In: Iglesias A, Takayama N, editors. Mathematical Software – ICMS 2006. Lect Notes Comp Sci. vol. 4151. Berlin: Springer; 2006. p. 144–57.
[17] Levandovskyy V, Studzinski G, Schnitzler B. Enhanced computations of Gröbner bases in free algebras as a new application of the letterplace paradigm. In: Kauers M, editor. Proc. Int. Symp. Symbolic and Algebraic Computation (ISSAC '13). New York: ACM; 2013. p. 259–66.
[18] Möller HM. A reduction strategy for the Taylor resolution. In: Caviness BF, editor. Proc. European Conf. Computer Algebra (EUROCAL '85). Lect Notes Comp Sci. vol. 162. Heidelberg: Springer; 1985. p. 526–34.
[19] Mora T. Gröbner bases for non-commutative polynomial rings. In: Calmet J, editor. Proc. Int. Conf. Algebraic Algorithms and Error-Correcting Codes (AAECC-3). Heidelberg: Springer; 1986. p. 353–62.
[20] Mora T. An introduction to commutative and non-commutative Gröbner bases. J Theor Comput Sci. 1994;134:131–73.
[21] Mora T. Solving Polynomial Equation Systems II – Macaulay's Paradigm and Gröbner Technology. Cambridge: Cambridge Univ. Press; 2005.
[22] Mora T. Solving Polynomial Equation Systems IV – Buchberger Theory and Beyond. Cambridge: Cambridge Univ. Press; 2016.
[23] Rosenberger G, Scheer M. Classification of the finite generalized tetrahedron groups. Contemp Math. 2002;296:207–29.
[24] Scala R, Levandovskyy V. Letterplace ideals and non-commutative Gröbner bases. J Symb Comput. 2004;44:1374–93.
[25] Xiu X. Non-commutative Gröbner bases and applications, dissertation, Universität Passau, 2012.

Martin Kreuzer and Gerhard Rosenberger

On the numbers of the form $x^2 + 11y^2$

Abstract: Numbers of the form $x^2 + Ny^2$ have been studied for a long time, in particular when N is one of Euler's convenient numbers. For the first *inconvenient* number $N = 11$, most attention has gone to prime numbers of the form $x^2 + 11y^2$. Here, we use the class group G_{11} of level 11 to study the set S_1 of *all* numbers n having a primitive representation $n = x^2 + 11y^2$. The four conjugacy classes of elliptic elements of order 2 in G_{11} give rise to a set $C = S_1 \cup S_2$ in which we have to distinguish S_1 from the set S_2 of numbers n having a primitive representation $n = 4x^2 + 11xy + 33y^2$. The set S_1 is decomposed into disjoint subsets corresponding to specific products of primes and cubic numbers. The primes and cubic numbers in S_1 and S_2 are described in detail.

Keywords: Class group, binary quadratic form, congruence subgroup, number field order

MSC 2010: Primary 11E25, Secondary 11F06, 20F05, 20F65

1 Introduction

In the paper [5], L. Euler considers *convenient numbers*, that is, numbers N for which a positive integer n has a unique representation of the form $n = x^2 + Ny^2$ with $\gcd(x^2, Ny^2) = 1$ if and only if n is a prime, a prime power, twice one of these, or a power of 2. The set of known convenient numbers, that is, the set $\{1, 2, 3, 4, 5, 6, 7, 8, 9, 10, 12, 13, 15, 16, \ldots, 1848\}$ consists of 65 numbers, and it is conjectured that these are all of them. When we look at this set, we see that 11 is the first *inconvenient* number. So, it is a natural question to ask which positive integers have a representation of the form $n = x^2 + 11y^2$ with $\gcd(x, 11y) = 1$, and when this representation is unique.

For the study of prime numbers of the form $p = x^2 + Ny^2$, there is a huge literature, and many deep results have been found. An excellent introduction to this topic is given in [4]. In particular, using these results, we can characterize prime numbers of the form $p = x^2 + 11y^2$ as in Theorem 5.1. A connection between these prime numbers and Tribonacci numbers is established in [6].

Acknowledgement: The authors thank Fairfield University, and in particular Professor Ben Fine, for their kind hospitality during part of the preparation of this paper.

Martin Kreuzer, Fakultät für Informatik und Mathematik, Universität Passau, D-94030 Passau, Germany, e-mail: martin.kreuzer@uni-passau.de
Gerhard Rosenberger, Fachbereich Mathematik, Universität Hamburg, D-20146 Hamburg, Germany, e-mail: gerhard.rosenberger@math.uni-hamburg.de

https://doi.org/10.1515/9783110638387-016

However, for general positive integers of the form $n = x^2 + 11y^2$, much less seems to be known. An algorithm for computing such a representation, if it exists, is contained in [10], but no characterization is given when such a representation exists. On the other hand, in [11], Satz 6.2 and Satz 7.2, the author proves several formulas for the number of such representations. However, these formulas are difficult to evaluate in general, and even depend on all solutions of $12n = x^2 + 11y^2$ satisfying certain extra conditions, so that they do not allow us to characterize numbers of the form $x^2 + 11y^2$ intrinsically. One reason for these difficulties may be that, since the quadratic form $x^2 + 11y^2$ has discriminant -44 and its form class number is $h(-44) = 3$, one should not expect simple congruence relations that decide whether n is or is not of the form $n = x^2 + 11y^2$.

In another vein, B. Fine proved in [7] Fermat's two-square theorem, that is, the characterization of primes representable by $x^2 + y^2$, using the modular group $\mathrm{PSL}_2(\mathbb{Z})$. The second author, together with G. Kern-Isberner, extended this method in [9] to all forms $x^2 + Ny^2$ such that $h(-4N) \leq 2$ and $N \neq 15$. In particular, they showed that, in these cases, one may in essence characterize numbers of the form x^2+Ny^2 by the conditions that $-N$ is a quadratic residue modulo n, that n is a quadratic residue modulo N, and possibly a simple congruence condition. This approach met with some obstacles for the convenient number $N = 15$, because the underlying class group G_{15} has a more complicated structure. Notwithstanding such impediments, we attempt to generalize these group theoretic methods further to the first *inconvenient* case $N = 11$, and the central ideal of the current paper is to follow the approach of [7] and [9] in order to deduce as many results about positive integers of the form $n = x^2 + 11y^2$ as possible.

Let us have a closer look at the contents of the paper. In Section 2, we lay the group theoretic foundation. Our main goal here is to introduce the class group G_{11} of level 11 and to give a detailed description of its structure. In particular, we prove that there are four conjugacy classes of elliptic elements of order 2, we provide concrete matrices t_1, t_2, t_3, t_4 representing these conjugacy classes, and we give an explicit presentation of G_{11} in terms of these elements (see Corollary 2.8).

This allows us in Section 3 to initiate the study of the numbers of the form $x^2 + 11y^2$ as follows: by conjugating the matrix $t_1 = \left(\begin{smallmatrix} 0 & 1 \\ -1 & 0 \end{smallmatrix}\right)$ with an element of G_{11}, we get a matrix whose top right entry is of the form $x^2 + 11y^2$, where $x, y \in \mathbb{Z}$, and $\gcd(x, 11y) = 1$. Conversely, given a number n such that $11 \nmid n$, and such that -11 is a quadratic residue modulo n, we construct an elliptic element $A_n(\ell)$ of order 2 in G_{11}. Then $A_n(\ell)$ is conjugated to exactly one of the matrices t_i, and n is of the form $n = x^2 + 11y^2$ if and only if $A_n(\ell)$ is conjugated to t_1. For $i = 1, \ldots, 4$, let S_i be the set of integers n for which the top right corner of $A_n(\ell)$ is n and $A_n(\ell)$ is conjugated to t_i. Thus we are interested in

$$S_1 = \{n = x^2 + 11y^2 \mid x, y \in \mathbb{Z}; \gcd(x, 11y) = 1\}$$

We prove that $S_2 = S_3$ and $S_4 = 2S_2$ (see Proposition 3.9) and that S_4 is easily discernible, since its elements n are exactly the ones satisfying $n \equiv 2 \pmod 4$. Consequently, the main task is to distinguish S_1 from

$$S_2 = \{n = 4x^2 + 22xy + 33y^2 \mid x, y \in \mathbb{Z}; \gcd(x, 11y) = 1\}$$

inside $C = S_1 \cup S_2$. We also show that the even numbers in C are precisely the fourfolds of the odd numbers in C (see Proposition 3.13) and that the set of odd numbers in C is multiplicatively closed (see Proposition 3.14).

In order to move on, we need to introduce some methods of algebraic number theory in Section 4. More precisely, since $x^2+11y^2 = (x+y\sqrt{-11})(x-y\sqrt{-11})$, we look at the algebraic number field $\mathbb{Q}(\sqrt{-11})$, its ring of integers $\mathbb{Z}[\omega]$, where $\omega = (-11 + \sqrt{-11})/2$, and its order $\mathbb{Z}[\sqrt{-11}]$ of conductor 2. We describe the structure of $\mathbb{Z}[\omega]$, and in particular which products of its elements are contained in $\mathbb{Z}[\sqrt{-11}]$ (see Proposition 4.3). By realizing the elements of S_1 and S_2 as the norms of elements in $\mathbb{Z}[\omega]$, we get basic properties of products of elements in S_1 and S_2 (see Corollary 4.4). Moreover, we describe how various prime numbers split in $\mathbb{Z}[\omega]$ (see Proposition 4.5).

Next, we turn our attention to special types of numbers in S_1 which occur in our main theorem: prime numbers and cubic numbers. In Section 5, we characterize prime numbers of the form $x^2 + 11y^2$. This case has been studied extensively before, so that it suffices to recall and simplify some results from [4] in order to get a good characterization (see Theorem 5.1). We also show that every prime is either in S_1 or in S_2 (see Proposition 5.2).

As for cubic numbers in $C = S_1 \cup S_2$, we prove in Section 6 that they are all odd and contained in $S_1 \setminus S_2$ (see Proposition 6.5.a). More precisely, the numbers m such that $m^3 \in S_1$ are of the form $m = p^\alpha \tilde{m}$, where p is a prime in S_2, where $\alpha \in \{0, 1, 2\}$, and where $\tilde{m}$ is of the form $\tilde{m} = x^2 + 11y^2$ (see Proposition 6.5.b).

Finally, in Section 7, we provide a detailed decomposition of the set S_1 in Theorem 7.4. It says that every number n with a primitive representation $n = x^2 + 11y^2$ such that $x, y \in \mathbb{Z}$ and $\gcd(x, 11y) = 1$ is of one of the following types:

(1) If n is even, it is of the form $n = 4\tilde{n}$ with an odd number $\tilde{n} \in S_2$.
(2) The number n is a product of powers of primes in S_1.
(3) The number n is a cubic number.
(4) The number n is an odd number in $S_1 \cap S_2$.

Here, only the second and third sets intersect nontrivially and in the obvious way. For the odd numbers in S_1 and S_2, we then remove cubic factors and go on to provide a detailed characterization when they are in $S_1 \setminus S_2$ or $S_2 \setminus S_1$ or $S_1 \cap S_2$ based on their prime factors (see Corollary 7.7).

Since we apply methods from a number of different areas, we tried to keep this paper as self-contained as possible. The knowledgeable readers may bear with us for including some down-to-earth proofs which could have been replaced by high-level references. Many characterizations and properties of the sets of numbers we study were found and checked using the computer algebra system ApCoCoA (see [1]).

2 The class group of level 11

In the following, we consider the subgroup G_{11} of $\mathrm{PSL}_2(\mathbb{R})$ consisting of the matrices of one of the types

$$U = \begin{pmatrix} a & b\sqrt{11} \\ c\sqrt{11} & d \end{pmatrix} \quad \text{with } a, b, c, d \in \mathbb{Z} \text{ and } ad - 11bc = 1,$$

$$V = \begin{pmatrix} a\sqrt{11} & b \\ c & d\sqrt{11} \end{pmatrix} \quad \text{with } a, b, c, d \in \mathbb{Z} \text{ and } 11ad - bc = 1,$$

where a matrix is identified with its negative. Equivalently, we consider a matrix of one of these types as a linear fractional transformation. This group is called the **class group of level 11** (see [8], pp. 314–316).

The matrices of type U form a normal subgroup H_{11} of index 2 in G_{11}, and we have $G_{11} = H_{11} \cup T \cdot H_{11}$, where $T = \left(\begin{smallmatrix} 0 & 1 \\ -1 & 0 \end{smallmatrix}\right)$. Since $T \cdot \left(\begin{smallmatrix} a & b\sqrt{11} \\ c\sqrt{11} & d \end{smallmatrix}\right) = \left(\begin{smallmatrix} c\sqrt{11} & d \\ -a & b\sqrt{11} \end{smallmatrix}\right)$, the matrices in $T \cdot H_{11}$ are precisely the matrices of type V. The group H_{11} can also be described as follows.

Remark 2.1. By conjugating G_{11} with the matrix $X = \left(\begin{smallmatrix} 1 & 0 \\ 0 & \sqrt{11} \end{smallmatrix}\right)$, we get the discrete group

$$G'_{11} = XG_{11}X^{-1} = H'_{11} \cup T' \cdot H'_{11}$$

where $H'_{11} = \{\left(\begin{smallmatrix} a & b \\ c' & d \end{smallmatrix}\right) \mid ad - bc = 1, c' \equiv 0 \pmod{11}\}$ and $T' = XTX^{-1} = \left(\begin{smallmatrix} 0 & 1/\sqrt{11} \\ -\sqrt{11} & 0 \end{smallmatrix}\right)$. Thus H'_{11} is the **Hecke congruence subgroup of level 11**

$$\Gamma_0(11) = \left\{\begin{pmatrix} a & b \\ c & d \end{pmatrix} \mid a, b, c, d \in \mathbb{Z} \text{ such that } ad - bc = 1 \text{ and } c \equiv 0 \pmod{11}\right\}$$

of the **(inhomogeneous) modular group** $\Gamma = \mathrm{PSL}_2(\mathbb{Z})$.

In the following, we first determine the structure of the group G'_{11} and then we translate everything back to the group G_{11}. In particular, we want to determine the conjugacy classes of the elliptic elements of order 2 in $G_{11} \subset \mathrm{PSL}_2(\mathbb{Z})$. Clearly, they have to be residue classes of matrices of the form V. Notice that a matrix V of this form satisfies $V^2 = -I_2$ if and only if $d = -a$. (Here, I_2 denotes the identity matrix of size 2×2, and we may use $-I_2$, since we work in $\mathrm{PSL}_2(\mathbb{Z})$.) It is known that the number of these conjugacy classes is $m(11) = 4$ (cf. [9], p. 152). Our goal in this section is to find explicit representatives of these conjugacy classes. Recall that the group

$$\Gamma(11) = \left\{\begin{pmatrix} a & b \\ c & d \end{pmatrix} \in \mathrm{PSL}_2(\mathbb{Z}) \mid \begin{pmatrix} a & b \\ c & d \end{pmatrix} \equiv \begin{pmatrix} 1 & 0 \\ 0 & 1 \end{pmatrix} \pmod{11}\right\}$$

is called the **(inhomogeneous) principal congruence subgroup of level 11** of the modular group $\Gamma = \mathrm{PSL}_2(\mathbb{Z})$.

Remark 2.2 (The index and cosets of $\Gamma_0(11)$). The principal congruence subgroup $\Gamma(11)$ of Γ is a normal subgroup, and it satisfies $\Gamma/\Gamma(11) \cong \mathrm{PSL}_2(\mathbb{F}_{11})$. Hence the index of $\Gamma(11)$ in Γ is $\frac{1}{2}11^3(1-\frac{1}{11^2}) = 660$.

(a) The group $\Gamma(11)$ is clearly a subgroup of $\Gamma_0(11)$. Under the hypothesis that $c \equiv 0 \pmod{11}$, the condition $ad - bc = 1$ implies that the residue classes of a and d in $\mathbb{F}_{11}$ are inverses of each other, and the residue class of b can be chosen freely. Hence the congruence $ad - bc \equiv 1 \pmod{11}$ has 110 incongruent solutions for (a, b, d). Since we identify a matrix with its negative, it follows that the index of $\Gamma(11)$ in $\Gamma_0(11)$ is 55. Altogether, we see that the index of $\Gamma_0(11)$ in Γ is 12.

(b) It is well known that, as a system of representatives of $\Gamma/\Gamma_0(11)$, we can use $\{S_{-5}, S_{-4}, \ldots, S_5, T\}$, where $S_i = S^i = \left(\begin{smallmatrix} 1 & 0 \\ -i & 1 \end{smallmatrix}\right)$ with $S = \left(\begin{smallmatrix} 1 & 0 \\ -1 & 1 \end{smallmatrix}\right)$, and where $T = \left(\begin{smallmatrix} 0 & 1 \\ -1 & 0 \end{smallmatrix}\right)$.

Next, we want to determine generators for $\Gamma_0(11)$ via geometric arguments. For this, it is more convenient to use the isomorphic group

$$\Gamma^0(11) = T \cdot \Gamma_0(11) \cdot T^{-1} = \{\left(\begin{smallmatrix} d & -c \\ -b & a \end{smallmatrix}\right) \mid ad - bc = 1, -c \equiv 0 \pmod{11}\}$$

The conjugation of the above decomposition of $\Gamma_0(11)$ yields

$$\Gamma^0(11) = \bigcup_{i=-5}^{5} \Gamma^0(11) \cdot U_i \cup \Gamma^0(11) \cdot T$$

where $U_i = U_1^i = \left(\begin{smallmatrix} 1 & i \\ 0 & 1 \end{smallmatrix}\right)$ and $U_1 = \left(\begin{smallmatrix} 1 & 1 \\ 0 & 1 \end{smallmatrix}\right)$.

Remark 2.3 (A fundamental domain for $\Gamma^0(11)$). Let $\mathbb{H} = \{\tau \in \mathbb{C} \mid \Im(\tau) > 0\}$ denote the upper half-plane, as usual. Recall that the fundamental domain of the modular group Γ is given by

$$D_\Gamma = \left\{\tau \in \mathbb{H} \mid |\Re(\tau)| < \frac{1}{2}, |\tau| > 1\right\} \cup \{i\infty\} \cup \left\{\tau \in \mathbb{H} \mid \Re(\tau) = -\frac{1}{2}, |\tau| \geq 1\right\}$$
$$\cup \left\{\tau \in \mathbb{H} \mid |\tau| = 1, -\frac{1}{2} \leq \Re(\tau) \leq 0\right\}$$

Then a fundamental domain of $\Gamma^0(11)$ is given by

$$D_{\Gamma^0(11)} = \bigcup_{i=-5}^{5} U_i(D_\Gamma) \cup T(D_\Gamma)$$

In order to get a geometric presentation of $\Gamma^0(11)$, we need to study the boundary correspondence for $D_{\Gamma^0(11)}$ next. For $i = -5, -4, \ldots, 5$, we let E_i be the circular arc of radius 1 centered at i, that is, we let $E_i = \{\tau \in \mathbb{H} \mid |\tau - i| = 1\}$.

Proposition 2.4. *Under the action of $\Gamma^0(11)$, there are precisely the following boundary correspondences:*

$$E_{-5} \leftrightarrow E_{-2}, \quad E_5 \leftrightarrow E_2, \quad E_{-4} \leftrightarrow E_3, \quad E_4 \leftrightarrow E_{-3}, \quad E_{-1} \leftrightarrow E_1$$

Proof. For $-5 \le i, j \le 5$, there are exactly two maps in Γ which transform E_i to E_j, namely U_1^{j-i} and $U_1^j T U_1^{-i}$. Since $U_1^{j-i} \notin \Gamma^0(11)$ for $j \neq i$, there remains at most the map $U_1^j T U_1^{-i} = \left(\begin{smallmatrix} -j & ij+1 \\ -1 & i \end{smallmatrix}\right)$. This is an element of $\Gamma^0(11)$ if and only if $ij + 1 \equiv 0 \pmod{11}$. Hence two circular arcs E_i and E_j are equivalent under $\Gamma^0(11)$ if and only if $ij + 1 \equiv 0 \pmod{11}$, and from this observation the claim follows readily. □

Now we can read off the structure of $\Gamma^0(11)$ and $\Gamma_0(11)$.

Corollary 2.5. *The groups $\Gamma^0(11)$ and $\Gamma_0(11)$ have genus 1 and are free groups of rank 3.*
(a) *The group $\Gamma^0(11)$ is freely generated by $\tilde{A} = \left(\begin{smallmatrix} 2 & 11 \\ -1 & -5 \end{smallmatrix}\right)$, $\tilde{B} = \left(\begin{smallmatrix} 3 & -11 \\ -1 & 4 \end{smallmatrix}\right)$, and $\tilde{P} = \left(\begin{smallmatrix} -1 & 0 \\ -1 & -1 \end{smallmatrix}\right)$.*
(b) *The group $\Gamma_0(11)$ is freely generated by $A = \left(\begin{smallmatrix} -5 & 1 \\ -11 & 2 \end{smallmatrix}\right)$, $B = \left(\begin{smallmatrix} 4 & 1 \\ 11 & 3 \end{smallmatrix}\right)$, and $P = \left(\begin{smallmatrix} -1 & 1 \\ 0 & -1 \end{smallmatrix}\right)$.*

Proof. To prove (a), it suffices to note that the boundary correspondences in Proposition 2.4 are generated by $\tilde{A} = U_1^{-2} T U_1^5$ and $\tilde{B} = U_1^{-3} T U_1^{-4}$ and $\tilde{P} = U_1 T U_1$. Claim (b) follows immediately from (a) by conjugating back. □

Notice that P is a parabolic element of $\Gamma_0(11)$. A further parabolic element can be found via $Q = P \cdot [A, B^{-1}] = \left(\begin{smallmatrix} -21 & -11 \\ 44 & 23 \end{smallmatrix}\right)$. Next, we determine the structure of the group G'_{11}.

Remark 2.6 (The signature of G'_{11}). From the preceding corollary and the fact that $H'_{11} = \Gamma_0(11)$ has index 12 in the modular group Γ, it follows that H'_{11} is a cofinite Fuchsian group.

In general, a cofinite Fuchsian group F has a presentation of the form

$$F = \Big\langle s_1, \dots, s_r, p_1, \dots, p_t, a_1, b_1, \dots, a_g, b_g \mid s_1^{m_1} = \dots = s_r^{m_r} = s_1 \cdots s_r \cdot p_1 \cdots p_r \cdot \prod_{i=1}^{g} [a_i, b_i] = 1 \Big\rangle$$

where $m_i \ge 2$, where the elements s_i represent the conjugacy classes of maximal elliptic cyclic subgroups, where the elements p_j represent the conjugacy classes of maximal parabolic cyclic subgroups, and where g is the genus of F (see, for instance, [2], Chapter 10). The group F can be described by the symbol $(g; m_1, \dots, m_r; t)$ which is called the **signature** of F. Moreover, the (finite) **hyperbolic area** for F is given by

$$\mu(F) = 2\pi\left(2g - 2 + t + \left(1 - \frac{1}{m_1}\right) + \dots + \left(1 - \frac{1}{m_r}\right)\right) > 0$$

A subgroup F' of F of finite index is also a cofinite Fuchsian group, and its hyperbolic area satisfies the Riemann–Hurwitz relation $\mu(F') = [F : F'] \cdot \mu(F)$.

In our setting, the corollary says that H'_{11} is a cofinite Fuchsian group with signature $(1; 0; 2)$ and we get $\mu(H'_{11}) = 2\pi(2-2+2) = 4\pi$. Clearly, the group G'_{11} is also a cofinite Fuchsian group. Let $(g; m_1, \dots, m_r; g)$ be its signature. Then $[G'_{11} : H'_{11}] = 2$ implies $t = 1$

and $m_1 = \cdots = m_r = 2$. Moreover, we get $\mu(G'_{11}) = 2\pi$, and hence $2g - 2 + \frac{r}{2} = 0$. This is possible only if $g = 0$ and $r = 4$, so that altogether we obtain the signature $(0; 2, 2, 2, 2; 1)$ for G'_{11}.

Using the information gathered above, we are ready to construct a nice presentation of G'_{11}.

Proposition 2.7. *In G'_{11}, consider the following elements:*

$$T_1 = \begin{pmatrix} 0 & 1/\sqrt{11} \\ -\sqrt{11} & 0 \end{pmatrix}, \quad T_2 = \begin{pmatrix} -\sqrt{11} & 4/\sqrt{11} \\ -3\sqrt{11} & \sqrt{11} \end{pmatrix}, \quad T_3 = \begin{pmatrix} \sqrt{11} & -3/\sqrt{11} \\ 4\sqrt{11} & -\sqrt{11} \end{pmatrix},$$

$$T_4 = \begin{pmatrix} -\sqrt{11} & -6/\sqrt{11} \\ 2\sqrt{11} & \sqrt{11} \end{pmatrix}, \quad \text{and} \quad P = \begin{pmatrix} -1 & 1 \\ 0 & -1 \end{pmatrix}$$

Then the group G'_{11} has the presentation

$$G'_{11} = \langle T_1, T_2, T_3, T_4, P \mid T_1^2 = T_2^2 = T_3^2 = T_4^2 = T_2T_3T_1T_4P = 1 \rangle$$

where T_1, T_2, T_3, T_4 are elliptic elements of order 2 representing the conjugacy classes of such elements, and where P is a parabolic element.

Proof. Clearly, the matrix T_1 agrees with the matrix T' given above, while $T_2 = BT'$ and $T_3 = A^{-1}BT'$ with A, B as in Corollary 2.5. Together with this corollary, we see that G'_{11} is generated by $\{A, B, P, T'\}$, and hence by $\{T', BT', A^{-1}BT', P\} = \{T_1, T_2, T_3, P\}$.

To find a representative of the fourth conjugacy class, we may look for an element T_4 of order 2 such that $T_4PT_4^{-1} = Q^{-1}$, and the given element T_4 satisfies this. Notice that $T_4 = -PT_2T_3T_1$ and that $T_i^2 = -I_2$, where I_2 is the identity matrix of size 2×2. Altogether, it follows that G'_{11} has the stated presentation.

Notice that this implies that G'_{11} is a free product of four cyclic groups of order 2, and that it has the signature $(0; 2, 2, 2, 2; 1)$. By [12], Satz 2.15, it follows that T_1, T_2, T_3, and T_4 represent different conjugacy classes, and that these are all conjugacy classes of elliptic elements of order 2. □

The last step is to translate the above presentation of G'_{11} to a presentation of G_{11} This is easily achieved by conjugating back using the matrix X of Remark 2.1.

Corollary 2.8. *The group G_{11} has a presentation*

$$G_{11} = \langle t_1, t_2, t_3, t_4, p \mid t_1^2 = t_2^2 = t_3^2 = t_4^2 = t_2t_3t_1t_4p = 1 \rangle$$

where $t_1 = X^{-1}T_1X = \left(\begin{smallmatrix} 0 & 1 \\ -1 & 0 \end{smallmatrix}\right)$, $t_2 = X^{-1}T_2X = \left(\begin{smallmatrix} -\sqrt{11} & 4 \\ -3 & \sqrt{11} \end{smallmatrix}\right)$, $t_3 = X^{-1}T_3X = \left(\begin{smallmatrix} \sqrt{11} & -3 \\ 4 & -\sqrt{11} \end{smallmatrix}\right)$, $t_4 = X^{-1}T_4X = \left(\begin{smallmatrix} -\sqrt{11} & -6 \\ 2 & \sqrt{11} \end{smallmatrix}\right)$, *and* $p = X^{-1}PX = \left(\begin{smallmatrix} -1 & \sqrt{11} \\ 0 & -1 \end{smallmatrix}\right)$.

Here, t_1, t_2, t_3, t_4 are elliptic elements of order 2 representing the conjugacy classes of such elements, and p is a parabolic element.

3 Representing numbers in the form $x^2 + 11y^2$

In this section, we want to represent numbers in the form $n = x^2 + 11y^2$ with $x, y \in \mathbb{Z}$ using the matrices in the class group of level 11. Let us begin with some easy observations.

Remark 3.1. In the following, we let $n \in \mathbb{Z}$.

(a) Suppose that n is divisible by 11, and write $n = 11\tilde{n}$ with $\tilde{n} \in \mathbb{Z}$. Then n is of the form $n = x^2 + 11y^2$ if and only of $\tilde{n}$ is of this form.
Namely, if $\tilde{n} = x^2 + 11y^2$, then $n = 11\tilde{n} = (11y)^2 + 11x^2$. Conversely, if $n = x^2 + 11y^2$ is divisible by 11, then x is divisible by 11 and we can write $x = 11\tilde{x}$ with $\tilde{x} \in \mathbb{Z}$. Consequently, we have $\tilde{n} = y^2 + 11\tilde{x}^2$. So, from now on we shall assume that n is not divisible by 11.

(b) Clearly, $n \equiv x^2 \pmod{11}$ says that n is a quadratic residue modulo 11. So, it suffices to consider numbers n which are quadratic residues modulo 11.

(c) If $n = x^2 + 11y^2$ and $\gcd(n, x, y) = 1$, we say that (n, x, y) is a **primitive representation** of n. Clearly, if $\gcd(n, x) > 1$ or $\gcd(n, y) > 1$ or $\gcd(x, y) > 1$ then the representation is not primitive. Moreover, it suffices to check whether n has a primitive representation, as all representations can be obtained by multiplying a primitive representation by a square number. To check whether a number n has a representation of the form $n = x^2 + 11y^2$, it suffices to check whether n, or a number n/s with a square number s dividing n, has a primitive representation. Therefore, we will be interested only in primitive representations.

(d) If n has a primitive representation of the form $n = x^2 + 11y^2$, then -11 is a quadratic residue modulo n. In effect, we have $x^2 + 11y^2 \equiv 0 \pmod{n}$ and $\gcd(y, n) = 1$. Hence y is a unit modulo n and $-11 \equiv (x/y)^2 \pmod{n}$ is a quadratic residue modulo n.

Altogether, we are led to define the following sets.

Notation 3.2. The set $\mathbb{D}$ of all positive integers n such that $11 \nmid n$, and such that -11 is a quadratic residue modulo n, is called the *domain* of our investigation. We have

$$\mathbb{D} = \{1, 2, 3, 4, 5, 6, 9, 10, 12, 15, 18, 20, 23, 25, 27, 30, 31, 36, 37, 45, 46, 47, 50, \dots\}$$

Moreover, the set of all numbers $n \in \mathbb{D}$ which have a primitive representation $n = x^2 + 11y^2$ with $x, y \in \mathbb{Z}$ is denoted by

$$\begin{aligned} S_1 &= \{n \in \mathbb{D} \mid n = x^2 + 11y^2 \text{ for some } x, y \in \mathbb{Z} \text{ with } \gcd(x, y) = 1\} \\ &= \{1, 12, 15, 20, 27, 36, 45, 47, 53, 60, 69, 75, 92, 93, 100, 103, 111, 115, 124, \dots\} \end{aligned}$$

Notice that, if -11 is a quadratic residue modulo n and n is a prime, then n is a quadratic residue modulo 11. In fact, for prime numbers n, both conditions are equivalent by the quadratic reciprocity theorem, since $11 \equiv 3 \pmod{4}$. The following construction is the key for representing numbers in the form $x^2 + 11y^2$.

Remark 3.3. Let $n \in \mathbb{D}$. Consider the following steps:

(a) Since -11 is a unit modulo n, we can calculate $b = (-11)^{-1} \pmod{n}$.

(b) Then b is a square modulo n, that is, we can find a number $\ell \in \mathbb{Z}$ such that $\ell^2 \equiv b \pmod{n}$.

(c) In particular, we have $(-11)\ell^2 \equiv 1 \pmod{n}$. Hence we find $q \in \mathbb{Z}$ such that $-11\ell^2 + nq = 1$.

(d) Now we form the matrix $A_n(\ell) = \begin{pmatrix} \ell\sqrt{11} & n \\ -q & -\ell\sqrt{11} \end{pmatrix}$.

In this way, we obtain a matrix $A_n(\ell) \in G_{11}$ which satisfies $A_n(\ell)^2 = -I_2$. Consequently, the matrix $A_n(\ell)$ is conjugate in G_{11} to exactly one of the matrices t_1, t_2, t_3, t_4 in Corollary 2.8.

Notice that, for a given number n, there exist infinitely many different matrices $A_n(\ell)$. In order to find out which of the matrices t_1, t_2, t_3, t_4 is conjugate to a given matrix $A_n(\ell)$, we first calculate the general shapes of the conjugates of these matrices.

Lemma 3.4. *Let the elements of G_{11} be described by the matrices of type $U = \begin{pmatrix} a & b\sqrt{11} \\ c\sqrt{11} & d \end{pmatrix}$ and $V = \begin{pmatrix} a\sqrt{11} & b \\ c & d\sqrt{11} \end{pmatrix}$ as at the beginning of Section 2. Then the following equalities hold:*

(a) $Ut_1U^{-1} = \begin{pmatrix} (-ac-bd)\sqrt{11} & a^2+11b^2 \\ -11c^2-d^2 & (ac+bd)\sqrt{11} \end{pmatrix}$

(b) $Vt_1V^{-1} = \begin{pmatrix} (-ac-bd)\sqrt{11} & 11a^2+b^2 \\ -c^2-11d^2 & (ac+bd)\sqrt{11} \end{pmatrix}$.

(c) $Ut_2U^{-1} = \begin{pmatrix} (-4ac-11bc-ad-3bd)\sqrt{11} & 4a^2+22ab+33b^2 \\ -44c^2-22cd-3d^2 & (4ac+11bc+ad+3bd)\sqrt{11} \end{pmatrix}$.

(d) $Vt_2V^{-1} = \begin{pmatrix} (-4ac-bc-11ad-3bd)\sqrt{11} & 44a^2+22ab+3b^2 \\ -4c^2-22cd-33d^2 & (4ac+bc+11ad+3bd)\sqrt{11} \end{pmatrix}$.

(e) $Ut_3U^{-1} = -\begin{pmatrix} (-3ac-11bc-ad-4bd)\sqrt{11} & 3a^2+22ab+44b^2 \\ -33c^2-22cd-4d^2 & (3ac+11bc+ad+4bd)\sqrt{11} \end{pmatrix}$.

(f) $Vt_3V^{-1} = -\begin{pmatrix} (-3ac-bc-11ad-4bd)\sqrt{11} & 33a^2+22ab+4b^2 \\ -3c^2-22cd-44d^2 & (3ac+bc+11ad+4bd)\sqrt{11} \end{pmatrix}$.

(g) $Ut_4U^{-1} = -\begin{pmatrix} (-6ac+11bc+ad-2bd)\sqrt{11} & 6a^2-22ab+22b^2 \\ -66c^2+22cd-2d^2 & (6ac-11bc-ad+2bd)\sqrt{11} \end{pmatrix}$.

(h) $Vt_4V^{-1} = -\begin{pmatrix} (-6ac+bc+11ad-2bd)\sqrt{11} & 66a^2-22ab+2b^2 \\ -6c^2+22cd-22d^2 & (6ac-bc-11ad+2bd)\sqrt{11} \end{pmatrix}$.

Proof. This follows by multiplying out the matrix products. □

The following proposition lies at the heart of our method.

Proposition 3.5. *For $n \in \mathbb{D}$, the following conditions are equivalent:*

(a) *The number n has a primitive representation $n = x^2 + 11y^2$ with $x, y \in \mathbb{Z}$.*

(b) *There exists a number $\ell \in \mathbb{Z}$ such that $\ell^2 \equiv (-11)^{-1} \pmod{n}$ and such that the matrix $A_n(\ell) = \begin{pmatrix} \ell\sqrt{11} & n \\ -q & -\ell\sqrt{11} \end{pmatrix}$ is conjugate to $t_1 = \begin{pmatrix} 0 & 1 \\ -1 & 0 \end{pmatrix}$ in G_{11}, where $q = (1 + 11\ell^2)/n$.*

Proof. First, we show that (a) implies (b). If n has a primitive representation $n = x^2 + 11y^2$, then we have $\gcd(x, -11y) = 1$ and we can find $c, d \in \mathbb{Z}$ such that $dx - 11cy = 1$. Furthermore, let $a = x$ and $b = y$. Then we let $\ell = -ac - bd$ and $q = 11c^2 + d^2$, and part (a) of the lemma yields

$$Ut_1U^{-1} = \begin{pmatrix} (-ac - bd)\sqrt{11} & a^2 + 11b^2 \\ -11c^2 - d^2 & (ac + bd)\sqrt{11} \end{pmatrix} = \begin{pmatrix} \ell\sqrt{11} & n \\ -q & -\ell\sqrt{11} \end{pmatrix} = A_n(\ell)$$

since it is easy to verify $nq = (a^2 + 11b^2)(11c^2 + d^2) = 1 + 11(-ac - bd)^2 = 1 + 11\ell^2$.

Conversely, if the matrix $A_n(\ell)$ is conjugate to t_1 then parts (a) and (b) of the lemma show that $n = a^2 + 11b^2$ resp. $n = 11a^2 + b^2$ has the desired representation. Notice that this representation is primitive, since we have $\det(U) = ad - 11bc = 1$ resp. $\det(V) = 11ad - bc = 1$. □

The following example shows that different choices of ℓ may yield matrices which are conjugate to different matrices t_i.

Example 3.6. Consider the number $n = 12$. It is a quadratic residue modulo 11, $-11 \equiv 1 \pmod{12}$ is a quadratic residue modulo n, and it has representations $n = x^2 + 11y^2$ with $x = y = 1$ as well as $n = 4a^2 + 22ab + 33b^2$ with $a = -5$ and $b = 2$. Let us examine the conjugacy classes of the following matrices $A_n(\ell)$:

(a) For $U = \begin{pmatrix} 1 & \sqrt{11} \\ \sqrt{11} & 12 \end{pmatrix}$, we see that $Ut_1U^{-1} = \begin{pmatrix} -13\sqrt{11} & 12 \\ -155 & 13\sqrt{11} \end{pmatrix} = A_n(\ell)$ for $\ell = -13$. Thus $A_n(-13)$ is conjugate to t_1 in G_{11}.

(b) For $U = \begin{pmatrix} -5 & 2\sqrt{11} \\ 2\sqrt{11} & -9 \end{pmatrix}$, we see that $Ut_2U^{-1} = \begin{pmatrix} 5\sqrt{11} & 12 \\ -23 & -5\sqrt{11} \end{pmatrix} = A_n(\ell)$ for $\ell = 5$. Thus $A_n(5)$ is conjugate to t_2 in G_{11}.

Indeed, both $\ell = -13$ and $\ell = 5$ satisfy $\ell^2 \equiv (-11)^{-1} \equiv 1 \pmod{12}$.

To get the dependency of the conjugacy class of $A_n(\ell)$ in ℓ under control, we may use the following proposition.

Proposition 3.7. *Let $n \in \mathbb{D}$, let $\ell \in \mathbb{Z}$ be chosen such that $\ell^2 \equiv (-11)^{-1} \pmod{n}$, and let $i \in \{1, \ldots, 4\}$ be such that $A_n(\ell)$ is in the conjugacy class of t_i in G_{11}.*

(a) *The matrices $A_n(\ell - n)$ and $A_n(\ell + n)$ are in the conjugacy class of t_i.*

(b) *The matrix $A_n(-\ell)$ is in the conjugacy class of t_1 if and only if $A_n(\ell)$ is in the conjugacy class of t_1.*

Proof. To prove (a), we conjugate a matrix of the form $A_n(\ell) = \begin{pmatrix} \ell\sqrt{11} & n \\ -q & -\ell\sqrt{11} \end{pmatrix}$ by $p^{tr} = \begin{pmatrix} -1 & 0 \\ \sqrt{11} & -1 \end{pmatrix}$ and obtain

$$p^{tr} \cdot A_n(\ell) \cdot (p^{tr})^{-1} = \begin{pmatrix} (\ell + n)\sqrt{11} & n \\ -22\ell - 11n - q & (-\ell - n)\sqrt{11} \end{pmatrix} = A_n(\ell + n)$$

as well as

$$(p^{\mathrm{tr}})^{-1} \cdot A_n(\ell) \cdot p^{\mathrm{tr}} = \begin{pmatrix} (\ell - n)\sqrt{11} & n \\ 22\ell - 11n - q & (-\ell + n)\sqrt{11} \end{pmatrix} = A_n(\ell - n)$$

To prove (b), we assume that $A_n(\ell)$ is in the conjugacy class of t_1. Observe that any matrix of type V is of the form $V = Ut_1$, so that we may assume $U^{-1}A_n(\ell)U = t_1$ for some $U \in H_{11}$. We write $U = \left(\begin{smallmatrix} a & b\sqrt{11} \\ c\sqrt{11} & d \end{smallmatrix}\right)$ with $ad - 11bc = 1$ and form $\widehat{U} = \left(\begin{smallmatrix} a & -b\sqrt{11} \\ -c\sqrt{11} & d \end{smallmatrix}\right)$. Then $Ut_1U^{-1} = A_n(\ell)$ implies $\widehat{U}t_1\widehat{U}^{-1} = A_n(-\ell)$ by Lemma 3.4.a, and the observation that $\widehat{U} \in H_{11}$, since $\det(\widehat{U}) = ad - 11bc = 1$, completes the proof. The converse implication follows analogously. □

This proposition allows us to characterize primes of the form $x^2 + 11y^2$ as follows.

Corollary 3.8. *For an odd prime number $p \in \mathbb{D}$, the following conditions are equivalent:*
(a) *The number p has a primitive representation $p = x^2 + 11y^2$ with $x, y \in \mathbb{N}$.*
(b) *There exists a number $\ell \in \mathbb{Z}$ such that $\ell^2 \equiv (-11)^{-1} \pmod p$ and such that $A_p(\ell)$ is in the conjugacy class of t_1 in G_{11}.*
(c) *For every number $\ell \in \mathbb{Z}$ such that $\ell^2 \equiv (-11)^{-1} \pmod p$, the matrix $A_p(\ell)$ is in the conjugacy class of t_1 in G_{11}.*

Proof. By Proposition 3.5, it suffices to prove that (b) implies (c). Using part (a) of the proposition, we may assume that $0 \le \ell < p$. Since p is a prime number, there are at most two square roots of $(-11)^{-1}$ in this interval, namely ℓ and $p - \ell$. Thus the claim follows from the proposition. □

The following proposition collects some properties of the numbers represented by the quadratic forms in the top right corners of the matrices in Lemma 3.4.

Proposition 3.9. *Consider the following six quadratic forms: $u_2 = 4x^2 + 22xy + 33y^2$, $v_2 = 44x^2 + 22xy + 3y^2$, $u_3 = 3x^2 + 22xy + 44y^2$, $v_3 = 33x^2 + 22xy + 4y^2$, $u_4 = 6x^2 - 22xy + 22y^2$, and $v_4 = 66x^2 - 22xy + 2y^2$.*
(a) *For $x, y \in \mathbb{Z}$ and $(x, y) \neq (0, 0)$, these quadratic forms represent positive integers.*
(b) *Let $S_2 = \{u_2(x, y) \mid x, y \in \mathbb{Z}; (x, y) \neq (0, 0); \gcd(x, 11y) = 1\}$ be the set of numbers represented by u_2, and let $S_2' = \{v_2(x, y) \mid x, y \in \mathbb{Z}; (x, y) \neq (0, 0); \gcd(11x, y) = 1\}$ be the set of numbers represented by v_2. Then we have $S_2' = S_2$.*
(c) *Both the set of numbers represented by u_3 and the set of numbers represented by v_3 agree with S_2.*
(d) *Let $S_4 = \{u_4(x, y) \mid x, y \in \mathbb{Z}; (x, y) \neq (0, 0); \gcd(x, 11y) = 1\}$ be the set of numbers represented by u_4. This set agrees with the set of numbers represented by v_4, and we have $S_4 = \{2n \mid n \in S_2\} \cup \{2\}$.*
(e) *The domain $\mathbb{D}$ is the disjoint union of $S_1 \cup S_2$ and S_4. More precisely, the numbers in S_4 are precisely those numbers n in $\mathbb{D}$ which satisfy $n \equiv 2 \pmod 4$.*

Proof. First, we prove (a). To show the claim for u_2, it suffices to prove $4x^2 - 22xy + 33y^2 > 0$ for $x, y \in \mathbb{R}_+$. Since the left-hand side is homogeneous, we may assume $y = 1$. The inequality $4x^2 - 22x + 33 > 0$ holds for all $x \in \mathbb{R}$, since it holds for $x \gg 0$ and the discriminant is negative. For the other quadratic forms, the claim follows similarly.

Next, we show (b) and (c). If we substitute $x' = x$ and $y' = -y$ in u_2, we see that u_2 and $u_2' = 4(x')^2 - 22x'y' + 33(y')^2$ represent the same numbers, since (x', y') is again coprime and different from $(0,0)$. Now we substitute $x = 3x' + 11y'$ and $y = x' + 4y'$ in u_2' and see that u_2' and $u_3 = v_2(y,x) = 3x^2 + 22xy + 44y^2$ represent the same numbers. Again we have $\gcd(x,y) = 1$ and $(x,y) \neq (0,0)$ because $\det\left(\begin{smallmatrix} 3 & 11 \\ 1 & 4 \end{smallmatrix}\right) = 1$. Finally, by interchanging x and y in $v_2(y,x)$, we get that v_2 represents the numbers in S_2' and that this set agrees with S_2. Consequently, the set of numbers represented by $v_3 = u_2(y,x)$ is also equal to S_2.

In order to prove the first claim in (d), we apply the invertible transformation $x = 11x' + 2y'$ and $y = 5x' + y'$ to u_4 and get $v_4(x', -y') = 66(x')^2 + 22x'y' + 2(y')^2$. By using $y' \mapsto -y'$, the claim follows.

It remains to show (e). Since every number in $\mathbb{D}$ is represented by one of the forms $x^2 + 11y^2$ or u_i or v_i with $i \in \{2,3,4\}$, it is in $S_1 \cup S_2 \cup S_4$ by the previous parts. Thus we need to show that $S_1 \cup S_2$ and S_4 are disjoint. This follows by viewing these numbers modulo 4. The elements of $S_1 \cup S_2$ have remainders in $\{0,1,3\}$, while the elements of S_4 are congruent to 2 modulo 4. □

Thus we can now write down a simple characterization of the numbers in $S_1 \cup S_2$.

Corollary 3.10. *For a number $n \in \mathbb{N}_+$, the following conditions are equivalent:*

(a) *The number n is of the form $n = x^2 + 11y^2$ or of the form $n = 4x^2 + 22xy + 33y^2$ with $x, y \in \mathbb{Z}$ and $\gcd(x, 11y) = 1$.*

(b) *We have $n \in \mathbb{D}$ and $n \not\equiv 2 \pmod 4$.*

The preceding proposition and its corollary suggest to introduce the following notation.

Notation 3.11. The set of all positive integers n of the form $n = 4x^2 + 22xy + 33y^2$ with $x, y \in \mathbb{Z}$ and $\gcd(x, 11y) = 1$ is denoted by

$$S_2 = \{3, 4, 5, 9, 12, 15, 20, 23, 25, 31, 36, 37, 45, 59, 60, 67, 69, 71, 75, 81, 89, 92, \dots\}$$

The union $S_1 \cup S_2$ is called the set of *candidate* numbers and denoted by

$$\begin{aligned} C &= S_1 \cup S_2 \\ &= \{1, 3, 4, 5, 9, 12, 15, 20, 23, 25, 27, 31, 36, 37, 45, 47, 53, 59, 60, 67, \dots\} \end{aligned}$$

Thus we are left with the task of distinguishing between the sets S_1 and S_2 inside the candidate set C. For further usage, let us describe the sets S_2 and C in different ways.

Proposition 3.12. *The set S_2 defined above is equal to each of the following sets:*

(a) $S_{2,a} = \{4x^2 + 22xy + 33y^2 \mid x, y \in \mathbb{Z}, \gcd(x, 11y) = 1\}$

(b) $S_{2,b} = \{3x^2 + 22xy + 44y^2 \mid x, y \in \mathbb{Z}, \gcd(x, 11y) = 1\}$

(c) $S_{2,c} = \{3x^2 + 2xy + 4y^2 \mid x, y \in \mathbb{Z}, \gcd(x, y) = 1, 11 \nmid (3x^2 + 2xy + 4y^2)\}$

Proof. The set $S_{2,a}$ agrees with the original definition of S_2 in Proposition 3.9. In the proof of this proposition, we showed that it agrees with the set $S_{2,b}$. Thus remains to prove that $S_{2,c}$ is the same set again.

Clearly, the set $S_{2,b}$ agrees with the set of numbers of the form $3x^2 - 22xy + 44y^2$ where $x, y \in \mathbb{Z}$ and $\gcd(x, 11y) = 1$. Now we use the coordinate transformation

$$\begin{pmatrix} \tilde{x} \\ \tilde{y} \end{pmatrix} = \begin{pmatrix} 1 & 4 \\ 0 & 1 \end{pmatrix} \cdot \begin{pmatrix} x \\ y \end{pmatrix}$$

and get the quadratic form $3\tilde{x}^2 + 2\tilde{x}\tilde{y} + 4\tilde{y}^2$. Since the condition $\gcd(x, 11y) = 1$ is equivalent to $11 \nmid (3x^2 - 22xy + 44y^2)$ and $\gcd(x, y) = 1$, and since the determinant of the transformation matrix is 1, it follows that we are looking for $\tilde{x}, \tilde{y}$ such that $\gcd(\tilde{x}, \tilde{y}) = 1$ and $11 \nmid (3\tilde{x}^2 + 2\tilde{x}\tilde{y} + 4\tilde{y}^2)$. Consequently, the set $S_{2,b}$ equals $S_{2,c}$. □

Proposition 3.13. *The set $C = S_1 \cup S_2$ is the disjoint union of the following two sets:*

$$C_1 = \{n \in \mathbb{Z} \mid n = x^2 + 11xy + 33y^2; x, y \in \mathbb{Z}; \gcd(x, 11y) = 1\}$$
$$C_2 = \{n \in \mathbb{Z} \mid n = x^2 + 11xy + 33y^2; x, y \in \mathbb{Z}; \gcd(x, 11y) = 2\}$$

Here, the numbers in C_1 are odd and the numbers in C_2 are divisible by 4. The numbers in C_1 will be called the odd candidates and the numbers in C_2 the even candidates.

Proof. First, we show that C_1 and C_2 are contained in C. For $n \in C_1$, we use Corollary 3.10. It is easy to check that n is odd if at least one of x, y is odd. Hence we have $n \not\equiv 2 \pmod 4$. It is also clear that $11 \nmid n$ and n is a quadratic residue modulo 11. Moreover, since the quadratic form $x^2 + 11xy + 33y^2$ has discriminant $-11 \equiv 1 \pmod 4$ and n has a primitive representation by this quadratic form, it follows that -11 is a quadratic residue modulo n (see, for instance, [3], Proposition 4.1).

Next, we show that C_2 is contained in C. Since $\gcd(x, 11y) = 2$, we can write $x = 2\tilde{x}$ and get $n = 4\tilde{x}^2 + 22\tilde{x}y + 33y^2$ with $\gcd(\tilde{x}, y) = 1$. Thus n is contained in S_2.

By the definition of C_1 and C_2, it is clear that these two sets are disjoint. Hence it remains to show that their union is C. The set S_1 is contained in $C_1 \cup C_2$, because for every number n in S_1 we know that -11 is a quadratic residue modulo n. If n is odd, this implies that -11 is a quadratic residue modulo $4n$, and [3], Proposition 4.1 yields that n is represented by the quadratic form $x^2 + 11xy + 33y^2$. On the other hand, if n is even, then $n = 4\tilde{n} = x^2 + 11y^2$ with odd integers x, y. Therefore, -11 is a square modulo $4\tilde{n}$, and [3], Proposition 4.1 shows that $\tilde{n}$ is represented by $x^2 + 11xy + 33y^2$. Thus $n = 4\tilde{n}$ is contained in C_2.

In order to prove that S_2 is contained in $C_1 \cup C_2$, we write $n \in S_2$ in the form $n = 4x^2+22xy+33y^2$ with $x, y \in \mathbb{Z}$ and $\gcd(x, 11y) = 1$. If y is odd, then $n = (2x)^2+11(2x)y+33y^2$ and $\gcd(2x, y) = 1$ imply $n \in C_1$. If y is even, then x has to be odd, and $n = (2x)^2+11(2x)y+33y^2$ together with $\gcd(2x, y) = 2$ shows $n \in C_2$. Altogether, we obtain $n \in C = S_1 \cup S_2$, and the proof is complete. □

Another useful property is the fact that C_1 is multiplicatively closed, as the following proposition shows.

Proposition 3.14. *For odd numbers $n_1, n_2 \in C$, we have $n_1 n_2 \in C$. In particular, the set of odd candidates C_1 is a multiplicative submonoid of $\mathbb{N}_+$.*

Proof. Using the description $C_1 = \{n \in \mathbb{N}_+ \mid 11 \nmid n; n \equiv 1 \pmod 2; (\frac{n}{11}) = 1; -11$ is a quadratic residue modulo $n\}$, which follows from Corollary 3.10, it suffices to show that the condition "-11 is a quadratic residue modulo n" is multiplicative.

Given a number $n = p_1^{\alpha_1} \cdots p_s^{\alpha_s}$ with pairwise distinct odd prime numbers p_i and with $\alpha_i \geq 1$, and such that -11 is a quadratic residue modulo n, we see that -11 is a quadratic residue modulo p_i for $i = 1, \dots, s$. It follows from Hensel's lemma that if $x^2 + 11 \equiv 0 \pmod{p_i}$ has a solution a with $2a \not\equiv 0 \pmod p$ then $x^2 + 11 \equiv 0 \pmod{p^\beta}$ has a solution for every $\beta \geq 1$. Via the Chinese remainder theorem we then conclude that -11 is a quadratic residue modulo every number $p_1^{\beta_1} \cdots p_s^{\beta_s}$ with $\beta_i \geq 1$.

By combining the prime factorizations of n_1 and n_2, these observations imply the claim. □

At this point, we need to insert further background material.

4 The ring of integers of $\mathbb{Q}(\sqrt{-11})$

In this section, we provide some basic material from algebraic number theory. Since we have $x^2 + 11y^2 = (x + y\sqrt{-11})(x - y\sqrt{-11})$, it is natural to study the quadratic field $\mathbb{Q}(\sqrt{-11})$ and its ring of integers. Let us collect some well-known facts.

Proposition 4.1. *Let $K = \mathbb{Q}(\sqrt{-11})$, and let $\mathcal{O}_K$ be the ring of integers of K.*

(a) *The ring of integers of K is given by $\mathcal{O}_K = \mathbb{Z}[(1 + \sqrt{-11})/2] = \mathbb{Z}[\omega]$, where $\omega = (-11 + \sqrt{-11})/2$. It is a free $\mathbb{Z}$-module of rank 2 with basis $\{1, \omega\}$.*
(b) *The minimal polynomial of ω is $\mu_\omega(x) = x^2 + 11x + 33$. In particular, we have $\mathcal{O}_K \cong \mathbb{Z}[x]/\langle x^2 + 11x + 33\rangle$.*
(c) *The Galois group of $K/\mathbb{Q}$ has two elements. The nontrivial automorphism maps $\sqrt{-11}$ to $-\sqrt{11}$ and ω to $\overline{\omega} = (-11 - \sqrt{-11})/2$.*
(d) *We have $\omega + \overline{\omega} = -11$ and $\omega \cdot \overline{\omega} = 33$.*
(e) *The ideal class number of $\mathbb{Z}[\omega]$ is 1, that is, this ring is a PID. In particular, it is a factorial ring.*
(f) *The unit group of $\mathcal{O}_K$ is $\{1, -1\}$.*

(g) *The norm map* $N_{\mathcal{O}_K} : \mathcal{O}_K \to \mathbb{Z}$ *is given by* $N_{\mathcal{O}_K}(a + b\omega) = (a + b\omega)(a + b\overline{\omega}) = a^2 - 11ab + 33b^2$ *for all* $a, b \in \mathbb{Z}$. *It turns the ring* $\mathbb{Z}[\omega]$ *into a Euclidean domain.*

Proof. Most results are well known an can be found in many textbooks on algebraic number fields or quadratic fields; see, for instance, [13]. Parts (b) and (d) can be verified easily by direct calculation. □

As the factorization $x^2 + 11y^2 = (x + y\sqrt{-11})(x - y\sqrt{-11})$ actually takes place in the ring $\mathbb{Z}[\sqrt{-11}]$, let us also introduce its properties and its relation to $\mathbb{Z}[\omega]$.

Remark 4.2. Let $K = \mathbb{Q}(\sqrt{-11})$ and $\mathcal{O}_K = \mathbb{Z}[\omega]$ with $\omega = (-11 + \sqrt{11})/2$.

(a) The ring $\mathcal{O} = \mathbb{Z}[\sqrt{-11}]$ is the order of conductor 2 in $\mathbb{Z}[\omega]$. In particular, we have $\mathcal{O} = \mathbb{Z} + 2\omega\mathbb{Z}$.

(b) The ring $\mathbb{Z}[\sqrt{-11}]$ is a free $\mathbb{Z}$-module with basis $\{1, \sqrt{-11}\}$, the ring $\mathbb{Z}[\omega]$ is a $\mathbb{Z}[\sqrt{-11}]$-module which is generated by $\{1, \omega\}$, and $2\omega \in \mathbb{Z}[\sqrt{-11}]$. Thus the elements of $\mathbb{Z}[\omega]$ are either in $\mathbb{Z}[\sqrt{-11}]$ or in $\omega + \mathbb{Z}[\sqrt{-11}]$.

Using this remark, we can analyze products of elements of $\mathbb{Z}[\omega]$ as follows.

Proposition 4.3. *Let* $a + b\omega, a' + b'\omega$ *be elements of* $\mathbb{Z}[\omega]$, *where* $a, a', b, b' \in \mathbb{Z}$ *and* $\omega = (-11 + \sqrt{-11})/2$.

(a) *The element* $a + b\omega$ *is contained in* $\mathbb{Z}[\sqrt{-11}]$ *if and only if* b *is even.*

(b) *For* $a + b\omega \in \mathbb{Z}[\sqrt{-11}]$ *and* $a' + b'\omega \notin \mathbb{Z}[\sqrt{-11}]$ *we have* $(a + b\omega)(a' + b'\omega) \in \mathbb{Z}[\sqrt{-11}]$ *if and only if* a *is even, that is, if and only if* $a + b\omega$ *is a multiple of 2.*

(c) *For* $a + b\omega, a' + b'\omega \notin \mathbb{Z}[\sqrt{-11}]$, *we have* $(a + b\omega)(a' + b'\omega) \in \mathbb{Z}[\sqrt{-11}]$ *if and only if* $a + a'$ *is odd.*

(d) *For every element* $a + b\omega \in \mathbb{Z}[\omega]$ *with* $a, b \in \mathbb{Z}$, *we have* $(a + b\omega)^3 \in \mathbb{Z}[\sqrt{-11}]$.

Proof. The first claim follows immediately from part (a) of the preceding remark. To prove (b) and (c), we calculate

$$\begin{aligned}(a + b\omega)(a' + b'\omega) &= aa' + (ab' + ba')\omega + bb'(-11\omega - 33)\\ &= (aa' - 33bb') + (ab' + ba' - 11bb')\omega\end{aligned}$$

Now the claims follow from (a) by analyzing the coefficient of ω in this product.

It remains to prove (c). We compute

$$\begin{aligned}(a + b\omega)^3 &= a^3 + 3a^2b\omega + 3ab^2\omega^2 + b^3\omega^3\\ &= (a^3 - 99ab^2 + 363b^3) + (3a^2b - 33ab^2 + 88b^3)\omega\end{aligned}$$

By (a), the claim now follows from $a^2b + ab^2 \equiv ab(a + b) \equiv 0 \pmod 2$. □

This proposition yields the following properties of products of elements in C.

Corollary 4.4. *In the setting of the proposition, let $\overline{S}_1$ be the set of all numbers of the form $c^2 n$ where $c \in \mathbb{Z}$ and $n \in S_1$. Then the following statements hold:*

(a) *The element $N_{\mathcal{O}_K}(a + b\omega) = (a + b\omega)(a + b\overline{\omega})$ is contained in C if and only if* $\gcd(a, 11b) = 1$.
(b) *The element $N_{\mathcal{O}_K}(a + b\omega) = (a + b\omega)(a + b\overline{\omega})$ is contained in S_1 if and only if b is even and* $\gcd(a, 11b) = 1$.
(c) *Assume that we have $a+b\omega, a'+b'\omega \in \mathbb{Z}[\sqrt{-11}]$. Then the element $N_{\mathcal{O}_K}((a+b\omega)(a'+b'\omega))$ is in $\overline{S}_1$. Consequently, the product of two numbers in S_1 is contained in $\overline{S}_1$.*
(d) *Assume that we have $a + b\omega \in \mathbb{Z}[\sqrt{-11}]$ and $a' + b'\omega \notin \mathbb{Z}[\sqrt{-11}]$. Then the element $N_{\mathcal{O}_K}((a+b\omega)(a'+b'\omega))$ is in $\overline{S}_1$ if and only if a is even, that is, if and only if 2 divides $a + b\omega$. Consequently, the product of a number n in S_1 and a number in $S_2 \setminus S_1$ is contained in $\overline{S}_1$ if and only if n is even.*
(e) *Assume that we have $a+b\omega, a'+b'\omega \notin \mathbb{Z}[\sqrt{-11}]$. Then the element $N_{\mathcal{O}_K}((a+b\omega)(a'+b'\omega))$ is in $\overline{S}_1$ if and only if $a + a'$ is an odd integer.*
(f) *A number $n \in C$ is contained in S_2 if and only if we have a representation $n = N_{\mathcal{O}_K}(a+b\omega)$ with $a, b \in \mathbb{Z}$ and* $\gcd(a, 11b) = 1$*, an even number a and an odd number b.*

Proof. Claim (a) follows from $N_{\mathcal{O}_K}(a + b\omega) = a^2 - 11ab + 33b^2$ and Corollary 3.10. To prove (b), we note that

$$N_{\mathcal{O}_K}(a + 2b'\omega) = a^2 - 22ab' + 132(b')^2 = (a - 11b')^2 + 11(b')^2$$

and apply part (a) of the proposition. Claims (c), (d), and (e) follow from the proposition and the fact that the norm map is multiplicative.

Thus we are left with proving (f). If $n \in S_2$, then we have $n = 4a^2 - 22ab + 33b^2$ for some $a, b \in \mathbb{Z}$ with $\gcd(a, 11b) = 1$. Thus we get $n = N_{\mathcal{O}_K}(2a + b\omega) = (2a+b\omega)(2a+b\overline{\omega}) = n$. Conversely, if $n = N_{\mathcal{O}_K}(a + b\omega)$ with $a = 2\tilde{a}$ then $n = 4\tilde{a}^2 - 22\tilde{a}b + 33b^3$ together with $\gcd(\tilde{a}, 11b) = \gcd(a, 11b) = 1$ shows $n \in S_2$. □

Finally, we collect results about the splitting of primes in the factorial ring $\mathbb{Z}[\omega]$.

Proposition 4.5. *Consider the ring of integers $\mathcal{O}_K = \mathbb{Z}[\omega]$ of the quadratic number field $K = \mathbb{Q}(\sqrt{-11})$, where $\omega = (-11 + \sqrt{-11})/2$.*

(a) *The only ramified prime in $\mathcal{O}_K$ is $p = 11$.*
(b) *The prime 2 is inert in $\mathcal{O}_K$*
(c) *An odd prime $p \neq 11$ splits in $\mathcal{O}_K$ if and only if -11 is a quadratic residue modulo p. In particular, all primes in $S_1 \cup S_2$ split in $\mathcal{O}_K$.*
(d) *Let p be an odd prime in $C = S_1 \cup S_2$. Then p is contained in S_1 if and only if the two factors of p are already contained in $\mathbb{Z}[\sqrt{-11}]$.*

Proof. The ramified primes in $\mathcal{O}_K$ are the ones which divide the discriminant, which in our case is $\operatorname{disc}(\mathcal{O}_K) = -11$. This proves (a).

To show (b), we note that the equation $x^2 + 11xy + 33 = 0$ has no solutions modulo 2. Hence 2 is inert in $\mathcal{O}_K$ by [13], Proposition 4.9.3. Claim (c) follows, for instance, from [13], Theorem 4.9.5.

Finally, we show (d). By the hypothesis $p \in C$, we know that p splits in $\mathcal{O}_K$. Since $\mathcal{O}_K$ is a factorial domain, this means that there exist $a, b \in \mathbb{Z}$ such that $p = (a + b\omega)(a + b\overline{\omega})$.

Thus, in $\mathcal{O}_K$, the ideal $\langle p \rangle$ is the intersection of the two maximal ideals $\mathfrak{m}_1 = \langle a + b\omega \rangle$ and $\mathfrak{m}_2 = \langle a + b\overline{\omega} \rangle$. When we intersect $\mathfrak{m}_1$ with $\mathbb{Z}[\sqrt{-11}]$, we get

$$\mathfrak{m}_1 \cap \mathbb{Z}[\sqrt{-11}] = \begin{cases} \langle 2a - 11b + b\sqrt{-11} \rangle & \text{if } b \text{ is odd,} \\ \langle a - 11\tilde{b} + \tilde{b}\sqrt{-11} \rangle & \text{if } b = 2\tilde{b} \text{ is even.} \end{cases}$$

An analogous result holds for $\mathfrak{m}_2 \cap \mathbb{Z}[\sqrt{-11}]$. Consequently, if $b = 2\tilde{b}$ is even, then

$$(\mathfrak{m}_1 \cap \mathbb{Z}[\sqrt{-11}])(\mathfrak{m}_2 \cap \mathbb{Z}[\sqrt{-11}]) = (2a - 11\tilde{b})^2 + 11\tilde{b}^2 = a^2 + 11ab + 33b^2 = p$$

shows that $p \in S_1$. On the other hand, if b is odd, then

$$(\mathfrak{m}_1 \cap \mathbb{Z}[\sqrt{-11}])(\mathfrak{m}_2 \cap \mathbb{Z}[\sqrt{-11}]) = (2a - 11b)^2 + 11b^2 = 4p$$

shows that p does not split in $\mathbb{Z}[\sqrt{-11}]$. Hence p splits in $\mathbb{Z}[\sqrt{-11}]$ if and only if b is even, that is, if and only if the two factors of p are already contained in $\mathbb{Z}[\sqrt{-11}]$. □

For powers of primes in S_1, this proposition implies the following result.

Corollary 4.6. *Given a prime number $p \in S_1$, we have $p^\alpha \in S_1$ for every $\alpha \geq 1$.*

Proof. According to the proposition, we have $p = q\bar{q}$ with a prime element q of $\mathbb{Z}[\omega]$ of the form $q = a + b\sqrt{-11}$, where $a, b \in \mathbb{Z}$ and $\gcd(a, 11b) = 1$. Then we get $p^\alpha = q^\alpha \bar{q}^\alpha$ for every $\alpha \geq 1$. Since we have $\gcd(q, \bar{q}) = \gcd(a + b\sqrt{-11}, 2a) = 1$, the elements q and $\bar{q}$ are coprime in $\mathcal{O}_K$. Hence, when we write $q^\alpha = \tilde{a} + \tilde{b}\sqrt{-11}$ with $\tilde{a}, \tilde{b} \in \mathbb{Z}$, we obtain $\gcd(\tilde{a}, 11\tilde{b}) = 1$ and, therefore, $p^\alpha \in S_1$. □

As a byproduct of the proof of Proposition 4.5, we see that $p \in S_2$ implies $p \notin S_1$. A better criterion for distinguishing the primes in S_1 and S_2 is coming up next.

5 Primes of the form $x^2 + 11y^2$

The next task is to characterize the prime numbers of the form $x^2 + 11y^2$. As for instance explained in [4], this problem has been explored intensively using class field theory and other methods of algebraic number theory. As mentioned in the preceding section, the form $x^2 + 11y^2$ splits in the ring $\mathcal{O} = \mathbb{Z}[\sqrt{-11}]$ which is the order of conductor 2 in the ring of integers $\mathcal{O}_K$ of the field $K = \mathbb{Q}(\sqrt{-11})$. Thus it is not a Dedekind domain. The discriminant of this order is -44. As a result of these studies, we have the following theorem.

Theorem 5.1 (Characterization of primes of the form $x^2 + 11y^2$). *A prime number $p \geq 13$ is of the form $p = x^2 + 11y^2$ with $x, y \in \mathbb{Z}$ if and only if the following two conditions are satisfied:*
(a) *The number -11 is a quadratic residue modulo p.*
(b) *The polynomial $f_{11}(x) = x^3 - 2x^2 + 2x - 2$ has a zero modulo p.*

Proof. According to [4], Theorem 9.2, we have to show that $f_{11}(x)$ is the minimal polynomial of a primitive element of the ring class field of $\mathbb{Z}[\sqrt{-11}]$. Notice that the discriminant of this order is -44, so that none the primes satisfying (a) divides it.

Since the ideal class number for the discriminant -44 is 3, the minimal polynomial of a primitive element of the ring class field of $\mathbb{Z}[\sqrt{-11}]$ has degree 3. According to [4], Theorem 12.2 and Theorem 12.24, the desired ring class field is $\mathbb{Q}(\gamma_2(\sqrt{-11})) = \mathbb{Q}(j(\sqrt{-11})) = \mathbb{Q}(\mathfrak{f}(\sqrt{-11})^2)$, where γ_2 is the cube root of the j-invariant and $\mathfrak{f}$ a Weber function (see [4], Section 12.B). The minimal polynomial of $\mathfrak{f}(\sqrt{-11})$ has been determined by Weber, and it is $f_{11}(x) = x^3 - 2x^2 + 2x - 2$ (see [14], Section 128, p. 475). Now it is easy to check that the degree of $\mathfrak{f}(\sqrt{-11})^2$ over $\mathbb{Q}$ is three, as well, so that the ring class field is also given by $\mathbb{Q}(\mathfrak{f}(\sqrt{-11}))$, and the claims are proved. □

Notice that the polynomial f_{11} actually splits into linear factors in $\mathbb{F}_p$, if p is a prime of the form $p = x^2 + 11y^2$.

To distinguish the prime numbers of the form $p = x^2+11y^2$ from the prime numbers of the form $p = 3x^2 + 2xy + 4y^2$, we need one further ingredient.

Proposition 5.2. *Let S_1 and S_2 be the sets defined in the preceding section. Then every prime number in $S_1 \cup S_2$ lies either in the set S_1 or in the set S_2, but not in both.*

Proof. Let p be a prime number in $S_1 \cup S_2$. When we construct the matrix $A_p(\ell)$, we need to find a number $\ell \in \{1, \ldots, p-1\}$ such that $\ell^2 \equiv -1/11 \pmod{p}$. Since p is a prime number and $-1/11$ is a quadratic residue modulo p, there are exactly two choices for ℓ, and if ℓ is one of them, then the other one is $p - \ell$. By Proposition 3.7, the matrix $A_p(\ell)$ is conjugate to t_1 if and only if $A_p(p-\ell)$ is conjugate to t_1 in G_{11}. By Corollary 3.8, this is equivalent to $p \in S_1$. Since $A_p(\ell)$ is in the conjugacy class of exactly one matrix t_i, the claim follows. □

As an immediate consequence of this proposition and the above theorem, we obtain the following characterization of primes of the form $p = 3x^2 + 2xy + 4y^2$.

Corollary 5.3. *A prime number $p \geq 3$ with $p \neq 11$ is of the form $p = 3x^2 + 2xy + 4y^2$ with $x, y \in \mathbb{Z}$ if and only if the following conditions hold:*
(a) *The number -11 is a quadratic residue modulo p.*
(b) *The polynomial $f_{11}(x) = x^3 - 2x^2 + 2x - 2$ is irreducible modulo p.*

Proof. By the proposition and the description of the set $S_1 \cup S_2$ in Corollary 3.10, the claim follows from the theorem. □

Finally, we point out that for the primes of the forms in the theorem and the corollary, we have the following uniqueness property.

Proposition 5.4. *Let $p \geq 3$ with $p \neq 11$ be a prime number such that -11 is a quadratic residue modulo p.*
(a) *If $f_{11}(x) = x^3 - 2x^2 + 2x - 2$ splits in $\mathbb{F}_p[x]$ into linear factors, then there exists a unique pair of numbers $(x, y) \in \mathbb{N}^2$ such that $p = x^2 + 11y^2$.*
(b) *If $f_{11}(x) = x^3 - 2x^2 + 2x - 2$ is irreducible in $\mathbb{F}_p[x]$, then there exists a unique pair of numbers $(x, y) \in \mathbb{N} \times \mathbb{Z}$ such that $p = 3x^2 + 2xy + 4y^2$.*

Proof. To show (a), we note that, by Theorem 5.1 and Proposition 3.5, there exists a number $\ell \in \{1, \ldots, p-1\}$ such that $A_p(\ell) = Ut_1U^{-1}$ for some matrix $U \in G_{11}$. Notice that we may assume $U \in H_{11}$ here, since $G_{11} = H_{11} \cup H_{11}t_1$ and since $(Ut_1)t_1(Ut_1)^{-1} = Ut_1U^{-1}$. Every matrix $U = \left(\begin{smallmatrix} a & b\sqrt{11} \\ c\sqrt{11} & d \end{smallmatrix}\right) \in H_{11}$ which satisfies $A_p(\ell) = Ut_1U^{-1}$ gives rise to a representation $p = a^2 + 11b^2$ in the top right position of the matrix. On the other hand, every such representation gives rise to a uniquely determined matrix U, because the entries c, d have to satisfy $\ell = -ac - bd$ and $\det(U) = ad - 11bc = 1$. Now the claim follows from the fact that the conjugating matrix U in $A_p(\ell) = Ut_1U^{-1}$ is uniquely determined, because H_{11} is a free group on three generators. Claim (b) follows in the same way. □

6 Cubic numbers of the form $x^2 + 11y^2$

In this section, we examine cubic numbers m^3 of the form $m^3 = x^2 + 11y^2$ with $m, x, y \in \mathbb{N}$. Recall the sets S_1 and S_2 defined in notation 3.2 and notation 3.11. The following easy observations will help us.

Lemma 6.1. *Every cubic number of the form $m^3 = x^2 + 11y^2$ with $x, y \in \mathbb{Z}$ and* $\gcd(x, 11y) = 1$ *is odd.*

Proof. Suppose that $m^3 = (2k)^3 = x^2 + 11y^2$ is even. Then we have $x^2 + 3y^2 \equiv 0 \pmod 4$, and since x and y are not both even, it follows that they are both odd. Then we have $x^2 \equiv y^2 \equiv 1 \pmod 8$, and hence $m^3 \equiv x^2 + 3y^2 \equiv 4 \pmod 8$, in contradiction to $m^3 \equiv 8k^3 \equiv 0 \pmod 8$. □

Lemma 6.2. *For every cubic number of the form $m^3 = x^2 + 11y^2$ with $x, y \in \mathbb{Z}$ and* $\gcd(x, 11y) = 1$, *the number m is contained in the set $S_1 \cup S_2$.*

Proof. Clearly, the condition $\gcd(x, 11y) = 1$ implies $\gcd(m, 11) = 1$, and the number m^3 is a quadratic residue modulo 11. The cube of a quadratic residue modulo 11 is a quadratic residue, while the cube of a quadratic nonresidue is a quadratic nonresidue. Hence m is a quadratic residue modulo 11. Given $a \in Z$ with $a^2 \equiv -11 \pmod{m^3}$, it follows that $a^2 \equiv -11 \pmod m$. Finally, we note that m is not in the set S_4 of Proposition 3.9, since it is an odd number, and thus $m \in S_1 \cup S_2$ by part (e) of that proposition. □

Lemma 6.3. *Let $\overline{S}_1$ be the set of all integers n such that $n = x^2 + 11y^2$ for some $x, y \in \mathbb{Z}$ and such that* $\gcd(n, 11) = 1$, *and let p_1, p_2 be prime numbers in C.*
(a) *If $p_1, p_2 \in S_1$ then we have $p_1^{\alpha_1} p_2^{\alpha_2} \in \overline{S}_1$ for all $\alpha_1, \alpha_2 \in \mathbb{N}$.*
(b) *If $p \in S_2$ then we have $p^2 \in S_2 \setminus S_1$.*
(c) *If $p_1 \in S_1$ and $p_2 \in S_2$ then we have $p_1^{\alpha_1} p_2^{\alpha_2} \in S_2 \setminus S_1$ for $\alpha_1, \alpha_2 \in \{1, 2\}$.*
(d) *If $p_1, p_2 \in S_2$ are distinct primes, then we have $p_1^{\alpha_1} p_2^{\alpha_2} \in S_1 \cap S_2$ for $\alpha_1, \alpha_2 \in \{1, 2\}$.*

Proof. To show (a), we use Proposition 4.5.d and write $p_i = q_i \bar{q}_i$ with prime elements q_i of $\mathbb{Z}[\omega]$ which are contained in $\mathbb{Z}[\sqrt{-11}]$ for $i = 1, 2$. Then we have $p_1^{\alpha_1} p_2^{\alpha_2} = (q_1^{\alpha_1} q_2^{\alpha_2})(\bar{q}_1^{\alpha_1} \bar{q}_2^{\alpha_2})$ with $q_1^{\alpha_1} q_2^{\alpha_2} \in \mathbb{Z}[\sqrt{-11}]$, and the claim follows.

Next, we prove (b). By Proposition 4.5.d, the number p is of the form $p = q\bar{q}$ with a prime element q of $\mathbb{Z}[\omega]$ which is not contained in $\mathbb{Z}[\sqrt{-11}]$. Then Proposition 4.3.a says that we have $q = a + b\omega$ with $a, b \in \mathbb{Z}$ and $\gcd(a, 11b) = 1$ and an odd number b. Thus we see that in

$$q^2 = (a^2 - 33b^2) + (2ab - 11b^2)\omega$$

the coefficient $2ab - 11b^2$ is an odd number, too, and Proposition 4.3.a yields $q^2 \notin S_1$. It remains to show that $p \in C = S_1 \cup S_2$, that is, we need to show that -11 is a quadratic residue modulo p^2. Let $k \in \mathbb{Z}$ such that $x^2 + 11 = pk$ for some $x \in \mathbb{Z}$. Notice that $2x$ is coprime to p. Then the equality $p^2k^2 = (x^2 - 11)^2 + 44x^2$ yields $p^2(k^2/(4x^2)) = ((x^2 - 11)/2x)^2 + 11$, and hence the desired result.

In order to show (c), we note that $p_1^2 \in S_1$ by Corollary 4.6 and $p_2^2 \in S_2 \setminus S_1$ by part (b). Hence Corollary 4.4.d yields the claim.

Finally, we prove (d). Using (b), we get $p_i^2 \in S_2$. By the definition of the set S_2, we then find representations $p_i^{\alpha_i} = 4a_i^2 - 22a_ib_i + 33b_i^2 = (2a_i + b_i\omega)(2a_i + b_i\overline{\omega})$ with $a_i, b_i \in \mathbb{Z}$ and $\gcd(a_i, 11b_i) = 1$. Now Proposition 4.3.c and Corollary 4.4 imply that we have $p_1^{\alpha_1} p_2^{\alpha_2} \notin \bar{S}_1$. Consequently, it remains to show that $p_i^2 \in C$ in order to finish the proof. This follows exactly as in the proof of (b). □

Lemma 6.4. *Let $a \in \mathbb{Z}$ be even and $b \in \mathbb{Z}$ be odd. Then* $\gcd(a + b\omega, a + b\overline{\omega}) = 1$ *holds in* $\mathbb{Z}[\omega]$.

Proof. Since $\mathbb{Z}[\omega]$ is an Euclidean domain, and since $\omega - \overline{\omega} = \sqrt{-11}$, we get $\gcd(a + b\omega, a + b\overline{\omega}) = \gcd(a + b\omega, b\sqrt{-11})$. By assumption, we have $\gcd(a, b) = 1$. Thus it suffices to consider $\gcd(a + b\omega, \sqrt{-11})$. Using the fact that b is odd, this gcd equals $\gcd(a+\omega, \sqrt{-11}) = \gcd(a+\omega, 2\omega+11) = \gcd(a+\omega, 11-2a)$. Now $11-2a$ is odd and $2 \in \mathbb{Z}[\omega]$ a prime, so that we may as well consider $\gcd(2a + 2\omega, 11 - 2a) = \gcd(11 + 2\omega, 11 - 2a) = \gcd(\sqrt{-11}, 11 - 2a)$. Thus the claim follows from $\gcd(11, 11 - 2a) = 1$. □

Finally, we can characterize cubic numbers of the form $x^2 + 11y^2$ as follows.

Proposition 6.5. *Let the sets S_1, S_2, and C be defined as in Section 3, let $\bar{S}_1 = \{x^2 + 11y^2 \mid x, y \in \mathbb{Z}; \gcd(x, 11) = 1\}$, and let S_1^{cube} be the set of cubic numbers in C. Then the following statements hold:*

(a) *Every cube of a number in C is contained in $S_1 \setminus S_2$, that is, we have $S_1^{\text{cube}} \subset S_1 \setminus S_2$.*
(b) *Every number m such that $m^3 \in S_1^{\text{cube}}$ is of the form $m = p^\alpha \tilde{m}$ with a prime number $p \in S_2$, with $0 \le \alpha \le 2$, and with $\tilde{m} \in \bar{S}_1$.*
(c) *The set S_1^{cube} is the multiplicative monoid generated by the cubes of the prime numbers in $S_1 \cup S_2$.*

Proof. First, we show (a). Given $m \in C$, we write down its prime factorization $m = p_1^{\alpha_1} \cdots p_s^{\alpha_s}$, where $p_i \in \mathbb{Z}$ is a prime number for $i = 1, \dots, s$. Since -11 is a quadratic residue modulo m, it is a quadratic residue modulo p_i, and by Proposition 4.5.c, we have a decomposition $p_i = q_i \bar{q}_i$ with conjugated prime elements $q_i, \bar{q}_i \in \mathbb{Z}[\omega]$. Then we obtain

$$m^3 = (p_1^{\alpha_1} \cdots p_s^{\alpha_s})^3 = (q_1^{\alpha_1} \cdots q_s^{\alpha_s})^3 (\bar{q}_1{}^{\alpha_1} \cdots \bar{q}_s^{\alpha_s})^3 = c\bar{c}$$

where $c = (q_1^{\alpha_1} \cdots q_s^{\alpha_s})^3$ and $\bar{c} = (\bar{q}_1^{\alpha_1} \cdots \bar{q}_s^{\alpha_s})^3$ are contained in $\mathbb{Z}[\sqrt{-11}]$ by Proposition 4.3. This means $m^3 \in S_1$.

Furthermore, assume that $m^3 = (a + b\omega)(a + b\bar{\omega})$ is a decomposition corresponding to a representation of m as an element of S_2, that is, such that $a \in \mathbb{Z}$ even, $b \in \mathbb{Z}$ odd, and $\gcd(a, 11b) = 1$. By Lemma 6.4, the elements $a + b\omega$ and $a + b\bar{\omega}$ of $\mathbb{Z}[\omega]$ are coprime. Hence, for every prime p of $\mathbb{Z}[\omega]$, the cube p^3 divides either $a + b\omega$ or $a + b\bar{\omega}$. Now $p^3 \in S_1$ and Corollary 4.4.b yield $p^3 = (\tilde{a} + \tilde{b}\omega)(\tilde{a} + \tilde{b}\bar{\omega})$ with $\tilde{a} \in \mathbb{Z}$ odd, $\tilde{b} \in \mathbb{Z}$ even, and $\gcd(\tilde{a}, 11\tilde{b}) = 1$. The formula given in the proof of Proposition 4.3 then implies that the product of such cubes of primes has an even coefficient of ω, in contradiction to our assumption. This proves $m^3 \notin S_2$.

Next, we prove (b). In view of Proposition 6.5.a and Corollary 4.6, we may assume that m is of the form $m = p_1^{\alpha_1} \cdots p_s^{\alpha_s}$ with pairwise distinct primes $p_i \in S_2$ and $\alpha_i \in \{1, 2\}$. Via Lemma 6.3.d, we then reduce m to a product of an element of S_1 and zero or one prime powers $p_i^{\alpha_i}$, depending on whether s is even or odd.

Finally, we note that (c) follows immediately from (b). □

The next example shows that it is possible that m^3 has a primitive representation of the form $x^2 + 11y^2$, while $m \in \bar{S}_1$ does not.

Example 6.6. The number $m = 675 = 25 \cdot 27$ is not contained in S_1, since the only representations $m = 20^2 + 11 \cdot 5^2 = 24^2 + 11 \cdot 3^2$ are not primitive. However, $675^3 = 12136^2 + 11 \cdot 3817^2$ is a primitive representation, and thus we have $675^3 \in S_1$.

7 Classifying numbers of the form $x^2 + 11y^2$

In this section, we prove several decompositions of the set S_1. To simplify the discussion, we introduce the following abbreviations.

Notation 7.1. Let the sets C, S_1, and S_2 be defined according to Notation 3.2 and Notation 3.11. For $i \in \{1,2\}$, we define

$$S_i^{\text{even}} = \{n \in S_i \mid n \text{ is even}\}$$
$$S_i^{\text{odd}} = \{n \in S_i \mid n \text{ is odd}\}$$
$$S_i^{\text{prim}} = \{n \in S_i \mid n \text{ is a prime}\}$$
$$S_1^{\text{cube}} = \{n \in S_1 \mid n \text{ is a cubic number}\}$$

Proposition 7.2. *The even numbers in S_1 satisfy $S_1^{\text{even}} = 4 \cdot S_2^{\text{odd}}$.*

Proof. For every number $n = 4\tilde{n}$ with an odd number $\tilde{n} \in S_2$, we use Corollary 4.4.b to write $\tilde{n} = (a+b\omega)(a+b\overline{\omega})$, where $a, b \in \mathbb{Z}$, where $\omega = (-11+\sqrt{-11})/2$, where $\gcd(a, 11b) = 1$, and where b is odd. Then we get $n = 4\tilde{n} = (2a - 11b + b\sqrt{-11}) \cdot (2a - 11b - b\sqrt{-11})$ and $\gcd(2a - 11b, b) = 1$, and hence $n \in S_1^{\text{even}}$.

Conversely, Proposition 3.13 shows that every even number $n \in S_1^{\text{even}}$ is of the form $n = 4\tilde{n}$ with an odd number $\tilde{n} \in C$. Using the fact that 2 is a prime in $\mathbb{Z}[\omega]$ by Proposition 4.5.a, we write $n = (2a + 2b\omega)(2a + 2b\overline{\omega})$ in $\mathbb{Z}[\omega]$, where $a, b \in \mathbb{Z}$. Hence we have $n = (2a - 11b + b\sqrt{-11})(2a - 11b - b\sqrt{-11})$, and $n \in S_1$ yields $\gcd(2a - 11b, b) = 1$. In particular, the number b is odd and $\tilde{n} = (a + b\omega)(a + b\overline{\omega})$ is contained in S_2^{odd} by Corollary 4.4.b. □

In view of this proposition, our task of decomposing S_1 is reduced to decomposing S_1^{odd} and S_2^{odd}. In order to move to the main theorem of this section, we need one final ingredient.

Lemma 7.3. *Let $p \in S_2^{\text{prim}}$ and $\alpha \geq 1$. Then we have $p^\alpha \in S_1$ if α is divisible by 3, and $p^\alpha \in S_2$ otherwise.*

Proof. To begin with, assume that $\alpha = 3\beta$ for some $\beta \geq 1$. Using Proposition 6.5.a, it follows that $p^\alpha = (p^\beta)^3 \in S_1$. Now assume that $\alpha = 3\beta + \gamma$ for some $\beta \geq 0$ and $\gamma \in \{1,2\}$. Then Lemma 6.3.b yields $p^\gamma \in S_2$, and together with $p^{3\beta} \in S_1$ we obtain $p^\alpha = p^{3\beta} \cdot p^\gamma \in S_2$ by applying Corollary 4.4.d. □

The following decomposition is one of the main results of this paper.

Theorem 7.4. *The set S_1 is the union*

$$S_1 = S_1^{\text{even}} \cup \langle S_1^{\text{prim}} \rangle \cup S_1^{\text{cube}} \cup (S_1^{\text{odd}} \cap S_2^{\text{odd}})$$

where $\langle S_1^{\text{prim}} \rangle$ is the multiplicative monoid generated by the prime numbers in S_1, and where the only nontrivial intersection is

$$\langle S_1^{\text{prim}} \rangle \cap S_1^{\text{cube}} = \{p_1^{3\alpha_1} \cdots p_s^{3\alpha_s} \mid p_i \in S_1^{\text{prim}}; \alpha_i \geq 1\}.$$

Proof. By Proposition 7.2, it suffices to prove that we have a union

$$S_1^{\mathrm{odd}} = \langle S_1^{\mathrm{prim}} \rangle \cup S_1^{\mathrm{cube}} \cup (S_1^{\mathrm{odd}} \cap S_2^{\mathrm{odd}})$$

where only the first two sets intersect nontrivially and in the obvious way. The second and the third set are disjoint by Proposition 6.5.a. To prove that the first and the third set are disjoint, we write $n \in \langle S_1^{\mathrm{prim}} \rangle$ in the form $n = p_1^{\alpha_1} \cdots p_s^{\alpha_s}$ with pairwise distinct primes $p_i \in S_1^{\mathrm{prim}}$ and $\alpha_i \geq 1$. By Proposition 4.5.c, we have $p_i = q_i \bar{q}_i$ with prime elements $q_i = a_i + b_i\omega \in \mathbb{Z}[\omega]$, and by Corollary 4.4.b, the numbers b_i are even. Thus the formula in the proof of Proposition 4.3 shows that any product of elements $q_i, \bar{q}_i$ will be of the form $a + b\omega$ with an even integer b. Hence Corollary 4.4.f implies that the norm of a product like this is not contained in S_2, and consequently $n \notin S_2$.

Altogether, it remains to show that every element n of S_1^{odd} is contained in one of these three sets. Let us write $n = p_1^{\alpha_1} \cdots p_s^{\alpha_s} \cdot q_1^{\beta_1} \cdots q_t^{\beta_t}$ with pairwise distinct odd primes $p_i \in S_1^{\mathrm{prim}}$, $q_i \in S_2^{\mathrm{prim}}$, and with $\alpha_i \geq 1$. We distinguish four cases.

Case 1: If all prime factors of n are from S_1, then n is contained in $\langle S_1^{\mathrm{prim}} \rangle$.

Case 2: If n has a prime factor from S_1^{prim} and one from S_2^{prim}, we split cubes off $q_j^{\beta_j}$ and write $q_j^{\beta_j} = q_j^{3\gamma_j} \cdot q_j^{\delta_j}$ with $\gamma_j \geq 0$ and $0 \leq \delta_j \leq 2$. Then Proposition 6.5.a yields $q_j^{3\gamma_j} \in S_1$ and we obtain a representation $n = \tilde{n} \cdot q_1^{\delta_1} \cdots q_t^{\delta_t}$ with $\tilde{n} \in S_1 \setminus \{1\}$. Since the elements $q_j^{\delta_j}$ are contained in $S_2 \setminus S_1$ by Lemma 6.3.b, they have representations $q_j^{\delta_j} = (a_j + b_j\omega)(a_j + b_j\overline{\omega})$ with even numbers a_j and odd numbers b_j (see Corollary 4.4.f). Now Proposition 4.3.c shows $q_1^{\delta_1} \cdots q_t^{\delta_t} \in S_2$, and then Corollary 4.4.d yields $n \in S_2$. In particular, we have $n \in S_1^{\mathrm{odd}} \cap S_2^{\mathrm{odd}}$.

Case 3: All prime factors of n are from S_2^{prim} and at least one exponent is ≥ 3. As in the proof of Case 2, it follows that $n \in S_1^{\mathrm{odd}} \cap S_2^{\mathrm{odd}}$.

Case 4: All prime factors of n are from S_2^{prim} and their exponents are 1 or 2. Then, as in the proof of Case 2, we get $q_j^{\delta_j} = (a_j + b_j\omega)(a_j + b_j\overline{\omega})$ with even numbers a_j and odd numbers b_j. Now we can use Corollary 4.4.e to conclude that $n \in S_2 \setminus S_1$. This contradicts the hypothesis, and thus Case 4 does not occur. □

Let us compare the content of this theorem to some formulas proven in [11].

Remark 7.5. Using the theory of modular functions, several formulas for the number $a(n, 11)$ of representations of a given number n of the form $n = x^2 + 11y^2$ are derived in [11], Satz 6.2 and Satz 7.2. They are based on Fourier expansions of theta series and count primitive as well as imprimitive representations. In particular, it is shown that

$$a(n, 11) = \frac{2}{3}\alpha(n, 11) + \frac{4}{3}\beta(n, 11)$$

where $\alpha(n, 11) = \sum_{d|n} (-1)^{(d-1)(n/d-1)} (\frac{d}{11})$ involves Legendre symbols, and where $\beta(n, 11) = \sum_{(x,y) \in M} \mathrm{sign}(xy)(\frac{-1}{|xy|})$ involves Jacobi symbols. Here, the sum extends over the set M of all pairs $(x, y) \in \mathbb{Z}^2 \setminus \{(0,0)\}$ such that $x \equiv y \equiv 1 \pmod 6$ and $x^2 + 11y^2 = 12n$. Clearly, this formula is hard to evaluate in general, because it involves finding particular solutions of the equation $x^2 + 11y^2 = 12n$.

In comparison, the above theorem describes the set of numbers n having a primitive representation of the form $n = x^2 + 11y^2$ as the disjoint union of several special cases which are studied in greater detail in other parts of this paper.

As a consequence of the proof of this theorem, we can decompose S_1^{odd} further. We shall use the following subsets.

Definition 7.6. A number $n \in C$ is called **cubically reduced** if there is no cubic number $c \in S_1^{\text{cube}} \setminus \{1\}$ such that $c \mid n$. For $i = 1, 2$, we denote the set of cubically reduced numbers in S_i^{odd} by S_i^{cro}.

With this terminology, we can decompose S_1^{odd} as follows.

Corollary 7.7. *In the above setting, the following claims hold:*

(a) *Every number $n \in S_1^{\text{odd}}$ has a uniquely determined decomposition $n = c\tilde{n}$ with $c \in S_1^{\text{cube}}$ and $\tilde{n} \in S_1^{\text{cro}}$.*
(b) *Every number $n \in S_1^{\text{cro}} \cup S_2^{\text{cro}}$ has a prime decomposition $n = p_1^{\alpha_1} \cdots p_s^{\alpha_s} \cdot q_1^{\beta_1} \cdots q_t^{\beta_t}$ where $p_i \in S_1^{\text{prim}}$, where $q_j \in S_2^{\text{prim}}$, and where $\alpha_i, \beta_j \in \{1, 2\}$.*
(c) *A number n as in (b) is contained in $S_1^{\text{cro}} \setminus S_2^{\text{cro}}$ if and only if $t = 0$.*
(d) *A number n as in (b) is contained in $S_2^{\text{cro}} \setminus S_1^{\text{cro}}$ if and only if $s = 0$.*
(e) *A number n as in (b) is contained in $S_1^{\text{cro}} \cap S_2^{\text{cro}}$ if and only if $s \geq 1$ and $t \geq 1$.*

Proof. For every number $n \in S_1^{\text{odd}}$ and every prime number p dividing n, we know that -11 is a quadratic residue modulo p, since it is a quadratic residue modulo n. Hence we have $p \in S_1^{\text{prim}} \cup S_2^{\text{prim}}$. Now claim (a) follows by simply writing $p^\alpha = p^{3\beta} \cdot p^\gamma$ with $\beta \geq 0$ and $\gamma \in \{0, 1, 2\}$ for every prime p dividing n with exponent α, and by using Proposition 4.3.c to conclude that $p^3 \in S_1^{\text{odd}}$. The remaining claims were shown in the proof of the theorem. □

To complete the discussion, we formulate a similar decomposition for the set S_2. Since the proof of the next proposition uses the same tools from Section 4 and proceeds in analogy to the proof of the above theorem and its corollary, we leave it to the interested reader.

Proposition 7.8. *For the set S_2, the following claims hold:*

(a) *We have*

$$S_2^{\text{even}} = S_1^{\text{even}} \cup 4S_1^{\text{cube}} \cup 4\langle S_1^{\text{prim}} \rangle = 4S_2^{\text{odd}} \cup 4S_1^{\text{cube}} \cup 4\langle S_1^{\text{prim}} \rangle$$

where the unions are disjoint except for the union of S_1^{cube} and $\langle S_1^{\text{prim}} \rangle$.
(b) *The set S_2^{odd} is the disjoint union of $S_1^{\text{odd}} \cap S_2^{\text{odd}}$ and the set of all numbers $p_1^{\alpha_1} \cdots p_s^{\alpha_s} \cdot q_1^{\beta_1} \cdots q_s^{\beta_s}$ such that $p_i \in S_1^{\text{prim}}$, $q_j \in S_2^{\text{prim}}$, and $\beta_j = 3\gamma_j + \delta_j$ with $\delta_j \in \{0, 1, 2\}$ satisfying $\delta_1 + \cdots + \delta_t \equiv 1 \pmod 2$.*

Bibliography

[1] The ApCoCoA Team. ApCoCoA: Applied Computations in Commutative Algebra. Available at apcocoa.uni-passau.de.

[2] Beardon AF. The Geometry of Discrete Groups. Heidelberg: Springer; 1983.

[3] Buell DA. Binary Quadratic Forms. New York: Springer; 1989.

[4] Cox DA. Primes of the Form $x^2 + ny^2$. 2nd ed. Hoboken: J. Wiley & Sons; 2013.

[5] Euler L. Illustratio paradoxi circa progressionem numerorum idoneorum sive congruorum. Nova Acta Acad Sci Imperialis Petropolitinae. 1806;15:29–32 (in Latin). English translation available at arXiv:math/0507352 [math.HO].

[6] Evink T, Helminck P. Tribonacci numbers and primes of the form $p = x^2 + 11y^2$. Preprint 2018. Available at arXiv:1801.04605 [math.NT].

[7] Fine B. A note on the two-square theorem. Can Math Bull. 1977;20:93–4.

[8] Fricke R. Lehrbuch der Algebra, Dritter Band: Algebraische Zahlen. Braunschweig: F. Vieweg und Sohn; 1928.

[9] Kern-Isberner G, Rosenberger G. A note on numbers of the form $x^2 + Ny^2$. Arch Math. 1984;43:148–56.

[10] Özgür NY. On numbers of the form $n = x^2 + Ny^2$ and the Hecke groups $H(\sqrt{N})$. J Number Theory. 2010;130:1274–81.

[11] Petersson H. Modulfunktionen und quadratische Formen. Berlin: Springer; 1982 (in German).

[12] Rosenberger G. Über Darstellungen von Elementen und Untergruppen in freien Produkten. In: Kim AC, Neuman BH, editors. Groups – Korea 1983, Proc. Conf Kyoungju. Korea. Lect Notes Math. vol. 1098. Berlin: Springer; 1984. p. 142–60 (in German).

[13] Trifković M. Algebraic Theory of Quadratic Numbers. New York: Springer; 2013.

[14] Weber H. Lehrbuch der Algebra, Dritter Band. 2nd ed. Braunschweig: F. Vieweg und Sohn; 1908 (in German).

Stephen Majewicz and Marcos Zyman

Separability properties of nilpotent $\mathbb{Q}[x]$-powered groups

Abstract: In this paper, we study conjugacy and subgroup separability properties in the class of nilpotent $\mathbb{Q}[x]$-powered groups. Many of the techniques used to study these properties in the context of ordinary nilpotent groups carry over naturally to this more general class. Among other results, we offer a generalization of a theorem due to G. Baumslag. The generalized version states that if G is a finitely $\mathbb{Q}[x]$-generated $\mathbb{Q}[x]$-torsion-free nilpotent $\mathbb{Q}[x]$-powered group and H is a $\mathbb{Q}[x]$-isolated subgroup of G, then for any prime $\pi \in \mathbb{Q}[x]$, $\bigcap_{i=1}^{\infty} G^{\pi^i} H = H$.

Keywords: Nilpotent group, nilpotent R-group, nilpotent R-powered group, nilpotent $\mathbb{Q}[x]$-group, nilpotent $\mathbb{Q}[x]$-powered group, subgroup separability, conjugacy separability

MSC 2010: Primary 20F18, 20F19, 20E26, Secondary 13C12, 13C13, 13G05

1 Introduction

Let $\mathcal{C}$ be a class of groups. Among the types of separability properties of groups, there are:

- *Conjugacy $\mathcal{C}$-separability*: A group G is *conjugacy $\mathcal{C}$-separable* if whenever g and h are not conjugate elements in G, there exists $K \in \mathcal{C}$ and a homomorphism φ from G onto K such that $\varphi(g)$ and $\varphi(h)$ are not conjugate elements in K.
- *Residually $\mathcal{C}$*: A group G is *residually $\mathcal{C}$* if for every nonidentity element g in G, there exists $K \in \mathcal{C}$ and a homomorphism φ from G onto K such that $\varphi(g)$ is not the identity.
  Equivalently, for some nonempty index set Λ, there exists a family $\{N_\lambda \mid \lambda \in \Lambda\}$ of normal subgroups of G such that $G/N_\lambda \in \mathcal{C}$ for all $\lambda \in \Lambda$ and $\bigcap_{\lambda \in \Lambda} N_\lambda = 1$.
- *Subgroup $\mathcal{C}$-separability*: A subgroup H of a group G is called *$\mathcal{C}$-separable* in G if, for every element $g \in G \setminus H$, there is a homomorphism φ of G onto $K \in \mathcal{C}$ such that $\varphi(g) \notin \varphi(H)$. The group G is said to be *separated in a group from $\mathcal{C}$*, and K is said to *separate g and H*.

Note: This paper is dedicated to Ben Fine and Anthony Gaglione on the occasion of their birthdays; and to Gilbert Baumslag (in memoriam).

Stephen Majewicz, Department of Mathematics, CUNY-Kingsborough Community College, Brooklyn, NY 11235, USA, e-mail: smajewicz@kbcc.cuny.edu
Marcos Zyman, Department of Mathematics, CUNY-Borough of Manhattan Community College, New York, NY 10007, USA, e-mail: mzyman@bmcc.cuny.edu

https://doi.org/10.1515/9783110638387-017

It is clear that G is residually $\mathcal{C}$ if and only if the identity subgroup of G is $\mathcal{C}$-separable. Furthermore, every conjugacy $\mathcal{C}$-separable group is residually $\mathcal{C}$.

Let $\mathcal{F}$ and $\mathcal{F}_p$ denote the classes of finite groups and finite p-groups (p is a prime), respectively. In the literature, there are several results pertaining to these properties for nilpotent groups. For instance:

(i) In [4], N. Blackburn proved that every finitely generated nilpotent group is conjugacy $\mathcal{F}$-separable.

(ii) In [8], E. A. Ivanova proved that if G is a finitely generated torsion-free nilpotent group which is conjugacy $\mathcal{F}_p$-separable for some prime p, then G must be abelian.

(iii) Let G be a finitely generated torsion-free nilpotent group, and let H be an isolated subgroup of G. In [3], G. Baumslag proved that for any prime p,

$$\bigcap_{i=1}^{\infty} G^{p^i} H = H.$$

Equivalently, H is $\mathcal{F}_p$-separable in G for any given prime p.
If H is the trivial subgroup in (iii), then we get a classical theorem of K. W. Gruenberg [6] which states that finitely generated torsion-free nilpotent groups are residually $\mathcal{F}_p$ for any prime p.

(iv) In [1], R. B. J. T. Allenby and R. J. Gregorac prove the following: Let G be a finitely generated nilpotent group and $H \leq G$. Suppose there is an element $g \in G \setminus H$ such that $g^n \notin H$ for any positive integer n. If p is any prime, then g and H can be separated in a finite p-group.

Our goal in this paper is to prove results similar to those mentioned above for nilpotent $\mathbb{Q}[x]$-powered groups. The ring $\mathbb{Q}[x]$ is an example of a *binomial ring*, an integral domain of characteristic zero with unity such that for any $r \in \mathbb{Q}[x]$ and positive integer k,

$$\binom{r}{k} = \frac{r(r-1)\cdots(r-k+1)}{k!} \in \mathbb{Q}[x].$$

In this paper, R will always be a binomial ring.

Definition 1.1 ([7]). Let G be a (locally) nilpotent group which comes equipped with an action by R,

$$G \times R \to G \quad \text{defined by } (g, \alpha) \mapsto g^{\alpha},$$

such that for all $g \in G$ and for all $\alpha \in R$, the element $g^{\alpha} \in G$ is uniquely determined. Then G is called a *nilpotent R-powered group* if the following axioms hold:

- $g^1 = g$, $g^{\alpha} g^{\beta} = g^{\alpha+\beta}$, and $(g^{\alpha})^{\beta} = g^{\alpha\beta}$ for all $g \in G$ and $\alpha, \beta \in R$.
- $(h^{-1}gh)^{\alpha} = h^{-1}g^{\alpha}h$ for all $g, h \in G$ and for all $\alpha \in R$.
- (Hall–Petresco axiom) If $\{g_1, \ldots, g_n\} \subset G$ and $\alpha \in R$, then

$$g_1^\alpha \cdots g_n^\alpha = \tau_1(\bar{g})^\alpha \tau_2(\bar{g})^{\binom{\alpha}{2}} \cdots \tau_{k-1}(\bar{g})^{\binom{\alpha}{k-1}} \tau_k(\bar{g})^{\binom{\alpha}{k}},$$

where k is the nilpotency class of the group generated by $\{g_1, g_2, \ldots, g_n\}$, $\bar{g} = (g_1, \ldots, g_n)$, and $\tau_i(\bar{g})$ is the ith *Hall–Petresco word.*

The Hall–Petresco words are obtained by letting $\alpha = 1, 2$, and so on. For instance,

$$\tau_1(\bar{g}) = g_1 \cdots g_n \text{ and } \tau_2(\overline{g}) = (g_1 \cdots g_n)^{-2} g_1^2 \cdots g_n^2.$$

R-modules and R-completions of finitely generated torsion-free nilpotent groups with respect to a Mal'cev basis are examples of nilpotent R-powered groups [7]. Categorical notions such as R-subgroup, R-homomorphism, etc. are defined in the obvious way [7].

Various theorems on residual properties of nilpotent R-powered groups have already been proven in some papers by the authors ([10], [13], [14]). A result on the conjugacy separability of nilpotent $\mathbb{Q}[x]$-powered groups can also be found in [10].

The definitions of separability carry over to nilpotent R-powered groups in a natural way. We merely state them here without giving additional background definitions. All necessary definitions, notation, and results on nilpotent R-powered groups will be provided in the next section.

Definition 1.2. Let $\mathcal{C}$ denote a class of nilpotent R-powered groups. A nilpotent R-powered group G is *conjugacy $\mathcal{C}$-separable* if whenever g and h are not conjugate elements in G, there exists $K \in \mathcal{C}$ and an R-homomorphism φ of G onto K such that $\varphi(g)$ and $\varphi(h)$ are not conjugate in K.

From this point on, $\mathcal{FT}$ and $\mathcal{FT}_\pi$ will denote the classes of nilpotent $\mathbb{Q}[x]$-powered groups of finite type and finite π-type for a prime $\pi \in \mathbb{Q}[x]$, respectively.

Theorem 1.1. *Let G be a finitely $\mathbb{Q}[x]$-generated nilpotent $\mathbb{Q}[x]$-powered group. Two elements of G are conjugate if and only if they are conjugate in every factor $\mathbb{Q}[x]$-group of G of finite type.*

This establishes that finitely $\mathbb{Q}[x]$-generated nilpotent $\mathbb{Q}[x]$-powered groups are conjugacy $\mathcal{FT}$-separable, thus proving an analogue of (i). Our proof mimics the one given by G. Baumslag for Theorem 1.3 of [3].

Our next theorem is similar to (ii).

Theorem 1.2. *Let G be a finitely $\mathbb{Q}[x]$-generated $\mathbb{Q}[x]$-torsion-free nilpotent $\mathbb{Q}[x]$-powered group. If G is conjugacy $\mathcal{FT}_\pi$-separable for some prime $\pi \in \mathbb{Q}[x]$, then G is an abelian $\mathbb{Q}[x]$-group.*

The proof mimics the one given by E. A. Ivanova in [8].

Definition 1.3. Let G be a nilpotent R-powered group, and let $\mathcal{C}$ be a class of nilpotent R-powered groups. An R-subgroup H of G is called *$\mathcal{C}$-separable* in G if, for every element $g \in G \setminus H$, there is an R-homomorphism φ of G onto $K \in \mathcal{C}$ such that $\varphi(g) \notin \varphi(H)$.

The group G is said to be *separated in a group from* $\mathcal{C}$, and K is said to *separate* g *and* H.

We prove the following analogue of (iii).

Theorem 1.3. *Let G be a finitely $\mathbb{Q}[x]$-generated $\mathbb{Q}[x]$-torsion-free nilpotent $\mathbb{Q}[x]$-powered group, and let H be a $\mathbb{Q}[x]$-isolated subgroup of G. For any prime $\pi \in \mathbb{Q}[x]$,*

$$\bigcap_{i=1}^{\infty} G^{\pi^i} H = H.$$

Equivalently, H is $\mathcal{FT}_\pi$-separable for any given prime $\pi \in \mathbb{Q}[x]$.

Another result on $\mathcal{FT}_\pi$-separability is the following.

Theorem 1.4. *Let G be a finitely $\mathbb{Q}[x]$-generated nilpotent $\mathbb{Q}[x]$-powered group and $H \leq_{\mathbb{Q}[x]} G$. Suppose there is an element $g \in G \setminus H$ such that $g^\lambda \notin H$ for any $0 \neq \lambda \in \mathbb{Q}[x]$. If π is any prime in $\mathbb{Q}[x]$, then g and H can be separated in a nilpotent $\mathbb{Q}[x]$-powered group of finite π-type.*

The proof is similar to the one given for Theorem 1 (a) in [1].

2 Preliminaries

In this section, we discuss some terminology and results on nilpotent R-powered groups which will be used in this paper. We remind the reader that R will always be a binomial ring. We will usually denote a group identity element, a multiplicative identity ring element, and the trivial subgroup of a group by 1.

Let G is a nilpotent R-powered group. If $[g,h] = 1$ for $g, h \in G$, the Hall–Petresco axiom implies that $(gh)^\alpha = g^\alpha h^\alpha$ for any $\alpha \in R$. In particular, every abelian R-group is a R-module.

Proposition 2.1. *Suppose G is a nilpotent R-powered group. Let $g, h \in G$ such that $[g,h] \in Z(G)$. Then $[g,h]^\alpha = [g, h^\alpha]$ for all $\alpha \in R$.*

Proof. Since $[g,h] \in Z(G)$, we have

$$\begin{aligned}[g^{-1}h^{-1}g, h] &= g^{-1}hgh^{-1}[g,h] \\ &= g^{-1}hg[g,h]h^{-1} \\ &= 1.\end{aligned}$$

By the previous paragraph, $(g^{-1}h^{-1}gh)^\alpha = (g^{-1}h^{-1}g)^\alpha h^\alpha$ for any $\alpha \in R$. Since $(g^{-1}h^{-1}g)^\alpha = g^{-1}h^{-\alpha}g$, the result follows. □

Let G be a nilpotent R-powered group. If H is an R-subgroup or a normal R-subgroup of G, then we write $H \leq_R G$ and $H \trianglelefteq_R G$, respectively. We say that H is *R-generated by* $X = \{x_1, \ldots, x_j\} \subseteq G$ if

$$H = \bigcap_{X \subset H_i \leq_R G} \{H_i\}.$$

In this case, we write $H = gp_R(x_1, \ldots, x_j)$.

If G is a nilpotent R-powered group and $N \trianglelefteq_R G$, then the R-action on G induces an R-action on G/N,

$$(gN)^\alpha = g^\alpha N \quad \text{for all } gN \in G/N \text{ and } \alpha \in R,$$

which turns G/N into a nilpotent R-powered group.

It is not hard to show using the Hall–Petresco axiom that if G is a nilpotent R-powered group with $H \leq_R G$ and $N \trianglelefteq_R G$, then $HN \leq_R G$ and $HN = gp_R(H, N)$.

The usual isomorphism theorems carry over to nilpotent R-powered groups in the obvious way. If G_1 and G_2 are nilpotent R-powered groups, then the kernel of an R-homomorphism $\varphi : G_1 \to G_2$, abbreviated as $\ker \varphi$, is a normal R-subgroup of G_1. If G_1 and G_2 happen to be R-isomorphic, then we write $G_1 \cong_R G_2$.

The upper and lower central subgroups of a nilpotent R-powered group G are R-subgroups of G. We denote them by $\zeta_i G$ and $\gamma_i G$, respectively. The center of G will be written as $Z(G)$.

Theorem 2.2 ([9]). *If R is noetherian, then every R-subgroup of a finitely R-generated nilpotent R-powered group is finitely R-generated.*

Definition 2.1 ([15]). If G is a nilpotent R-powered group, then $g \in G$ is an *R-torsion element* if $g^\alpha = 1$ for some $0 \neq \alpha \in R$. If every element of G is R-torsion, then G is called an *R-torsion group*. We say that G is *R-torsion-free* if the only R-torsion element of G is 1.

Definition 2.2 ([10]). Let $\pi \in R$ be a prime. A finitely R-generated nilpotent R-powered group G is of:

(1) *finite type* if it is an R-torsion group.
(2) *finite π-type* if for each $g \in G$, there exists a positive integer k such that $g^{\pi^k} = 1$.

The next theorem and lemma are proven by the authors in [11].

Theorem 2.3. *Let R be noetherian and consider the short exact sequence*

$$1 \to H \to G \to G/H \to 1$$

in the category of nilpotent R-powered groups. Then G is of finite type if and only if H and G/H are both of finite type.

Lemma 2.4. *Let R be a PID. If G is a finitely R-generated nilpotent R-powered group and Z(G) is of finite type, then G is of finite type.*

The next lemma will be needed later.

Lemma 2.5. *Let R be a PID. If G is a finitely R-generated nilpotent R-powered group and is not of finite type, then $Z(G)$ has an element which is not R-torsion.*

Proof. Assume, on the contrary, that $Z(G)$ is an R-torsion group. Since G is finitely R-generated, so is $Z(G)$ by Theorem 2.2. Thus, $Z(G)$ is of finite type. By Lemma 2.4, G is also of finite type. □

If G is a nilpotent R-powered group and $\beta \in R$, then we define

$$G^\beta = gp_R(g^\beta \mid g \in G).$$

It is clear that $G^\beta \trianglelefteq_R G$.

Theorem 2.6. [12] *If R contains $\mathbb{Q}$ and G is a nilpotent R-powered group, then every element of G^β is a βth power of an element of G for any $\beta \in R$.*

The notion of the exponent of a group has been generalized for the category of nilpotent R-powered groups and will be used in the proof of Theorem 1.4.

Definition 2.3. [11] Let R be a PID, and let G be a nilpotent R-powered group of finite type. We say that G has *exponent* $\pi_1\pi_2\cdots\pi_n$, where the π_i's are primes (not necessarily distinct) in R, if

$$G^{\pi_1\pi_2\cdots\pi_n} = 1 \quad \text{but } G^{\pi_1\pi_2\cdots\widehat{\pi_i}\cdots\pi_n} \neq 1 \text{ for any } 1 \leq i \leq n.$$

Here, $\pi_1\pi_2\cdots\widehat{\pi_i}\cdots\pi_n$ stands for the omission of π_i in the product $\pi_1\pi_2\cdots\pi_n$.

The R-isolator of a nilpotent R-powered group plays an important role in what follows.

Definition 2.4. Let G be a nilpotent R-powered group.
(i) An R-subgroup H of G is *R-isolated* in G if $g \in G$ and $g^\lambda \in H$ for some $0 \neq \lambda \in R$ imply $g \in H$.
(ii) The *R-isolator* of $S \subseteq G$, denoted by $I(S, G)$, is the intersection of all R-isolated subgroups of G containing S.

Clearly, $I(S, G)$ is an R-isolated subgroup of G. The next theorem describes the R-isolator of an R-subgroup and can be found in [15].

Theorem 2.7. [15] *If G is a nilpotent R-powered group and $H \leq_R G$, then*

$$I(H, G) = \{g \in G \mid g^\lambda \in H \quad \text{for some } 0 \neq \lambda \in R\}.$$

Note that $I(1, G)$ is the R-subgroup of G consisting of the R-torsion elements of G. This is referred to as the *R-torsion subgroup* of G and denoted by $\tau(G)$. It is clear that $\tau(G) \trianglelefteq_R G$ and $G/\tau(G)$ is R-torsion-free.

We record a simple lemma which will be used in what follows.

Lemma 2.8. *Let G be a nilpotent R-powered group and $H \trianglelefteq_R G$. Then G/H is R-torsion-free if and only if H is R-isolated in G.*

The theory of extraction of roots appears in several papers by the authors (see [11], [12], [13], and [14]).

Definition 2.5. If G is a nilpotent R-powered group and $h^\alpha = g$ for some $g, h \in G$ and $\alpha \in R$, then h is called an *αth root* of g.

Theorem 2.9. [15] *Let G be a nilpotent R-powered group. Then G is R-torsion-free if and only if every $1 \neq g \in G$ has at most one αth root for every $0 \neq \alpha \in R$; that is, if $g, h \in G$ and $g^\alpha = h^\alpha$ for some $0 \neq \alpha \in R$, then $g = h$.*

Corollary 2.10. *If G is an R-torsion-free nilpotent R-powered group, then $Z(G)$ is R-isolated.*

Proof. If $g^\alpha \in Z(G)$ for some $g \in G$ and $\alpha \in R$, then $h^{-1}g^\alpha h = g^\alpha$ for any $h \in G$. Thus, $(h^{-1}gh)^\alpha = g^\alpha$. By Theorem 2.9, $h^{-1}gh = g$, and thus, $g \in Z(G)$. □

Another result that we will rely on is the following.

Lemma 2.11. *Let G be a nontrivial finitely $\mathbb{Q}[x]$-generated $\mathbb{Q}[x]$-torsion-free nilpotent $\mathbb{Q}[x]$-powered group. For any $1 \neq g \in G$ and prime $\pi \in \mathbb{Q}[x]$, there exists $n \in \mathbb{N}$ such that g has no π^nth root in G.*

Proof. Assume, on the contrary, that g has π^nth roots in G for every $n \in \mathbb{N}$. By Theorem 2.9, these π^nth roots must be unique. Thus, we let

$$H = gp_{\mathbb{Q}[x]}(g_1, g_2, \ldots), \quad \text{where } g = g_1^\pi \text{ and } g_j = g_{j+1}^\pi \text{ for all } j \in \mathbb{N}.$$

By Theorem 2.2, H is finitely $\mathbb{Q}[x]$-generated. Consequently, there exists $k \in \mathbb{N}$ such that $H = gp_{\mathbb{Q}[x]}(g_1, g_2, \ldots, g_k)$. By construction of the g_j's, this means that $H = gp_{\mathbb{Q}[x]}(g_k)$.

Now, H should contain all π^nth roots of g. However, this is not the case since g_k, the π^kth root of g, has no πth root in H. For if g_k had a πth root in H, then there would exist $\mu \in \mathbb{Q}[x]$ such that $(g_k^\mu)^\pi = g_k$; that is, $g_k^{\mu\pi-1} = 1$. Since H is $\mathbb{Q}[x]$-torsion-free and $g_k \neq 1$, $\mu\pi - 1 = 0$. It follows that π must be a unit. However, π is a prime by hypothesis, a contradiction. □

Next, we discuss some results on the R-torsion-free rank of a finitely R-generated nilpotent R-powered group, where R is noetherian (see [15]). First, we recall the notion of torsion-free rank for a module.

Definition 2.6. Let D be an integral domain with fraction field K, and let M be a D-module. The *torsion-free rank* of M is the K-dimension of the K-vector space $K \otimes_D M$, denoted by $\dim_K(K \otimes_D M)$.

If M is a finitely generated D-module, then $K \otimes_D M$ is finite dimensional as a K-vector space and $\dim_K(K \otimes_D M)$ is the maximal number of D-linearly independent elements in M, that is, the maximal rank of a free D-submodule of M (see Theorem 6.11 in [5] for a proof). In the literature, torsion-free rank is usually referred to simply as rank. In particular, if D is a PID and

$$M = \underbrace{D \oplus \cdots \oplus D}_{m \text{ summands}} \oplus T,$$

where T is the torsion part of M, then the (torsion-free) rank of M is m.

Definition 2.7. Let G be a finitely R-generated nilpotent R-powered group, where R is noetherian, and let K be the fraction field of R. If

$$G = G_1 \trianglerighteq_R G_2 \trianglerighteq_R \cdots \trianglerighteq_R G_{n+1} = 1$$

is a subnormal R-series, where G_i/G_{i+1} is an R-module for $i = 1, 2, \ldots, n$, then the *R-torsion-free rank* of G is

$$\sum_{i=1}^{n} \dim_K(K \otimes_R G_i/G_{i+1}).$$

It follows from Theorem 2.2 that each quotient G_i/G_{i+1} is a finitely generated R-module of finite rank. Thus, the above sum is well-defined and G has finite R-torsion-free rank. An application of the Jordan–Hölder theorem shows that the R-torsion-free rank of a nilpotent R-powered group is independent of the subnormal R-series chosen [15].

The next theorem and corollary are extensions of known facts about the rank of a finitely generated R-module over a PID.

Theorem 2.12. *Let R be a PID, and suppose G is a finitely R-generated nilpotent R-powered group with R-torsion-free rank n.*
(i) *If $H \leq_R G$, then the R-torsion-free rank of H is at most n.*
(ii) *If $N \trianglelefteq_R G$ and N has R-torsion-free rank m, then G/N has R-torsion-free rank $n - m$.*

Proof. (i) Suppose G has nilpotency class c, and let

$$G = \gamma_1 G \geq \gamma_2 G \geq \cdots \geq \gamma_{c+1} G = 1$$

be the lower central series of G. Assume that $\gamma_i G/\gamma_{i+1} G$ has R-torsion-free rank n_i for $i = 1, 2, \ldots, c$. Since G has R-torsion-free rank n, we have $n_1 + \cdots + n_c = n$. Put $H_i = H \cap \gamma_i G$ for $i = 1, 2, \ldots, c$. Then

$$H = H_1 \geq_R H_2 \geq_R \cdots \geq_R H_c \geq_R H_{c+1} = 1$$

is a central R-series for H. Now,

$$\frac{H_i}{H_{i+1}} = \frac{H \cap \gamma_i G}{H \cap \gamma_{i+1} G} = \frac{H \cap \gamma_i G}{(H \cap \gamma_i G) \cap \gamma_{i+1} G} \cong_R \frac{\gamma_{i+1} G(H \cap \gamma_i G)}{\gamma_{i+1} G}$$

for $i = 1, 2, \ldots, c$. Therefore, H_i/H_{i+1} is R-isomorphic to an R-subgroup of $\gamma_i G/\gamma_{i+1} G$. Since $\gamma_i G/\gamma_{i+1} G$ is an R-module, the R-torsion-free rank of H_i/H_{i+1} is at most n_i. Hence, the R-torsion-free rank of H is at most n.

(ii) Let

$$1 = N_0 \trianglelefteq_R N_1 \trianglelefteq_R \cdots \trianglelefteq_R N_k = N$$

and

$$1 = G_0/N \trianglelefteq_R G_1/N \trianglelefteq_R \cdots \trianglelefteq_R G_l/N = G/N$$

be subnormal R-series of N and G/N, respectively. Clearly, G has a subnormal R-series

$$1 = N_0 \trianglelefteq_R N_1 \trianglelefteq_R \cdots \trianglelefteq_R N_k \trianglelefteq_R G_1 \trianglelefteq_R \cdots \trianglelefteq_R G_l = G.$$

Since $(G_{i+1}/N)/(G_i/N) \cong_R G_{i+1}/G_i$ for $i = 0, 1, \ldots, l-1$, the R-torsion-free rank of G is the sum of the R-torsion-free ranks of N and G/N. □

The next corollary follows immediately from Theorem 2.12(ii).

Corollary 2.13. *Let R be a PID, and let G be a finitely R-generated R-torsion-free nilpotent R-powered group. If $N \trianglelefteq_R G$, then G and N have the same R-torsion-free rank if and only if G/N is an R-torsion group.*

Corollary 2.14. *Let R be a PID, and let G be a finitely R-generated R-torsion-free nilpotent R-powered group. If the R-torsion-free rank of G is 1, then G is an R-module.*

Proof. It suffices to show that G is abelian. Since $Z(G)$ is R-torsion-free, its R-torsion-free rank equals some $m > 0$. By Theorem 2.12(i), m must equal 1. Hence, $G/Z(G)$ is an R-torsion group by Corollary 2.13. Now, $Z(G)$ is R-isolated by Corollary 2.10. By Lemma 2.8, $G/Z(G)$ must be trivial. Thus, G is abelian. □

3 Proofs of theorems

We use the following notation: If g and h are (are not) conjugate group elements, then we write $g \sim h$ ($g \nsim h$).

Proof of Theorem 1.1. The proof is by induction on the $\mathbb{Q}[x]$-torsion-free rank r of G. If $r = 0$, then G is of finite type and there is nothing to prove.

Suppose that $r > 0$. Let g and h be elements of G which are conjugate in every factor $\mathbb{Q}[x]$-group of finite type. We claim that $g \sim h$. Assume, on the contrary, that $g \nsim h$. Since $r > 0$, G is not of finite type. By Lemma 2.5, there exists $a \in Z(G)$ that is not a $\mathbb{Q}[x]$-torsion element. Let

$$S = \{1, f \in \mathbb{Q}[x] \mid f \text{ is monic}\},$$

and for each $\sigma \in S$, define

$$H_\sigma = gp_{\mathbb{Q}[x]}(a^\sigma).$$

For each $\sigma \in S$, observe that $H_\sigma \trianglelefteq_{\mathbb{Q}[x]} G$. By Theorem 2.12, the $\mathbb{Q}[x]$-torsion-free rank of each G/H_σ is less than r.

Suppose that $gH_\sigma \nsim hH_\sigma$ in G/H_σ for some $\sigma \in S$. By induction and a $\mathbb{Q}[x]$-isomorphism theorem, there exists a normal $\mathbb{Q}[x]$-subgroup N_σ/H_σ of G/H_σ such that

$$(G/H_\sigma)/(N_\sigma/H_\sigma) \cong_{\mathbb{Q}[x]} G/N_\sigma$$

is of finite type and the images of g and h under the natural $\mathbb{Q}[x]$-homomorphism from G onto G/N_σ are not conjugate in G/N_σ. This contradicts the assumption that the images of g and h are conjugate in every factor $\mathbb{Q}[x]$-group of G of finite type. Therefore, $gH_\sigma \sim hH_\sigma$ for all $\sigma \in S$. In particular, $gH_1 \sim hH_1$ in G/H_1. Thus, since $a \in Z(G)$, we can find a nonzero element $t \in \mathbb{Q}[x]$ such that $h \sim ga^t$. Since $gH_\sigma \sim hH_\sigma$, we have

$$gH_\sigma \sim ga^t H_\sigma \text{ for each } \sigma \in S.$$

Thus, for each $\sigma \in S$, we can find $y_\sigma \in G$ and $u_\sigma \in \mathbb{Q}[x]$ such that

$$y_\sigma^{-1} g y_\sigma = ga^t (a^\sigma)^{u_\sigma}. \tag{1}$$

Let Y be the set of those $y_\sigma \in G$ arising in (1), and put $L = gp_{\mathbb{Q}[x]}(g, a, Y)$. Clearly, $[y, g] \in gp_{\mathbb{Q}[x]}(a)$ for all $y \in Y$ by (1). Moreover, $gp_{\mathbb{Q}[x]}(a) \leq_{\mathbb{Q}[x]} Z(L)$ since $a \in Z(G)$. It follows from the commutator calculus and the axioms that $[l, g] \in gp_{\mathbb{Q}[x]}(a)$ for all $l \in L$. Thus, we obtain a $\mathbb{Q}[x]$-homomorphism

$$\varphi : L \to gp_{\mathbb{Q}[x]}(a) \text{ defined by } \varphi(l) = [l, g].$$

Since $\mathbb{Q}[x]$ is a PID, $L/\ker\varphi$ is $\mathbb{Q}[x]$-cyclic. Thus, there exists an element $b \in L \setminus \ker\varphi$ and $\alpha \in \mathbb{Q}[x]$ such that

$$L = gp_{\mathbb{Q}[x]}(\ker\varphi, b) \quad \text{and} \quad [b, g] = a^\alpha. \tag{2}$$

Note that $\alpha \neq 0$ because $b \notin \operatorname{Ker}\varphi$. In fact, b and α can be chosen so that $\alpha \in S$.

Next, we find all of the conjugates of g in L. By (2), such a conjugate has the form $(b^\mu k)^{-1}g(b^\mu k)$, where $k \in \ker\varphi$ and $\mu \in \mathbb{Q}[x]$. Since $[g,b] = a^{-\alpha} \in Z(L)$ and $\ker\varphi$ is the centralizer of g in L,

$$\begin{aligned}(b^\mu k)^{-1}g(b^\mu k) &= k^{-1}b^{-\mu}gb^\mu k = k^{-1}g[g,b^\mu]k\\ &= k^{-1}g[g,b]^\mu k = k^{-1}gk[g,b]^\mu\\ &= ga^{-\alpha\mu}\end{aligned}$$

by Proposition 2.1. It follows that the set of conjugates of g in L is

$$C = \{ga^{\alpha\mu} \mid \mu \in \mathbb{Q}[x]\}. \tag{3}$$

Now, $g \nsim ga^t$ in L because $g \nsim ga^t$ in G. Thus $ga^t \notin C$, and so, $\alpha\mu \neq t$ for all $\mu \in \mathbb{Q}[x]$. On the other hand, (1) yields

$$y_\alpha^{-1}gy_\alpha = ga^t(a^\alpha)^{u_\alpha}$$

because $\alpha \in S$. This means that $ga^t(a^\alpha)^{u_\alpha} \in C$, so $t+\alpha u_\alpha = \alpha\lambda$ for some $\lambda \in \mathbb{Q}[x]$. Hence, $t = \alpha(\lambda - u_\alpha)$, a contradiction since $\lambda - u_\alpha \in \mathbb{Q}[x]$. Theorem 1.1 is proven. □

Proof of Theorem 1.2. Suppose that G has nilpotency class c, and assume that G is not abelian. In this case, $c > 1$. Thus, there exist elements $z_2 \in \zeta_2G \setminus Z(G)$ and $h \in G$ such that $[z_2,h] = z_1$ for some $1 \neq z_1 \in Z(G)$. Let α be a prime in $\mathbb{Q}[x]$ different from π. Since $\mathbb{Q}[x]$ is noetherian and G is finitely $\mathbb{Q}[x]$-generated, $Z(G)$ is also finitely $\mathbb{Q}[x]$-generated by Theorem 2.2. Hence, $Z(G)$ is a finitely $\mathbb{Q}[x]$-generated $\mathbb{Q}[x]$-torsion-free abelian $\mathbb{Q}[x]$-group. By Lemma 2.11, there exists $n \in \mathbb{N}$ such that z_1 has no α^nth root in $Z(G)$. We claim that $z_2^{\alpha^n} \nsim z_2^{\alpha^n}z_1$ in G, whereas $\varphi(z_2^{\alpha^n}) \sim \varphi(z_2^{\alpha^n}z_1)$ in $\varphi(G)$ for every $\mathbb{Q}[x]$-homomorphism φ from G onto a nilpotent $\mathbb{Q}[x]$-powered group of finite π-type. Once this is shown, the theorem will be proved.

First, we show that $z_2^{\alpha^n} \nsim z_2^{\alpha^n}z_1$ in G. If this were not the case, then there would exist $g \in G$ such that $g^{-1}z_2^{\alpha^n}g = z_2^{\alpha^n}z_1$. Now, since $z_2 \in \zeta_2G \setminus Z(G)$, there exists a central element z such that $[z_2,g] = z$. Thus,

$$z_2^{\alpha^n}z_1 = g^{-1}z_2^{\alpha^n}g = (g^{-1}z_2g)^{\alpha^n} = (z_2z)^{\alpha^n} = z_2^{\alpha^n}z^{\alpha^n}.$$

Therefore, $z_1 = z^{\alpha^n}$. This contradicts the fact that z_1 has no α^nth root in $Z(G)$.

Next, we show that the images of $z_2^{\alpha^n}$ and $z_2^{\alpha^n}z_1$ are conjugate under every $\mathbb{Q}[x]$-homomorphism from G onto an arbitrary nilpotent $\mathbb{Q}[x]$-powered group of finite π-type. Let $N \trianglelefteq_{\mathbb{Q}[x]} G$ such that G/N is of finite π-type and $(z_1N)^{\pi^m} = N$ for some nonnegative integer m; that is, $z_1^{\pi^m} \in N$. Since α and π are distinct primes, α^n and π^m are coprime. Thus, there exist $\beta,\gamma \in \mathbb{Q}[x]$ such that $\beta\alpha^n + \gamma\pi^m = 1$. Now, $[z_2,h] = z_1$ and $z_1 \in Z(G)$ imply that $h^{-\beta}z_2h^\beta = z_2z_1^\beta$ by Proposition 2.1. Therefore,

$$\begin{aligned} z_2^{\alpha^n} z_1 &= z_2^{\alpha^n} z_1^{\beta\alpha^n+\gamma\pi^m} = (z_2 z_1^{\beta})^{\alpha^n} z_1^{\gamma\pi^m} \\ &= (h^{-\beta} z_2 h^{\beta})^{\alpha^n} z_1^{\gamma\pi^m} = h^{-\beta} z_2^{\alpha^n} h^{\beta} (z_1^{\pi^m})^{\gamma}. \end{aligned}$$

Since $z_1^{\pi^m} \in N$, we have $z_2^{\alpha^n} z_1 N = h^{-\beta} z_2^{\alpha^n} h^{\beta} N$ in G/N. And so, $z_2^{\alpha^n} z_1 N \sim z_2^{\alpha^n} N$ in G/N. This completes the proof of Theorem 1.2. □

Proof of Theorem 1.3. This proof invokes the following lemma.

Lemma 3.1. *Let G be a finitely $\mathbb{Q}[x]$-generated $\mathbb{Q}[x]$-torsion-free nilpotent $\mathbb{Q}[x]$-powered group. Suppose that $Y \leq_{\mathbb{Q}[x]} Z(G)$, and let H be a $\mathbb{Q}[x]$-isolated subgroup of G. If $I(HY, G) = G$, then $H \trianglelefteq_{\mathbb{Q}[x]} G$.*

Proof. The proof is by induction on the $\mathbb{Q}[x]$-torsion-free rank r of G. If $r = 0$, then G is a $\mathbb{Q}[x]$-torsion group, contrary to assumption. If $r = 1$, then G is a $\mathbb{Q}[x]$-module by Corollary 2.14 and the result is trivial. Suppose then that $r > 1$. We first prove that $Z(H) \leq_{\mathbb{Q}[x]} Z(G)$. By Theorem 2.9, every element of G has at most one αth root for every $0 \neq \alpha \in \mathbb{Q}[x]$. It follows that the centralizer of every nonempty subset of G is $\mathbb{Q}[x]$-isolated in G (see Lemma 4.8 in [13]). In particular, $Z(H)$ is $\mathbb{Q}[x]$-isolated in G because $Z(H) = H \cap C_H(G)$, where $C_H(G)$ denotes the centralizer of H in G, and H is $\mathbb{Q}[x]$-isolated in G by hypothesis. We will use these observations freely in the remainder of the proof.

Pick any element $g \in G = I(HY, G)$. By Theorem 2.7, there exists $0 \neq \alpha \in \mathbb{Q}[x]$ such that $g^{\alpha} \in HY$. We assert that g^{α} centralizes $Z(H)$. Suppose that $h \in Z(H)$. Since $g^{\alpha} \in HY \leq_{\mathbb{Q}[x]} HZ(G)$, there exist elements $k \in H$ and $z \in Z(G)$ such that $g^{\alpha} = kz$. Now, $[h, k] = 1$ because $h \in Z(H)$, and so $[h, g^{\alpha}] = 1$ because $z \in Z(G)$. Thus, g^{α} centralizes $Z(H)$ as asserted. Therefore, g also centralizes $Z(H)$ because the centralizer of $Z(H)$ is $\mathbb{Q}[x]$-isolated in G. Since g is an arbitrary element of G, we have that $Z(H) \leq_{\mathbb{Q}[x]} Z(G)$ as claimed.

The fact that $Z(H) \leq_{\mathbb{Q}[x]} Z(G)$ immediately gives $Z(H) \trianglelefteq_{\mathbb{Q}[x]} G$. Clearly, $G/Z(H)$ is finitely $\mathbb{Q}[x]$-generated. It is also $\mathbb{Q}[x]$-torsion-free by Lemma 2.8. Since G has $\mathbb{Q}[x]$-torsion-free rank r, $G/Z(H)$ has $\mathbb{Q}[x]$-torsion-free rank less than r by Theorem 2.12(i) and Corollary 2.13. We check that the remaining hypotheses of the lemma hold for $G/Z(H)$. Indeed, $YZ(H)/Z(H) \leq_{\mathbb{Q}[x]} Z(G/Z(H))$ since $Y \leq_{\mathbb{Q}[x]} Z(G)$. Moreover, since H is $\mathbb{Q}[x]$-isolated in G, it is easy to see that $H/Z(H)$ must be $\mathbb{Q}[x]$-isolated in $G/Z(H)$. It remains to show that $G/Z(H) \leq_{\mathbb{Q}[x]} I(HY/Z(H), G/Z(H))$ (the reverse inclusion is trivially true). We begin by noting that $Z(H) \trianglelefteq_{\mathbb{Q}[x]} HY$ since $Y \leq_{\mathbb{Q}[x]} Z(G)$. Suppose $gZ(H) \in G/Z(H)$ for some $g \in G$. Since $G = I(HY, G)$, $g^{\beta} \in HY$ for some $0 \neq \beta \in \mathbb{Q}[x]$. Thus, $g^{\beta}Z(H) \in HY/Z(H)$; that is, $(gZ(H))^{\beta} \in HY/Z(H)$. And so, $gZ(H) \in I(HY/Z(H), G/Z(H))$ as required. By induction, $H/Z(H) \trianglelefteq_{\mathbb{Q}[x]} G/Z(H)$, and consequently, $H \trianglelefteq_{\mathbb{Q}[x]} G$. □

We now prove Theorem 1.3 by induction on the nilpotency class c of G. If $c = 1$, then G is a free $\mathbb{Q}[x]$-module of finite rank. By Lemma 2.8, G/H is $\mathbb{Q}[x]$-torsion-free, and thus, a free $\mathbb{Q}[x]$-module as well. Hence, $G = H \times K$ for some $K \leq_{\mathbb{Q}[x]} G$.

We claim that

$$G^{\pi^i}H = H \times K^{\pi^i} \quad (i = 1, 2, \ldots).$$

To begin with, observe that $HK^{\pi^i} = H \times K^{\pi^i}$ because $H \cap K^{\pi^i} = 1$. We show that $HK^{\pi^i} = G^{\pi^i}H$. Clearly, $HK^{\pi^i} \leq_{\mathbb{Q}[x]} G^{\pi^i}H$. We assert that $G^{\pi^i}H \leq_{\mathbb{Q}[x]} HK^{\pi^i}$. If $g^{\pi^i}h$ is a generator of $G^{\pi^i}H$, then

$$g^{\pi^i}h = (h_1k_1)^{\pi^i}h = h_1^{\pi^i}hk_1^{\pi^i}$$

for some $h_1 \in H$ and $k_1 \in K$. Thus, $g^{\pi^i}h \in HK^{\pi^i}$, proving the assertion and the claim.

Now, K is finitely $\mathbb{Q}[x]$-generated by Theorem 2.2, as well as $\mathbb{Q}[x]$-torsion-free. Thus, $\bigcap_{i=1}^{\infty} K^{\pi^i} = 1$ by Corollary 5.3 in [13]. And so,

$$\bigcap_{i=1}^{\infty} G^{\pi^i}H = \bigcap_{i=1}^{\infty}(H \times K^{\pi^i}) = H \times \left(\bigcap_{i=1}^{\infty} K^{\pi^i}\right) = H.$$

This proves the theorem when $c = 1$.

Suppose that $c > 1$. Put

$$Z = Z(G), \quad I = I(HZ, G), \text{ and } L = \bigcap_{i=1}^{\infty} G^{\pi^i}H.$$

We need to show that $L = H$. By Corollary 3.3 in [13], G/Z is $\mathbb{Q}[x]$-torsion-free because G is $\mathbb{Q}[x]$-torsion-free. Since I is $\mathbb{Q}[x]$-isolated in G, I/Z is $\mathbb{Q}[x]$-isolated in G/Z. By induction,

$$\bigcap_{i=1}^{\infty}\left(\frac{G}{Z}\right)^{\pi^i} \cdot \frac{I}{Z} = \frac{I}{Z}. \tag{4}$$

We claim that $\bigcap_{i=1}^{\infty} G^{\pi^i}I = I$. Clearly, $\bigcap_{i=1}^{\infty} G^{\pi^i}I \geq_{\mathbb{Q}[x]} I$. Let $x \in \bigcap_{i=1}^{\infty} G^{\pi^i}I$. Then

$$xZ \in \frac{G^{\pi^i}I}{Z} = \frac{G^{\pi^i}Z}{Z} \cdot \frac{I}{Z} = \left(\frac{G}{Z}\right)^{\pi^i} \cdot \frac{I}{Z}$$

for all $i \geq 1$. It follows from (4) that $x \in I$, establishing our claim. Now, $\bigcap_{i=1}^{\infty} G^{\pi^i}H \leq_{\mathbb{Q}[x]} I$. It follows that $H \leq_{\mathbb{Q}[x]} L \leq_{\mathbb{Q}[x]} I$. Now, I is finitely $\mathbb{Q}[x]$-generated by Theorem 2.2 and $\mathbb{Q}[x]$-torsion-free, and H is $\mathbb{Q}[x]$-isolated in I. By Lemma 3.1, $H \trianglelefteq_{\mathbb{Q}[x]} I$. Moreover, I/H is $\mathbb{Q}[x]$-torsion-free by Lemma 2.8 because H is $\mathbb{Q}[x]$-isolated in I.

We show that $L = H$. Consider the factor $\mathbb{Q}[x]$-group L/H. Let $\ell H \in L/H$, and let i be a positive integer. By Theorem 2.6, every element of G^{π^i} is a π^ith power of an element of G. Thus, there exist $h \in H$ and $g_i \in G$ such that

$$\ell = g_i^{\pi^i}h.$$

Hence, $g_i^{\pi^i} = \ell h^{-1} \in I$. Since I is $\mathbb{Q}[x]$-isolated in G, $g_i \in I$. This implies that $\ell H = (g_i H)^{\pi^i}$ with $g_i H \in I/H$. It follows that ℓH has a π^ith root in I/H for every $i = 1, 2, \ldots$. If $\ell H \neq H$, this would contradict Lemma 2.11. Consequently, $\ell H = H$, and thus, $L = H$ as desired. This completes the proof of Theorem 1.3. □

Proof of Theorem 1.4. Let c be the nilpotency class of G. We can assume, without loss of generality, that G is $\mathbb{Q}[x]$-torsion-free. To see why, put $T = \tau(G)$ and fix a prime $\pi \in \mathbb{Q}[x]$. Since T is a (normal) $\mathbb{Q}[x]$-subgroup of G, it is finitely $\mathbb{Q}[x]$-generated by Theorem 2.2, and thus, of finite type. Let $\sigma \in \mathbb{Q}[x]$ be the exponent of T, and assume that $(gT)^{\alpha} \in HT/T$ for some $0 \neq \alpha \in \mathbb{Q}[x]$. There exist $h \in H$ and $t \in T$ such that $g^{\alpha} = ht$. We claim that there exist nonzero distinct elements β and μ of $\mathbb{Q}[x]$ and $t' \in T$ such that $g^{\alpha\beta} = h^{\beta}t'$ and $g^{\alpha\mu} = h^{\mu}t'$. Choose any $0 \neq \beta \in \mathbb{Q}[x]$, and let $\mu = \beta + \sigma$. Put $\tau_i = \tau_i(h, t)$ for $i \in \mathbb{N}$. Since $T \trianglelefteq_{\mathbb{Q}[x]} G$, $\tau_i \in T$ for $i \geq 2$ (see the proof of Theorem 4.16 in [2] for details). By the Hall–Petresco axiom,

$$\begin{aligned} g^{\alpha\beta} &= h^{\beta}t^{\beta}\tau_c^{-\binom{\beta}{c}} \cdots \tau_2^{-\binom{\beta}{2}} \\ &= h^{\beta}t' \end{aligned}$$

with $t' \in T$ and

$$\begin{aligned} g^{\alpha\mu} &= h^{\mu}t^{\mu}\tau_c^{-\binom{\mu}{c}} \cdots \tau_2^{-\binom{\mu}{2}} \\ &= h^{\mu}t^{\beta+\sigma}\tau_c^{-\binom{\beta+\sigma}{c}} \cdots \tau_2^{-\binom{\beta+\sigma}{2}}. \end{aligned}$$

Since T has exponent σ, $t^{\beta+\sigma} = t^{\beta}$ and $\tau_i^{-\binom{\beta+\sigma}{i}} = \tau_i^{-\binom{\beta}{i}}$ for $2 \leq i \leq c$. This proves the claim.

Now, $\alpha\beta - \alpha\mu \neq 0$ and $g^{\alpha\beta-\alpha\mu} = h^{\beta-\mu} \in H$, contrary to the hypothesis. This means that $(gT)^{\alpha} \notin HT/T$ for any $0 \neq \alpha \in \mathbb{Q}[x]$. Hence, $g^{\alpha} \notin HT$, and thus, $g \notin T$. Consequently, we might as well assume that G is $\mathbb{Q}[x]$-torsion-free as mentioned in the beginning of the proof.

Since $I(H, G)$ is a $\mathbb{Q}[x]$-isolated subgroup of G,

$$\bigcap_{i=1}^{\infty} G^{\pi^i} I(H, G) = I(H, G)$$

by Theorem 1.3. Now,

$$g \notin \bigcap_{i=1}^{\infty} G^{\pi^i} I(H, G)$$

because $g \notin I(H, G)$ by hypothesis. Furthermore,

$$\bigcap_{i=1}^{\infty} G^{\pi^i} H \subseteq \bigcap_{i=1}^{\infty} G^{\pi^i} I(H, G)$$

since $H \leq_{\mathbb{Q}[x]} I(H, G)$. And so, there exists an integer j such that G/G^{π^j} is a nilpotent $\mathbb{Q}[x]$-powered group of finite π-type separating g and H. □

Bibliography

[1] Allenby RBJT, Gregorac RJ. Residual Properties of Nilpotent and Supersolvable Groups. J Algebra. 1972;23:565–73.

[2] Clement AE, Majewicz S, Zyman M. The Theory of Nilpotent Groups. Basel: Birkhauser; 2017.

[3] Baumslag G. Lecture Notes on Nilpotent Groups. CBMS Regional Conference Series. vol. 2. Providence: Amer. Math. Soc.; 1971.

[4] Blackburn N. Conjugacy in nilpotent groups. Proc Am Math Soc. 1965;16:143–8.

[5] Conrad K. http://www.math.uconn.edu/~kconrad/blurbs/linmultialg/tensorprod.pdf.

[6] Gruenberg KW. Residual properties of infinite soluble groups. Proc Lond Math Soc (3). 1957;7:29–62.

[7] Hall P. The Edmonton notes on nilpotent groups. Queen Mary College Mathematics Notes, Mathematics. College, London: Department, Queen Mary; 1969.

[8] Ivanova EA. On the Conjugacy Separability in the Class of Finite P-Groups of Finitely Generated Nilpotent Groups. arXiv:math.GR/0408393 v1 28 Aug 2004.

[9] Kargapolov MI, Remeslennikov VN, Romanovskii NS, Roman'kov VA, Churkin VA. Algorithmic problems for σ-power groups. J Algebra Logic. 1969;8(6):364–73. (Translated).

[10] Majewicz S. Nilpotent $\mathbb{Q}[x]$-powered groups. Amer Math Soc Contemporary Math. vol. 421. 2006.

[11] Majewicz S, Zyman M. Power-commutative nilpotent R-powered groups. Groups Complex Cryptol. 2009;1(2):297–309.

[12] Majewicz S, Zyman M. On the extraction of roots in exponential A-groups. Groups Geom Dyn. 2010;4:835–46.

[13] Majewicz S, Zyman M. On the extraction of roots in exponential A-groups II. Commun Algebra. 2012;40:64–86.

[14] Majewicz S, Zyman M. Localization of nilpotent R-powered groups. J Group Theory. 2012;15:119–35.

[15] Warfield RB Jr. Nilpotent Groups. Lecture Notes in Mathematics. vol. 513. Berlin: Springer; 1976.

Barry B. Mittag
Infinite nested radicals

Abstract: We prove that each positive integer n can be represented as an integer infinite nested radical and provide an easy method to determine the convergence. The method is straightforward and involves looking at integral quadratic forms. After we proved this, we discovered that the result appears in a somewhat different form in Ramanujan's Notebooks [1, 2]. However, our proof is so straightforward and the result so striking we think it should be published.

Keywords: Infinite nested radicals

MSC 2010: 11A99

1 Introduction

In this work, we will be concerned with simplifying radicals of the form

$$\sqrt{13+2\sqrt{15+3\sqrt{17+4\sqrt{19+5\sqrt{21+6\sqrt{23+\cdots}}}}}}.$$

The above radical will turn out to be 5. We will generalize this result and show that any natural number has an **infinite nested radical** (I. N. R.) representation.

The elements of our common number systems, the integers, the rationals, and the real numbers can be represented in a variety of ways. For example, a real number has a decimal expansion as well as an expansion as a continued fraction and a rational number has another representation within the p-adic numbers for each prime p. In this note, we look at representations of integers as I. N. R. s.

If $a_1, a_2, \ldots, a_n, \ldots$ is a sequence of real numbers, then an I. N. R. is a number

$$\sqrt{a_1+\sqrt{a_2+\sqrt{a_3+\sqrt{a_4+\cdots}}}}.$$

Putting aside for the time being questions of convergence, we say a real number $a \in \mathbb{R}$ is represented by an infinite nested radical if there exists a sequence (a_i) whose I. N. R. as defined above converges to a.

I. N. R. s play a role in several areas of mathematics. For example, if we start with the sequence $a_i = 1$ for all i, we get the infinite nested radical

$$\sqrt{1+\sqrt{1+\sqrt{1+\sqrt{1+\cdots}}}}.$$

Barry B. Mittag, 143 Federal St, Fairfield, CT 06825, USA, e-mail: barry.mittag@gmail.com

https://doi.org/10.1515/9783110638387-018

If we let x be the value of this, we must have $x = \sqrt{1+x}$, and hence $x^2 = 1+x$ and so $x = \frac{1+\sqrt{5}}{2}$, the golden section. What will be surprising is that we can start with integer sequences and get integer results, in fact, such an integer infinite nested radical for each integer.

We discovered this result by examining integer infinite nested radicals formed from certain quadratic forms. After discovering the striking main theorem, we found out that the result appears in a somewhat different form in Ramanujan's Notebooks [1, 2]. After giving the main theorem, we give some generalizations.

Our main result is the following: Given any natural number k, there exists an integer sequence $m_1, m_2, m_3 \ldots$, such that

$$k = \sqrt{m_1 + \sqrt{m_2 + \sqrt{m_3 + \cdots}}}.$$

For a given k, the elements m_i in this sequence can be found by examining various representations of the quadratic form $f(n) = n(n+k)$.

We give an example first and then generalize the example to give a proof of the theorem.

We first define the integer quadratic form $f(n) = n(n+4)$ and show how to construct an integer sequence from this that produces an infinite nested radical equal to 5.

2 Initial results

First, for $n \in \mathbb{N}$, we define

$$f(n) = n(n+4).$$

We note that in much of our work $n = 1$. We have:

$$f(n+1) = (n+1)(n+5) = n^2 + 6n + 5, \tag{1}$$

$$f(n+2) = (n+2)(n+6) = n^2 + 8n + 12, \tag{2}$$

$$f(n+3) = (n+3)(n+7) = n^2 + 10n + 21, \tag{3}$$

$$f(n+4) = (n+4)(n+8) = n^2 + 12n + 32, \tag{4}$$

$$\vdots$$

From the above:

$$f(n) = n(n+4) \tag{5}$$

$$= n\sqrt{(n+4)^2} \tag{6}$$

$$= n\sqrt{n^2 + 8n + 16} \tag{7}$$

$$= n\sqrt{2n + 11 + n^2 + 6n + 5} \tag{8}$$
$$= n\sqrt{2n + 11 + (n + 1)(n + 5)} \tag{9}$$
$$= n\sqrt{2n + 11 + f(n + 1)}. \tag{10}$$

We can continue in this vein:

$$f(n + 1) = (n + 1)(n + 5) \tag{11}$$
$$= (n + 1)\sqrt{(n + 5)^2} \tag{12}$$
$$= (n + 1)\sqrt{n^2 + 10n + 25} \tag{13}$$
$$= (n + 1)\sqrt{2(n + 1) + 11 + n^2 + 8n + 12} \tag{14}$$
$$= (n + 1)\sqrt{2(n + 1) + 11 + (n + 2)(n + 6)} \tag{15}$$
$$= (n + 1)\sqrt{2(n + 1) + 11 + f(n + 2)}. \tag{16}$$

Substituting (16) into (10), we get

$$f(n) = n\sqrt{2n + 11 + (n + 1)\sqrt{2(n + 1) + 11 + f(n + 2)}}.$$

If we expand $f(n + 2)$, we get

$$f(n + 2) = (n + 2)\sqrt{2(n + 2) + 11 + f(n + 3)},$$

so that

$$f(n) = n\sqrt{2n + 11 + (n + 1)\sqrt{2(n + 1) + 11 + (n + 2)\sqrt{2(n + 2) + 11 + f(n + 3)}}}.$$

Note, by induction this will continue in this manner forever

$$f(n + k) = (n + k)(n + k + 4) \tag{17}$$
$$= (n + k)\sqrt{(n + k + 4)^2} \tag{18}$$
$$= (n + k)\sqrt{(n + k)^2 + 8(n + k) + 16} \tag{19}$$
$$= (n + k)\sqrt{2(n + k) + 11 + (n + k)^2 + 6(n + k) + 5} \tag{20}$$
$$= (n + k)\sqrt{2(n + k) + 11 + (n + k + 1)(n + k + 5)} \tag{21}$$
$$= (n + k)\sqrt{2(n + k) + 11 + f(n + k + 1)}. \tag{22}$$

Thus continuing the substitution process we get the infinite nested radical:

$$f(n) = n\sqrt{2n + 11 + (n + 1)\sqrt{2(n + 1) + 11 + (n + 2)\sqrt{2(n + 2) + 11 + (n + 3)\sqrt{2(n + 3) + 11 + (n + 4)\sqrt{\cdots}}}}} \tag{23}$$

Now, by definition $f(1) = 1(1+4) = 5$ and by expansion

$$f(1) = \sqrt{13 + 2\sqrt{15 + 3\sqrt{17 + 4\sqrt{19 + 5\sqrt{21 + 6\sqrt{23 + \cdots}}}}}}. \quad (24)$$

Hence our particular I. N. R. is equal to 5. We can numerically compute various truncated nested radicals. The first 8 truncated radicals produce the following sequence of numbers (rounded to 4 decimal places) 3.6056, 4.5548, 4.8439, 4.9417, 4.9908, 4.9962, 4.9984, 4.9993.

3 Generalization

Now letting $f_{k+1}(n) = n(n+k)$ we can, in a similar manner get the following results:

$$\begin{aligned}
f_3(1) &= 1(1+2) = 3 = \sqrt{1 + 2\sqrt{1 + 3\sqrt{1 + 4\sqrt{1 + 5\sqrt{1 + \cdots}}}}},\\
f_4(1) &= 1(1+3) = 4 = \sqrt{6 + 2\sqrt{7 + 3\sqrt{8 + 4\sqrt{9 + 5\sqrt{10 + \cdots}}}}},\\
f_5(1) &= 1(1+4) = 5 = \sqrt{13 + 2\sqrt{15 + 3\sqrt{17 + 4\sqrt{19 + 5\sqrt{21 + \cdots}}}}},\\
f_6(1) &= 1(1+5) = 6 = \sqrt{22 + 2\sqrt{25 + 3\sqrt{28 + 4\sqrt{31 + 5\sqrt{34 + \cdots}}}}}. \quad (25)
\end{aligned}$$

The key observation that leads to these results is that for any $k \in \mathbb{N}$ we can generalize the result in (10) as follows:

$$\begin{aligned}
f_{k+1}(n) &= n(n+k) && (26)\\
&= n\sqrt{(n+k)^2} && (27)\\
&= n\sqrt{n^2 + 2kn + k^2} && (28)\\
&= n\sqrt{n^2 + (k - 2 + k + 2)n + (k^2 - k - 1) + k + 1} && (29)\\
&= n\sqrt{(k-2)n + (k^2 - k - 1) + n^2 + (k+2)n + (k+1)} && (30)\\
&= n\sqrt{(k-2)n + (k^2 - k - 1) + (n+1)(n+k+1)} && (31)\\
&= n\sqrt{(k-2)n + (k^2 - k - 1) + f_{k+1}(n+1)}. && (32)
\end{aligned}$$

So that the coefficient 2 and the constant term 11 in (10) are replaced by the coefficient $(k-2)$ and constant term (k^2-k-1). Writing a general nested radical:

$$f_{k+1}(n) = n\sqrt{(k-2)n + (k^2-k-1) + (n+1)\sqrt{(k-2)(n+1) + (k^2-k-1) + (n+2)\sqrt{\ldots}}},$$

and setting $n = 1$ we arrive at

$$k+1 = f_{k+1}(1) = \sqrt{(k^2-3) + 2\sqrt{(k^2-3) + (k-2) + 3\sqrt{(k^2-3) + 2(k-2) + 4\sqrt{\ldots}}}}.$$

With this, we can now easily write any integer as an infinite nested radical. For example, to write 20 as a nested radical let $k = 19$ so that

$$k^2 - 3 = 358,$$

$$k - 2 = 17$$

and

$$20 = f_{20}(1) \tag{33}$$

$$= \sqrt{358 + f_{20}(2)} \tag{34}$$

$$= \sqrt{358 + 2\sqrt{375 + f_{20}(3)}} \tag{35}$$

$$= \sqrt{358 + 2\sqrt{375 + 3\sqrt{392 + f_{20}(4)}}} \tag{36}$$

$$= \sqrt{358 + 2\sqrt{375 + 3\sqrt{392 + 4\sqrt{409 + f_{20}(5)}}}} \tag{37}$$

$$= \sqrt{358 + 2\sqrt{375 + 3\sqrt{392 + 4\sqrt{409 + 5\sqrt{426 + \cdots}}}}}. \tag{38}$$

Moreover, each subsequent term of the nested radical is determined by a closed-form formula

$$i\sqrt{(k^2-3) + (k-2)(i-1) + (i+1)\sqrt{\ldots}}, \tag{39}$$

where i is determined by the cluster level. For instance, the 50th term of the radical for 20 is

$$50\sqrt{358 + 17 \cdot 49 + 51\sqrt{\ldots}} = 50\sqrt{1191 + 51\sqrt{\ldots}}. \tag{40}$$

It would be worthwhile to consider allowing n to vary. This is the subject of future work. We conclude the paper with the exploration of how the truncated value and relative error behaves for various integers. To obtain the results, we use 8 terms in the truncation. Note that as the integer increases the relative error decreases. This is particularly interesting since the number of terms used in each truncation was kept fixed. The results are presented in Table 1.

Table 1: Numerical Results.

Integer	Truncated Nested Radical	Relative Error
3	2.962723005	0.0124256650
4	3.990708050	0.0023229875
5	4.996759071	0.0006481858
6	5.998643512	0.0002260813
10	9.999892377	0.0000107623
15	14.99998706	0.0000008627
20	19.99999726	0.0000001370
101	101.000000000	0.0000000000

Bibliography

[1] Ramanujan S. Notebooks (2 volumes). Bombay: Tata Institute of Fundamental Research; 1957.
[2] Ramanujan S. The Lost Notebook and Other Unpublished Papers. New Delhi: Narosa; 1988.

Mohammad Reza R. Moghaddam, Gerhard Rosenberger, and Mohammad Amin Rostamyari

Commutative transitivity property in groups and Lie algebras

Abstract: In 2007, a general notion of χ-transitive groups was introduced by C. Delizia, P. Moravec, and C. Nicotera, where χ is a class of groups. In 2013, L. Ciobanu, B. Fine, and G. Rosenberger studied the relationship among the notions of conjugately separated abelian, commutative transitive (CT) and fully residually χ-groups.

We introduce and discuss the concept of n-commutative transitive groups (n-CT), and under some conditions, it is shown the three notions n-commutative transitivity, conjugately separated n-central, and fully residually χ-groups are equivalent. Also the notion of 2-Engel transitive group (2-ET) will be discussed and give its relationship with conjugately separated 2-Engel group and fully residually χ-groups.

Finally, the concept of commutative transitive Lie algebras are also discussed.

Keywords: CT groups, 2-ET groups, CT Lie algebras

MSC 2010: Primary 20E05, 20E08, 15A27, Secondary 20F70, 16U80

1 Introduction and preliminaries

Residual properties have played a major role in infinite group theory. Let χ be a class of groups. Then a group G is *residually* χ if given any nontrivial element $g \in G$ there is a homomorphism $\phi : G \to H$, where H is a group in χ such that $\phi(g) \neq 1$. A group G is said to be *fully residually* χ if given finitely many nontrivial elements $g_1, \dots, g_n$ in G there is a homomorphism $\phi : G \to H$, where H is a group in χ such that $\phi(g_i) \neq 1$ for all $i = 1, \dots, n$. Fully residually free groups have played a crucial role in the study of equations and first-order formulas over free groups.

Definition 1.1. A subgroup H of a group G is called conjugately separated, if $H \cap H^x = 1$, for all $x \in G \setminus H$.

Note: In Memoriam: Gilbert Baumslag.

Mohammad Reza R. Moghaddam, Department of Mathematics, Khayyam University, Mashhad, Iran; and Department of Pure Mathematics, Centre of Excellence in Analysis on Algebraic Structures (CEAAS), Ferdowsi University of Mashhad, P. O. Box 1159, Mashhad, 91775, Iran, e-mail: rezam@ferdowsi.um.ac.ir
Gerhard Rosenberger, Department of Mathematics, Uni-Hamburg, 20146 Hamburg, Germany, e-mail: gerhard.rosenberger@math.uni-hamburg.de
Mohammad Amin Rostamyari, Department of Mathematics, Khayyam University, Mashhad, Iran, e-mail: m.a.rostamyari@khayyam.ac.ir

https://doi.org/10.1515/9783110638387-019

It is clear that the intersection of a family of conjugately separated subgroups is again conjugately separated.

Definition 1.2. Given a nonzero integer n, a group G is said to be *n-central* if $[x, y^n] = 1$, for all $x, y \in G$. Hence n-central property is equivalent to $G/Z(G)$ having finite exponent dividing n.

Here, $[x, y] = x^{-1}y^{-1}xy$ is the usual commutator of the elements x and y of the group G and $x^y = y^{-1}xy$ is the conjugate of x by y.

Also a group G is called *conjugately separated n-central* or CSC^n-group, if all of its maximal n-central subgroups are conjugately separated.

A group G is said to be a CSA-group, if all of its maximal abelian subgroups are conjugately separated and it is commutative transitive or CT group, if commutativity is transitive on the set of nontrivial elements of G. L. Ciobanu et al. [3] examined the relationships among the classes of non-abelian CSA, CT and fully residually χ-groups.

We will extend the definition of CT groups and introduce the notion of *n-commutative transitive* groups, then we study its relationship with CSC^n and fully residually χ-groups.

Definition 1.3. A group G is *n-commutative transitive* (henceforth n-CT), if $[x, y^n] = 1$ and $[y, z^n] = 1$ imply that $[x, z^n] = 1$, for any nontrivial elements x, y, z in G and $n \geq 1$.

Clearly n-CT groups are the usual CT groups, for $n = 1$. Also 1-central groups are abelian and CSC^1 groups are CSA.

We remark that B. Fine, A. M. Gaglione, A. G. Myasnikov, and D. Spellman [4] introduced a different concept of n-commutative transitivity. They call a group G commutative transitive of level n, if centralizers of elements not in $Z_n(G)$, and the nth-term of the upper central series of G, are abelian.

In [1], Benjamin Baumslag introduced the notion of fully residually free groups and proved that this property is equivalent to being residually free and commutative transitive.

We show that under some conditions the three notions n-commutative transitivity, CSC^n, and fully residually χ-groups are equivalent for many important classes of groups, including those of free products of cyclic groups not containing the infinite dihedral group, torsion-free hyperbolic groups and one-relator groups with only odd torsion.

So, the main theorem of this section reads as follows.

Theorem 1.4. *Let χ be a class of groups such that each of its n-noncentral group is* CSC^n *and G be n-noncentral residually χ-group. Then the following statements are equivalent:*

(i) *G is fully residually χ;*
(ii) *G is* CSC^n*;*
(iii) *G is n-CT.*

Proof. See [8], Theorem 2.1. □

Corollary 1.5. *If G is a residually free group. Then G is fully residually free if and only if G is n-commutative transitive.*

2 2-Engel transitive groups

A group G is called *conjugately separated* 2-*Engel* group (henceforth CSE^2-group), if all of its maximal 2-Engel subgroups are conjugately separated.

In the following, we discuss the notion of 2-Engel transitive group and then give its relationship with CSE^2-group and fully residually χ-groups.

Definition 2.1.

(a) A group G is 2-*Engel transitive* (henceforth 2-ET), whenever $[x,y,y] = 1$ and $[y,z,z] = 1$ imply that $[x,z,z] = 1$, for every nontrivial elements x,y,z in G.

(b) For any given element x of G, we call

$$E_G^2(x) = \{y \in G : [x,y,y] = 1, [y,x,x] = 1\}$$

to be the set of 2-*Engelizer* of x in G. The family of all 2-Engelizers in G is denoted by $E^2(G)$ and $|E^2(G)|$ denotes the number of distinct 2-Engelizers in G.

As an example consider $Q_{16} = \langle a,b : a^8 = 1, a^4 = b^4, b^{-1}ab = a^{-1}\rangle$, the Quaternion group of order 16 and take the element b in Q_{16}. Then one can easily check that the 2-Engelizer set of b is as follows:

$$E_{Q_{16}}^2(b) = \{1, a^2, a^4, a^6, b, a^2b, a^4b, a^6b\}.$$

In general, the 2-Engelizer of each nontrivial element of an arbitrary group G does not form a subgroup. The following example shows our claim.

Example 2.2. Let G be a finitely presented group of the following form:

$$\begin{aligned} G = \langle a_1, a_2, a_3, a_4 : a_3^3 = a_4^3 = 1,\ [a_1,a_2] = 1,\ [a_1,a_3] = a_4, \\ [a_1,a_4] = 1,\ [a_2,a_3] = 1,\ [a_2,a_4] = a_2,\ [a_3,a_4] = 1\rangle. \end{aligned}$$

It is clear that G is an infinite group. One can easily check that G is not 2-ET, as $[a_2,a_1,a_1] = 1$ and $[a_1,a_4,a_4] = 1$, while $[a_2,a_4,a_4] = a_2$. Moreover, $E_G^2(a_1)$ is not a subgroup of G, since it is easily calculated that $a_2, a_3 \in E_G^2(a_1)$ while $a_2a_3 \notin E_G^2(a_1)$.

Now, we give a condition under which the 2-Engelizer of each nontrivial element of a group G forms a subgroup.

Theorem 2.3. *Let G be an arbitrary group, then the set of each 2-Engelizer of a nontrivial element in G forms a subgroup if and only if the group $x^{E_G^2(x)}$ is abelian for all nontrivial elements x of G.*

Proof. See [9], Theorem 2.5. □

Baumslag's theorem is also true in the case of 2-Engel transitive groups (see [9] for more information).

Theorem 2.4. *Let G be a residually free group. Then G is fully residually free if and only if G is 2-Engel transitive.*

Proof. See [9], Theorem 2.12. □

3 Commutative transitive Lie algebras

In this section, we study Lie algebras L, in which every centralizer of non-zero elements of L is abelian. Such Lie algebras are equivalent to commutative transitive Lie algebras (see Lemma 3.2 below). The concept of commutative transitive groups was first introduced and studied by Weisner [11] in 1925.

Definition 3.1. A Lie algebra L is *commutative transitive* (henceforth CT), if $[x,y] = 0$ and $[y,z] = 0$ imply that $[x,z] = 0$, for any nonzero elements x, y, z in L.

The property of CT is clearly subalgebra closed, yet it is not quotient closed, as every free Lie algebra is CT (see [6], Example 4.4 for more detail).

The Frattini subalgebra $\Phi(L)$ of a Lie algebra L, is the intersection of all maximal subalgebras of L or it is L itself, when L has no maximal subalgebras (see also [7]).

Here, we study the concept of commutative transitive Lie algebras and among other results, their relationships with fully residually free Lie algebras are established.

Here, we give some basic notion and then state our main results of this section.

Lemma 3.2. *For any Lie algebra L, the following statements are equivalent:*
(i) *L is CT Lie algebra;*
(ii) *the centralizers of nonzero elements of L are abelian.*

Proof. (i) ⇒ (ii) Let L be a CT Lie algebra. For any nonzero element $x \in L$, if $y, z \in C_L(x)$, we have $[y,x] = 0$ and $[x,z] = 0$. The definition of CT implies that $[y,z] = 0$. Hence $C_L(x)$ is abelian.

(ii) ⇒ (i) Assume x, y, z are nonzero elements of L, with $[x,y] = 0$ and $[y,z] = 0$. Obviously, $x, z \in C_L(y)$, by the assumption $C_L(y)$ is abelian and hence $[x,z] = 0$. Thus L is commutative transitive. □

The proof of the following lemma is a routine argument by using Zorn's lemma.

Lemma 3.3. *Every abelian subalgebra K of a given Lie algebra L is contained in a maximal abelian subalgebra.*

The following fact is needed for proving our main results.

Proposition 3.4. *Let L be a non-abelian Lie algebra with $\Phi(L) \neq 0$. Then L has only one abelian maximal subalgebra.*

Proof. Let L be a non-abelian Lie algebra with nonzero Frattini subalgebra, $\Phi(L)$. Without loss of generality, we may assume that M_1 and M_2 are maximal abelian subalgebras with $M_1 \neq M_2$. Assume there exists an element $m_1 \in M_1 \setminus M_2$, then clearly

$$M_2 \subseteq M_2 \oplus \langle m_1 \rangle \subseteq L.$$

If $M_2 \oplus \langle m_1 \rangle = L$, then L is abelian which contradicts our assumption. Hence m_1 must be in M_2 and so $M_1 = M_2$. □

Using the above proposition and Lemma 3.3, we obtain the following useful result.

Theorem 3.5 ([10], Theorem 3.5). *Every non-abelian Lie algebra L with non-zero Frattini subalgebra is CT.*

In 2010, Klep and Moravec [6] classified all finite dimensional commutative transitive Lie algebras over an algebraically closed field of characteristic 0. They proved that these Lie algebras are either simple or soluble, where the only simple such Lie algebra is sl_2. Also, they showed that in the soluble case, Lie algebras are either abelian or a one-dimensional split extension of abelian Lie algebra (see [6] for more information).

Now, using Theorem 3.5 one can easily see that every non-abelian Lie algebra with $\Phi(L) \neq 0$ is either simple or soluble. One notes that all the results on CT Lie algebras in [6], carried out the assumption of nontriviality of Frattini subalgebras.

In the following, we focus on non-abelian CT Lie algebras and give some structural results.

Theorem 3.6. *The center of a non-abelian CT Lie algebra is trivial.*

Proof. Assume L is a non-abelian Lie algebra with nonzero center and z is a nonzero element in $Z(L)$. Clearly, for nonzero elements, $x, y \in L$,

$$[x, z] = 0, \quad [z, y] = 0,$$

then the definition of CT Lie algebras implies that $[x, y] = 0$. Hence L is abelian Lie algebra and this contradiction gives the result. □

A *derivation* of a Lie algebra L over a fixed field F is an F-bilinear transformation $d : L \longrightarrow L$ such that

$$d([x, y]) = [d(x), y] + [x, d(y)],$$

for all $x, y \in L$. We denote by $\mathrm{Der}(L)$ the vector space of derivations of L, which forms a Lie algebra with respect to the bracket of linear transformations, called the *derivation algebra* of L. Clearly, the space

$$\mathrm{ad}_L = \{\mathrm{ad}_x \mid x \in L\}$$

of *inner derivations* is an ideal of $\mathrm{Der}(L)$.

Theorem 3.7. *Let L be a non-abelian CT Lie algebra, then* $Z(\mathrm{Der}(L)) = 0$.

Proof. See [10], Theorem 3.7. □

Let χ be a class of Lie algebras. Then a Lie algebra L is *residually* χ if for every nonzero element $x \in L$, there exists a homomorphism $\phi : L \to K$, where K is a χ-Lie algebra such that $\phi(x) \neq 0$. Also a Lie algebra L is *fully residually* χ, if for finitely many nonzero elements $x_1, \ldots, x_n$ in L there exists a homomorphism $\phi : L \to K$, where K is a χ-Lie algebra such that $\phi(x_i) \neq 0$, for all $i = 1, \ldots, n$.

The following technical lemma is useful to give our final result.

Lemma 3.8 (Bokut and Kukin [2], Lemma 4.16.2). *A Lie algebra L is fully residually free if and only if, for every two linearly independent elements* x_1 *and* x_2 *in L, there exists a homomorphism* ϕ *from the Lie algebra L into a free Lie algebra* $\mathcal{F}$ *such that the elements* $\phi(x_1)$ *and* $\phi(x_2)$ *are linearly independent in* $\mathcal{F}$.

Now, using the above lemma we give the following result concerning free Lie algebras.

Theorem 3.9. *Let L be a residually free Lie algebra. Then L is fully residually free, if and only if L is CT.*

Proof. Without loss of generality, we may assume that L is non-abelian Lie algebra, and so assume L is a non-abelian residually free CT Lie algebra over a fixed field F. Then we show that for given nonzero linearly independent elements x_1 and x_2 in L, there exists a homomorphism $\phi : L \to \mathcal{F}$ such that $\mathcal{F}$ is a free Lie algebra, and $\phi(x_1)$ and $\phi(x_2)$ are linearly independent. Hence Lemma 3.8 implies that L is fully residually free.

For every nonzero element x_1 in L, there exits a homomorphism $\phi : L \to \mathcal{F}$ such that $\phi(x_1) \neq 0$, as by the assumption L is residually free Lie algebra. On the other hand, Theorem 3.6 implies that $Z(L) = 0$. Hence $[x_1, x_2] \neq 0$, for some x_2 in L. So x_1 and x_2 are linearly independent in L. Clearly, $[\phi(x_1), \phi(x_2)] \neq 0$, as $\mathcal{F}$ is free Lie algebra. Then $\phi(x_1)$ and $\phi(x_2)$ are linearly independent, and hence L is fully residually free.

Conversely, let L be a fully residually free Lie algebra such that $[x_1, x_2] = 0$ and $[x_2, x_3] = 0$, for any nonzero elements x_1, x_2, and x_3 in L. Assume that L is not CT and $x_4 = [x_1, x_3] \neq 0$, then there exits a homomorphism $\phi : L \to \mathcal{F}$, where $\mathcal{F}$ is a free Lie algebra and $\phi(x_i) \neq 0$ for $i = 1, 2, 3, 4$, as by the assumption L is fully residually free Lie algebra. Hence, $\phi(x_4) = [\phi(x_1), \phi(x_3)] \neq 0$. Now, to prove our claim it is enough to show that either $[\phi(x_1), \phi(x_2)] \neq 0$ or $[\phi(x_2), \phi(x_3)] \neq 0$. But both of which contradict the assumptions $[x_1, x_2] = 0$ and $[x_2, x_3] = 0$, respectively. Thus $[x_1, x_3] = 0$ and L is CT Lie algebra. □

Analogously, as in the case of residually free groups we may embed non-abelian residually free Lie algebras in ultrapowers of free Lie algebras (nonstandard free Lie algebras).

If we consider the arguments in chapter 6 of [5], then we get that a non-abelian residually free Lie algebra is fully residually free if and only if it is universally free.

Bibliography

[1] Baumslag B. Residually free groups. Proc Lond Math Soc. 1967;17(3):402–18.

[2] Bokut LA, Kukin GP. Algorithmic and Combinatorial Algebra. Norwell: Kluwer Academic Publisher; 1994.

[3] Ciobanu L, Fine B, Rosenberger G. Classes of groups generalizing a theorem of Benjamin Baumslag. Commun Algebra. 2016;44(2):656–67.

[4] Fine B, Gaglione AM, Myasnikov AG, Spellman D. Discriminating Groups. J Group Theory. 2001;4:463–74.

[5] Fine B, Gaglione AM, Myasnikov AG, Rosenberger G, Spellman D. The Elementary Theory of Groups: A Guide through the proofs of the Tarski Conjectures. 1st ed., Berlin: De Gruyter; 2014.

[6] Klep I, Moravec P. Lie algebras with abelian centralizers. Algebra Colloq. 2010;17(4):629–36.

[7] Marshall E. The Frattini subalgebra of a Lie algebra. J Lond Math Soc. 1967;42:416–22.

[8] Moghaddam MRR, Rosenberger G, Rostamyari MA. Some properties of n-commutative transitive groups. Submitted.

[9] Moghaddam MRR, Rostamyari MA. 2-Engelizer subgroup of a 2-Engel transitive groups. Bull Korean Math Soc. 2016;53(3):657–65.

[10] Saffarnia S, Moghaddam MRR, Rostamyari MA. Centralizers in Lie algebras. Indian J Pure Appl Math. 2018;49(1):39–49.

[11] Weisner L. Groups in which the normaliser of every element except identity is abelian. Bull Am Math Soc. 1925;31:413–6.

Leonard Wienke

Simplicial subdivisions and the chromatic number of a group

Abstract: We give an overview on simplicial subdivision theory and its role in current research. We then specialize to chromatic subdivisions and define the notion of chromatic groups. We state conjectures and open problems concerning the topics involved.

Keywords: Combinatorial algebraic topology, stellar subdivision, chromatic subdivision, PL topology, fundamental group, distributed computing

MSC 2010: Primary 05C15, 05E45, 57Q25, Secondary 57M05, 68Q85

Besides its role in classical topology, simplicial and in particular stellar subdivision theory has many applications, for instance, in mathematical physics (Turaev–Viro invariants, PL manifolds), in theoretical computer science (distributed protocol complexes) as well as open problems to the theory itself. In this short note, we give an overview on the topic and on our research motivated by so-called protocol complexes in the distributed computing context as well as on the structure theory of subdivisions. We end with an application to combinatorial group theory.

1 Simplicial subdivisions

To set the board, we recall some concepts from combinatorial algebraic topology and fix notations. If $v_0, v_1, \dots, v_n \in \mathbb{R}^d$, $n \leq d$, are affinely independent, we call the convex hull $\sigma = \operatorname{conv}(v_0, v_1, \dots, v_n)$ an n-dimensional simplex or just an *n-simplex* with *vertices* $v_0, v_1, \dots, v_n$. We let $\Delta^n = \operatorname{conv}(e_1, e_2, \dots, e_{n+1})$ denote the *n-dimensional standard simplex*. For any $S \subseteq \{0, 1, \dots, n\}$, we call $\tau = \operatorname{conv}(\{v_s \mid s \in S\})$ a $(|S| - 1)$-*face* of σ. A *geometric simplicial complex* K (in $\mathbb{R}^d$) is then a finite collection of simplices such that any face of $\sigma \in K$ is also in K and for any $\sigma, \tau \in K$ their intersection $\sigma \cap \tau$ is a face of each of them. We let $V(K)$ denote the set of vertices of K. In contrast, an *abstract simplicial complex* is a finite set $V(A)$ together with a collection A of subsets of $V(A)$ such that for all $X \in A$ and $Y \subseteq X$ also $Y \in A$. We adopt the custom to demand

Note: Dedicated to Ben Fine on the occasion of his 70th birthday.
Acknowledgement: The author would like to thank Gerhard Rosenberger for inspiring this note. He is also grateful to Dmitry Feichtner-Kozlov for his helpful comments.

Leonard Wienke, Institute for Algebra, Geometry, Topology and their Applications, Department of Mathematics and Computer Science, University of Bremen, MZH, Bibliothekstraße 5, 28359 Bremen, Germany, e-mail: wienke@uni-bremen.de

https://doi.org/10.1515/9783110638387-020

that $a \in A$ for all $a \in V(A)$ and call $V(A)$ the set vertices of A. An element $\sigma \in A$ has *dimension* $(|\sigma| - 1)$ and is called a $(|\sigma| - 1)$-*simplex*. A k-*face* of σ is a subset $\tau \subseteq \sigma$ of dimension k. For a simplicial complex L (geometric or abstract), we set $L(n)$ to be the subcomplex of simplices of dimension $\leq n$, the n-*skeleton*. We will use the short-hand notation $v_1 \dots v_n$ to denote the simplex $\{v_1, \dots, v_n\}$ in the abstract or $\operatorname{conv}(v_1, \dots, v_n)$ in the geometric setting.

A main motivation for abstract simplicial complexes in combinatorial algebraic topology is the observation that notions of interest such as homology and fundamental groups do not depend on the actual geometric embedding of the simplicial complex but only on the combinatorial incidence structure of simplices which suits algorithmic computations. The information on the geometric embedding is exactly what we forget going from the geometric to the abstract setting.

Given a geometric simplicial complex K, we can associate to it an abstract simplicial complex $U(K) = A$ with the same vertices as K, that is $V(A) = V(K)$, and a simplex $\{v_0, v_1, \dots, v_n\}$ for each $\operatorname{conv}(v_0, v_1, \dots, v_n) \in K$. Vice versa, there are many possibilities to geometrically realize an abstract simplicial complex A, that is, to find a K such that $U(K) = A$. A simple idea is to embed the vertices in general position. More explicitly, if $V(A) = \{v_0, v_1, \dots, v_d\}$ take Δ^d and let K be its subcomplex according to the simplices of A. Already the basic example of Δ^2 shows that this embedding dimension is not minimal. In general, any d-dimensional simplicial complex A can be emdedded into $\mathbb{R}^{2d+1}$ and this bound is sharp in the sense that there exist d-dimensional complexes that cannot be embedded into $\mathbb{R}^{2d}$. The characterization of simplicial complexes that embed in a certain dimension and the hardness of these questions is a recent field of research.

To each geometric complex K, let $|K| = \bigcup_{\sigma \in K} \sigma$ be its *polyhedron* which can be topologized with the usual subspace topology of $\mathbb{R}^d$. For any abstract simplicial complex A, we define $|A|$ to be the polyhedron of any geometric simplicial complex K with $U(K) = A$. As a fact, all such K will be simplicially isomorphic and hence all polyhedra of A are piecewise linear (PL) isomorphic.

We recall basic operations on simplicial complexes: For a simplicial complex L, we denote by $\operatorname{star}_L(\sigma) = \{\tau \in L \mid \sigma \cup \tau \in L\}$ the *(closed) star of σ (in L)* which is again a simplicial complex. For example, the star of a maximal simplex α contains exactly α itself and its subsets. We define the *link of σ (in L)* to be $\operatorname{link}_L(\sigma) = \{\tau \in L \mid \sigma \cap \tau = \emptyset, \sigma \cup \tau \in L\}$ which is the simplicial subcomplex of L that consists of all simplices of $\operatorname{star}_L(\sigma)$ that do not have common vertices with σ. Hence, the link of a maximal simplex is empty. See Figure 1 for a further example of a star and a link. Finally, for simplicial complexes M and N with disjoint vertex sets, we define their join to be the simplicial complex $M * N$ with vertices $V(M * N) = V(M) \cup V(N)$ and simplices $\alpha \cup \beta$ for all $\alpha \in M$ and $\beta \in N$. We adopt the shorthand notation $a * N$ for $\{\emptyset, a\} * N$ for a simplicial complex that just consists of one vertex and the empty set. We remark that such a join of a simplicial complex L with a point is its *cone*, $\operatorname{cone}(L)$, and the join with the simplicial complex S^0 consisting of two isolated vertices is its *suspension*, susp(L). All three concepts are compared by the simplicial isomorphism $\text{link}_L(\sigma) * \sigma \cong \text{star}_L(\sigma)$.

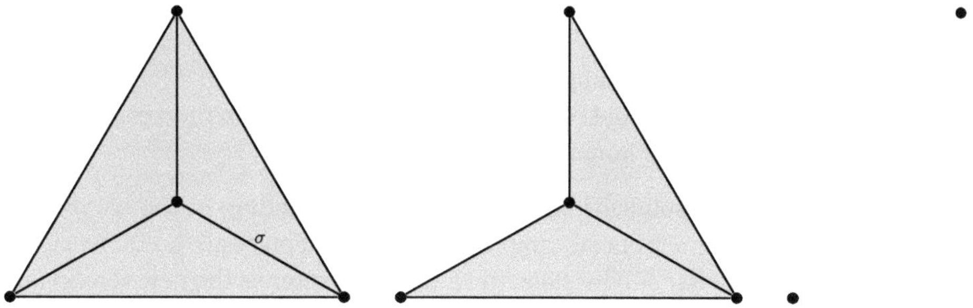

Figure 1: The edge σ, its star and its link.

Another notion from topology that is of importance for this note is that of a simplicial subdivision. The intuition behind that is to subdivide k-dimensional simplices into smaller k-dimensional simplices in a coherent way.

Definition 1.1. Let K_1 and K_2 be geometric simplicial complexes. Then K_2 is a subdivision of K_1 if $|K_1| = |K_2|$ and each simplex of K_1 is the union of finitely many simplices of K_2.

Subdivisions are of particular interest for PL topology where two polyhedra $|K_1|$ and $|K_2|$ are PL isomorphic if and only if K_1 and K_2 have simplicially isomorphic subdivisions. The complexes are then said to be *combinatorially equivalent*. A stronger version of this statement will be recalled at the end of this section with Theorem 1.4. However, one should remark at this point that being PL isomorphic, i. e. sharing a common subdivision, is a stronger notion for complexes than being homeomorphic. The so-called hauptvermutung of combinatorial topology conjectured that homeomorphic complexes are combinatorially equivalent but [6] provides a famous counterexample. For further details on this historic incident see also [7].

For abstract simplicial complexes A_1 and A_2, we define A_2 to be a subdivision of A_1 if there are geometric simplicial complexes K_1 and K_2 such that $U(K_1) = A_1$, $U(K_2) = A_2$ and K_2 subdivides K_1. Although it is possible to further characterize the subdivision property for abstract simplicial complexes, it is an open problem to get rid of the geometric embedding and to obtain a purely combinatorial characterization of subdivisions. This situation imposes the need to find sophisticated strategies in order to prove the subdivision property in the abstract setting. We proceed here with basic examples of subdivisions.

Example 1.2 (Stellar subdivision). The intuition for the stellar subdivision of a maximal simplex σ in a simplicial complex L is to replace its star by a new vertex x and cone over the boundary of σ with apex x. For nonmaximal simplices, one also has to consider the link of σ. The procedure can uniformly be described for geometric and abstract simplicial complexes by the formula

$$\mathrm{stel}_{L,x}(\sigma) = (L \setminus \mathrm{star}_L(\sigma)) \cup (x * \partial\sigma * \mathrm{link}_L(\sigma))$$

The next example has many applications in classical topology. It is for instance an important ingredient in the proof of the simplicial approximation theorem that compares simplicial and singular homology.

Example 1.3 (Barycentric subdivision). A suitable understanding of the barycentric subdivision of geometric simplicial complexes is that we replace all top-dimensional simplices with their stellar subdivisions (take the barycenter as the new vertex) and proceed for lower dimensions in the same fashion. The abstract definition is as follows: For an abstract simplicial complex A, we define its barycentric subdivision $\mathrm{bary}(A)$ to be $\mathrm{bary}(A) = \{\{\sigma_1, \ldots, \sigma_t\} \mid \sigma_1 \supset \sigma_2 \supset \cdots \supset \sigma_t, \sigma_i \in A \setminus \{\emptyset\}, t \geq 1\} \cup \{\emptyset\}$.

We would like to point out that the barycentric subdivision has an instructive interpretation as a functor: Send the abstract simplicial complex A to its poset of simplices partially ordered by inclusion and map this poset back to the abstract simplicial complex whose simplices consist of the flags in the poset. By construction, the barycentric subdivision is a sequence of stellar subdivisions and Figure 2 shows the subdivision process for the standard simplex Δ^2. For a general proof see [4].

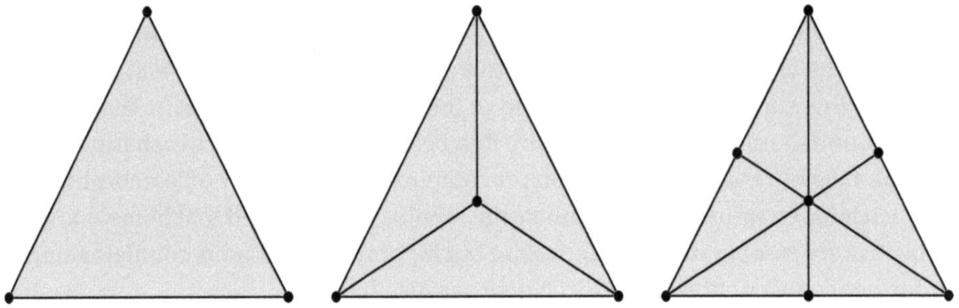

Figure 2: The barycentric subdivision of Δ^2 as a sequence of stellar ones.

We end this section with a famous theorem from [1] to which we will refer again later.

Theorem 1.4 (Alexander). *Let K_1 and K_2 be n-dimensional simplicial complexes such that $|K_1|$ and $|K_2|$ are piecewise linearly isomorphic. Then K_1 and K_2 can be transformed into each other by stellar subdivisions and their inverses.*

We mention here the classic and still unsolved conjecture by Alexander that the simplicially isomorphic subdivisions of PL isomorphic complexes can be obtained by a sequence of stellar subdivisions without using inverses.

2 Chromatic subdivisions

Subdivisions have recently become important in the context of distributed computing where protocol complexes model the possible execution threads of processes performing a distributed protocol; see [3] for details. In particular, the notion of chromatic complexes comes natural as the color represents a process number and multiple executions of a distributed protocol yield chromatic subdivisions of the protocol complex.

Definition 2.1. An *m-labeling* of a simplicial complex L is a map $h\colon V(L) \to C$ where C is a set of cardinality m. The map h is called an *m-coloring* if it is a coloring of the underlying graph of L. We then write $h = \chi$. A simplicial complex together with an m-coloring (L,χ) is called an m-chromatic complex. If (L_1,χ_1) and (L_2,χ_2) are m-chromatic complexes where L_2 is a subdivision of L_1 then L_2 is called an *m-chromatic subdivision* of L_1 if $\chi_2(v) = \chi_1(v)$ for all $v \in V(L_1)$.

We recall Examples 1.2 and 1.3 and see that the underlying graph of $\mathrm{stel}_{\Delta^2,x}(\Delta^2)$ is the complete graph K_4 on four vertices, and hence $\mathrm{stel}_{\Delta^2,x}(\Delta^2)$ cannot obtain a 3-coloring. Albeit $\mathrm{bary}(\Delta^2)$ is 3-colorable as a graph it is not a chromatic subdivision: If we fix the colors of the vertices of Δ^2, the barycenter in the top-dimension being adjacent to all three of them requires a fourth color. Hence both subdivisions are not chromatic. In general, neither the stellar nor barycentric subdivision of a simplex will be chromatic. For the stellar subdivision this follows from the construction via coning, and inductively for the barycentric subdivision.

In this section, we will demonstrate how to make the stellar subdivisions of Δ^1 and Δ^2 chromatic in a minimal way with respect to adding new vertices. We will obtain the well-known Schlegel diagrams of d-cross polytopes, see Definition 2.2 or equivalently the basic chromatic subdivision popular in distributed computing. In the following, let E be the simplicial complex that consists of the edge ab and its subsets together with the coloring $\chi_E(a) = 1$ and $\chi_E(b) = 2$. We calculate

$$\begin{aligned} E' = \mathrm{stel}_{E,d}(ab) &= E \setminus \mathrm{star}_E(ab) \cup (d * \partial ab * \mathrm{link}_E(ab)) \\ &= \emptyset \cup (\{\emptyset, d\} * \{\emptyset, a, b\} * \{\emptyset\}) \\ &= \{\emptyset, a, b, d, ad, bd\} \end{aligned}$$

and further stellar subdivide the edge bd,

$$\begin{aligned} E'' = \mathrm{stel}_{E',e}(bd) &= E' \setminus \mathrm{star}_{E'}(bd) \cup (e * \partial bd * \mathrm{link}_{E'}(bd)) \\ &= \{a, ad\} \cup (\{\emptyset, e\} * \{\emptyset, b, d\} * \{\emptyset\}) \\ &= \{\emptyset, a, b, d, e, ad, be, de\}. \end{aligned}$$

As discussed above, E' is not 2-colorable. However, setting $\chi_{E''}(a) = \chi_{E''}(e) = 1$ and $\chi_{E''}(d) = \chi_{E''}(b) = 2$ constitutes a 2-coloring for E'' that restricts to the coloring of E. Hence E'' is a chromatic subdivision of E.

As an aside, let $K = \text{cone}(E)$ be the simplicial complex that consists of the 2-simplex abc and its subsets together with the coloring $\chi_K(a) = 1$, $\chi_K(b) = 2$ and $\chi_K(c) = 3$. We just record the results of stellar subdividing the edge $ab \in K$ twice:

$$\tilde{K} = \{\emptyset, a, b, c, d, e, ac, ad, bc, be, cd, ce, de, acd, cde, bce\}.$$

We note that $\tilde{K}$ is the chromatic subdivision that equals the cone over E' - a standard fact of subdivision theory. We now calculate the top-dimensional stellar subdivision K' of K:

$$\begin{aligned}\text{stel}_{K,d}(abc) &= K \setminus \text{star}_K(abc) \cup (d * \partial abc * \text{link}_K(abc))\\ &= \emptyset \cup (\{\emptyset, d\} * \{\emptyset, a, b, c, ab, ac, bc\} * \{\emptyset\})\\ &= \{\emptyset, a, b, c, d, ab, ac, ad, bc, bd, cd, abd, acd, bcd\}\end{aligned}$$

and further stellar subdivide the edge bd to obtain K'' as

$$\begin{aligned}\text{stel}_{K',e}(bd) &= K' \setminus \text{star}_{K'}(bd) \cup (e * \partial bd * \text{link}_K(bd))\\ &= \{ac, acd\} \cup (\{\emptyset, e\} * \{\emptyset, b, d\} * \{\emptyset, a, c\})\\ &= \{\emptyset, a, b, c, d, e, ab, ac, ad, ae, bc, be, cd, ce, de, abe, acd, ade, bce, cde\}\end{aligned}$$

and finally stellar subdivide the edge ae to obtain $K''' = \text{stel}_{K'',f}(ae)$ as

$$\{\emptyset, a, \ldots, f, ab, ac, ad, af, bc, be, bf, cd, ce, de, df, ef, abf, bce, acd, adf, bef, cde, def\}.$$

Now define a coloring via $\chi_{K'''}(a) = \chi_K(a) = 1 = \chi_{K'''}(e)$, $\chi_{K'''}(b) = \chi_K(b) = 2 = \chi_{K'''}(d)$ and $\chi_{K'''}(c) = \chi_K(c) = 3 = \chi_{K'''}(f)$. Altogether we have obtained a chromatic subdivision $(K''', \chi_{K'''})$ of (K, χ_K). Figure 3 displays the chromatic subdivisions we have calculated so far.

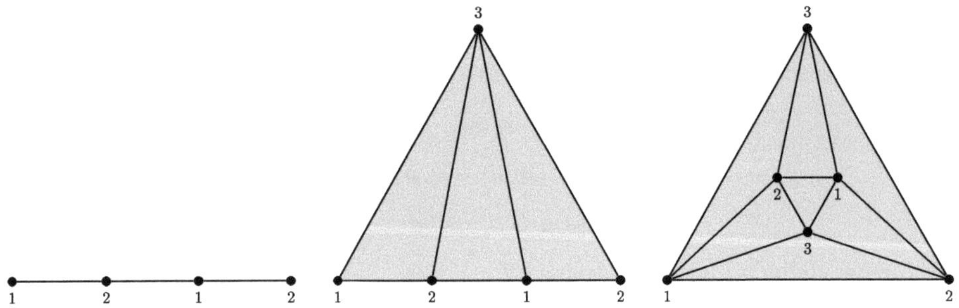

Figure 3: The chromatic subdivisions E'', $\tilde{K}$ and K'''.

The reader will have noticed that the construction of the chromatic subdivisions above involved choices. Any sequence of subdividing Δ^1 twice will yield isomorphic complexes, however, different sequences of (three) stellar subdivisions of Δ^2 are not necessarily isomorphic nor chromatic which raises the question as to how they are related.

Next, we will point out that the subdivisions obtained above are no new objects. Another fruitful way to describe the above simplicial complexes is to recognize them as (cones of) Schlegel diagrams of d-cross polytopes for $d = 1, 2$.

Definition 2.2. Let $P_d = \operatorname{conv}(e_1, \ldots, e_{d+1}, -e_1, \ldots, -e_{d+1})$ be the *d-cross polytope*. Let $\operatorname{Sch}(P_d)$ be its *Schlegel diagram*, that is, the perspective projection centered over any d-face of P_d.

We observe that P_d is simplicial, and that the Schlegel diagram is invariant of the d-face we choose. Furthermore, immediately from its construction we know that the Schlegel diagram is a subdivision. In our example above, it turns out that we have simplicial isomorphisms $E'' \cong \operatorname{Sch}(P_1)$, $\tilde{K} \cong \operatorname{cone}(\operatorname{Sch}(P_1))$ and $K''' \cong \operatorname{Sch}(P_2)$.

Remarkably, the above complexes have been identified in theoretical computer science and given the name *basic chromatic subdivisions*, see for example [2]. The overall philosophy—in comparison to the stellar subdivision of a simplex—is not to replace a simplex by a single vertex but by a duplicate of the simplex itself, that is, we replace Δ^n by $\partial\Delta^n * \Delta^n$; see Definition 2.3 for a generalization to simplicial complexes. In total, there are three different perspectives on the above subdivisions: (1) as a sequence of stellar subdivisions, (2) as Schlegel diagrams of d-cross polytopes and (3) as the basic chromatic subdivision. While the last two perspectives naturally generalize to higher dimensions, this is not obvious for (1) due to the choices indicated above. We suggest the following definition.

Definition 2.3. For a simplicial complex K and an n-simplex $\sigma \in K$, we define

$$\operatorname{stel}^{\chi}_{K,\tilde{\sigma}}(\sigma) = (K \setminus \operatorname{star}_K(\sigma)) \cup (\tilde{\sigma} * \partial\sigma * \operatorname{link}_K(\sigma))$$

to be its *chromatic stellar subdivision.*

Analoguosly to the stellar subdivision, $\operatorname{stel}^{\chi}_{K,\tilde{\sigma}}(\sigma)$ becomes $\tilde{\sigma} * \partial\sigma$ if σ is a maximal simplex. We remark that identifying $\operatorname{stel}^{\chi}_{K,\tilde{\sigma}}(\sigma)$ with the corresponding Schlegel diagram in particular shows that it is indeed a subdivision. Phrasing a new notion for a known object hardly matters if it does not provide new insight. However, perspective (1) reveals combinatorial features of chromatic stellar subdivisions that match those of the nonchromatic analogues. Thus we are led to the following conjecture as a chromatic version of Theorem 1.4.

Conjecture 2.4. Let K_1 and K_2 be two n-dimensional chromatic complexes such that $|K_1|$ and $|K_2|$ are piecewise linearly isomorphic. Then K_1 and K_2 can be transformed into each other by chromatic stellar subdivisions and their inverses.

As a special case, our conjecture predicts that any chromatic subdivision of Δ^n will be related to it via chromatic stellar subdivisions and their inverses. We conclude our detour with an application of Conjecture 2.4. We recall the *standard chromatic subdivision* $\chi(\Delta^2)$ (see [3]) of Δ^2 and construct it in such a way that $\chi(\Delta^2)$ is the resulting complex if we mimic the construction of the barycentric subdivision in the chromatic context. Figure 4 shows how the standard chromatic subdivision can be obtained via a sequence of chromatic stellar subdivisions.

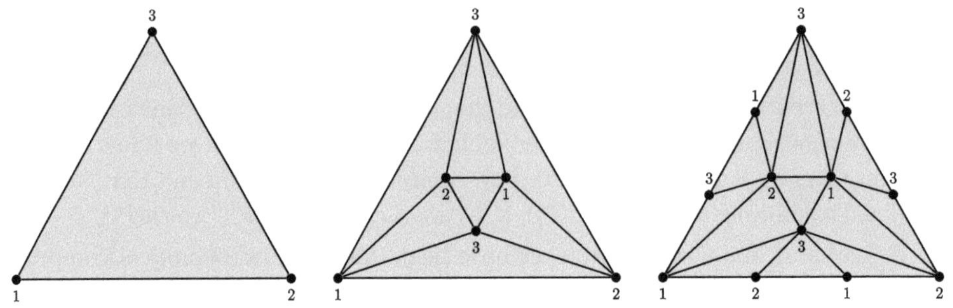

Figure 4: The standard chromatic subdivision $\chi(\Delta^2)$ as a sequence of stellar chromatic subdivisions.

The standard chromatic subdivision $\chi(\Delta^n)$ has rich applications in distributed computing as the protocol complex of the layered immediate snapshot communication model. A rigorous proof of the subdivision property of $\chi(\Delta^n)$ for all $n \geq 1$ has been given in [5], using Schlegel diagrams. We suggest to further develop chromatic stellar subdivision theory. In particular, a functorial interpretation (and relations between subdivisions being natural transformations) might provide insight into the universal property of $\chi(\Delta^n)$ and help to determine its role in combinatorial algebraic topology.

3 The chromatic number of a group

We restrict ourselves to finitely presented groups in this section. We recall that any group G is the fundamental group of a path connected 2-dimensional simplicial complex K, that is $\pi_1(K) \cong G$. This gives rise to the following definition.

Definition 3.1. For a group G let $\chi(G) = \min\{\chi(K) \mid \pi_1(K) \cong G\}$, where K denotes a path connected 2-dimensional simplicial complex and $\chi(K)$ denotes the chromatic number of its underlying graph, be the *chromatic number* of G.

A first observation is that the trivial group $\{e\}$ has chromatic number $\chi(\{e\}) = 1$. We next consider the integers.

Proposition 3.2. *Let $G = \mathbb{Z}$. Then $\chi(G) = 2$.*

Proof. Let K be the simplicial complex that consists of a hollow square. Then $\chi(K) = 2$ and $\pi_1(K) \cong \pi_1(S^1) \cong \mathbb{Z}$, hence $\chi(\mathbb{Z}) \leq 2$. As $\mathbb{Z}$ has at least one generator (we could have also consider presentations like $\langle a, b \mid b = 1 \rangle$), any simplicial complex K' such that $\pi_1(K') \cong \mathbb{Z}$ must contain a circle, hence $\chi(\mathbb{Z}) \geq 2$. □

A first structure theorem on the chromatic number is given by the following.

Proposition 3.3. *Let $G \neq \{e\}$ and $H \neq \{e\}$ be groups. Then $\chi(G * H) = \max\{\chi(G), \chi(H)\}$.*

Proof. As G and H have at least one generator, their corresponding 2-dimensional complexes K_1 and K_2 with $\pi_1(K_1) \cong G$ and $\pi_1(K_2) \cong H$ have at least chromatic number 2. Now connect both complexes via an edge attached to vertices of different colors and observe that this process does not increase the chromatic number. As this construction has the homotopy type of $K_1 \vee K_2$ and $\pi_1(K_1 \vee K_2) \cong \pi_1(K_1) * \pi_1(K_2) \cong G * H$, the resulting complex has chromatic number $\max\{\chi(G), \chi(H)\}$ as desired. □

If we combine the above two propositions, we get the following corollary.

Corollary 3.4. *Let F_n denote the free group on n generators, $n > 0$. Then $\chi(F_n) = 2$.*

Our next example shows that the chromatic number of a group detects relators.

Proposition 3.5. *Let G be a nonfree group presented by n generators, $n > 0$, $x_1, \ldots, x_n$ and one defining relator r. Then $\chi(G) = 3$.*

The proof of this proposition makes use of the following lemma.

Lemma 3.6. *Let K be the simplicial complex consisting of the 2-simplices acd, bcd and their faces. Let $h\colon V(K) \to C$ be a 3-labeling such that $h(c) = h(d)$ and $h(a), h(c), h(b)$ are pairwise distinct. Then there exists a subdivision K'' of K and a 3-coloring $\chi\colon V(K'') \to C$ such that $h(x) = \chi(x)$ for all $x \in \{a, b, c, d\}$.*

Proof. Let $K' = \mathrm{stel}_{K,e}(cd)$ and $K'' = \mathrm{stel}_{K',f}(be)$. Then K'' consists of the 2-simplices $ade, ace, cef, bcf, bdf, def$ and their subsets and

$$\chi(x) = \begin{cases} h(x) & \text{for } x \in \{a, b, c, d\}, \\ h(b) & \text{for } x = e, \\ h(a) & \text{for } x = f, \end{cases}$$

defines a 3-coloring with the desired properties. □

Observe that the proof mimics the construction of the chromatic stellar subdivision of Δ^2 in Section 2 and see Figure 5 for an illustration. We now continue with the proof of Proposition 3.5.

Proof. Let the group G be presented by generators $x_1, \ldots, x_n$ and the relator r. To any generator x_i, associate a hollow 2-simplex a_i with vertices $v_{i,1}, v_{i,2}$ and $v_{i,3}$ and coloring $f(v_{i,j}) = j$ for all $i = 1, \ldots, n$. Identify $v_{i,1}$ for all $i = 1, \ldots, n$ to obtain a wedge of circles. Now take the simplicial disc containing the 2-simplex abc and its faces. Perform $3(n-1)$ stellar subdivisions of the edge ab and label the $3n$ vertices according to the sequence induced by r with the convention that $x_i^{\varepsilon} = a_{i,1}a_{i,2}a_{i,3}$ if $\varepsilon = 1$ and $x_i^{\varepsilon} = a_{i,1}a_{i,3}a_{i,2}$ if $\varepsilon = -1$. Now subdivide repeatedly by means of Lemma 3.6 to extend this labeling to a 3-coloring, glue the boundary of the subdivided disc to the wedge of spheres according

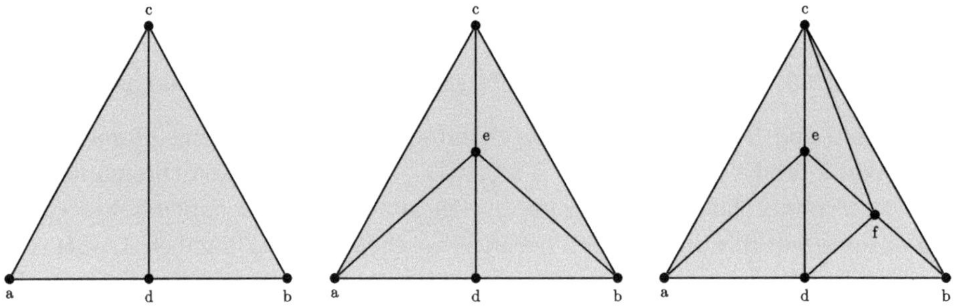

Figure 5: The stellar subdivision sequence from the proof of Lemma 3.6.

to r, and call this complex K. As $\pi_1(K) \cong G$ we obtain $\chi(G) \leq 3$ and due to the fact that a relator requires a disc we also have $\chi(G) \geq 3$. □

Note that Proposition 3.5 yields in particular the chromatic number of surface groups, that is, fundamental groups of compact surfaces, and nontrivial finite cyclic groups. We end this note with another conjecture.

Conjecture 3.7. Let G be a group. Then $\chi(G) \leq 3$.

We phrase an equivalent conjecture: Any 2-dimensional simplicial complex K can be 3-colored after a possible subdivision. We think that this conjecture can be proven by a case-by-case study of the incidence structure of 2-dimensional simplicial complexes with successive application of Lemma 3.6.

Bibliography

[1] Alexander JW. The combinatorial theory of complexes. Ann Math. 1930;31(2):292–320.

[2] Castañeda A, Rajsbaum S. New combinatorial topology bounds for renaming: The upper bound. Journal of the ACM (JACM); 2012;59(1).

[3] Herlihy M, Kozlov D, Rajsbaum S. Distributed Computing through Combinatorial Topology. Amsterdam: Elsevier; 2014. 336 pp.

[4] Kozlov D. Combinatorial Algebraic Topology. Algorithms and Computation in Mathematics. vol. 21. Berlin Heidelberg: Springer; 2008. XX, 390 pp., 115 illus.

[5] Kozlov DN. Chromatic subdivision of a simplicial complex. Homol Homotopy Appl. 2012;14(2):197–209.

[6] Milnor J. Two complexes which are homeomorphic but combinatorially distinct. Ann Math. 1961;74(3):575–90.

[7] Ranicki AA (ed). The Hauptvermutung Book. K-Monograph in Mathematics. vol. 1. Kluwer; 1996.

De Gruyter Proceedings in Mathematics

Galina Filipuk, Alberto Lastra, Sławomir Michalik, Yoshitsugu Takei, Henryk Żołądek (Eds.)
Complex Differential and Difference Equations. Proceedings of the School and Conference held at Będlewo, Poland, September 2-15, 2018
ISBN 978-3-11-060952-3, e-ISBN 978-3-11-061142-7

Mohammad Ashraf, Vincenzo De Filippis, Syed Tariq Rizvi (Eds.)
Algebra and Its Applications. Proceedings of the International Conference held at Aligarh Muslim University, 2016, 2018
ISBN 978-3-11-054092-5, e-ISBN 978-3-11-054240-0

Alexander Katz (Ed.)
Topological Algebras and their Applications. Proceedings of the 8th International Conference on Topological Algebras and their Applications, 2014, 2018
ISBN 978-3-11-041433-2, e-ISBN 978-3-11-041355-7

J. Pintz, A. Biró, K. Győry, G. Harcos, M. Simonovits, J. Szabados (Eds.)
Number Theory, Analysis, and Combinatorics: Proceedings of the Paul Turán Memorial Conference held August 22-26, 2011 in Budapest, 2013
ISBN 978-3-11-028237-5, e-ISBN 978-3-11-028242-9

I. Assani (Ed.)
Ergodic Theory and Dynamical Systems: Proceedings of the Ergodic Theory workshops at University of North Carolina at Chapel Hill, 2011-2012, 2013
ISBN 978-3-11-029813-0, e-ISBN 978-3-11-029820-8

B. Landman, M. B. Nathanson, J. Nešetril, R. J. Nowakowski, C. Pomerance, A. Robertson (Eds.)
Combinatorial Number Theory: Proceedings of the "Integers Conference 2011", Carrollton, Georgia, USA, October 26-29, 2011, 2013
ISBN 978-3-11-028048-7, e-ISBN 978-3-11-028061-6

C. Francisco, L. C. Klingler, S. Sather-Wagstaff, J. C. Vassilev (Eds.)
Progress in Commutative Algebra 1: Combinatorics and Homology, 2012
ISBN 978-3-11-025034-3, e-ISBN 978-3-11-025040-4

C. Francisco, L. C. Klingler, S. Sather-Wagstaff, J. C. Vassilev (Eds.)
Progress in Commutative Algebra 2: Closures, Finiteness and Factorization, 2012
ISBN 978-3-11-027859-0, e-ISBN 978-3-11-027860-6

Bei Fragen zur Produktsicherheit wenden Sie sich bitte an:
If you have any questions regarding product safety,
please contact:

Walter de Gruyter GmbH
Genthiner Straße 13
10785 Berlin
productsafety@degruyterbrill.com